Applications
on Advanced
Architecture
Computers

Applications on Advanced Architecture Computers

Edited by Greg Astfalk

Hewlett-Packard Company
Richardson, Texas

SOFTWARE · ENVIRONMENTS · TOOLS

Society for Industrial and Applied Mathematics
Philadelphia

Library of Congress Cataloging-in-Publication Data

Applications on advanced architecture computers / edited by Greg
    Astfalk.
        p. cm.
    Includes bibliographical references and index.
    ISBN 0-89871-368-4 (pbk.)
    1. Computer architecture. 2. Parallel computers. 3. Application
software. I. Astfalk, Greg.
    QA76.9.A73A66 1996
    502'.85'435--dc20   96-31473

The royalties from the sales of this book are being placed in a fund to help students attend SIAM meetings and other SIAM-related activities. This fund is administered by SIAM and qualified individuals are encouraged to write directly to SIAM for guidelines.

siam is a registered trademark.

Contributing Authors

Jan Almlöf
University of Minnesota
Department of Chemistry
Minneapolis, MN 55455 USA
612–624–6524
612–626–7541 (FAX)
almlof@chemsun.chem.umn.edu

Greg Astfalk
Hewlett–Packard Company
Convex Technology Center
PO Box 833851
Richardson, TX 75083–3851 USA
214–497–4787
214–497–4441 (FAX)
astfalk@rsn.hp.com

Clive F. Baillie
University of Colorado
Computer Science Department
Campus Box 430
Boulder, CO 80309 USA
303–492–7852
303–492–2844 (FAX)
clive@cs.colorado.edu

David M. Beazley
University of Utah
Department of Computer Science
Salt Lake City, UT 84112 USA
801–581–7977
801–581–5843 (FAX)
beazley@cs.utah.edu

Isabel M. Beichl
National Institute of Standards &
Technology
Gaithersburg, MD 20879 USA
301–975–3821
301–990–4127 (FAX)
isabel@cam.nist.gov

Pierre Bessiere
CNRS - LEIBNIZ
46 ave Felix Viallet
38031 Grenoble, France
+33–76–57–46–73
+33–76–57–46–02 (FAX)
Pierre.Bessiere@imag.fr

Robert Bixby
Rice University
Department of Computational &
Applied Mathematics
6100 S. Main St.
Houston, TX 77005 USA
713–527–6073
713–285–5318 (FAX)
bixby@rice.edu

Rob Bjornson
Scientific Computing Associates
265 Church St.
New Haven, CT 06510 USA
203–777–7442
203–776–4074 (FAX)
bjornson@sca.com

Kevin Burrage
University of Queensland
Department of Mathematics
Brisbane 4072, Australia
+61–07–33654387
+61–07–33651477 (FAX)
kb@flash.maths.uq.oz.au

Eduardo F. D'Azevedo
Oak Ridge National Laboratory
Mathematical Sciences Section
PO Box 2008, Bldg 6012
Oak Ridge, TN 37831–6367 USA
423–576–7925
423–574–0680 (FAX)
e6d@ornl.gov

John Dennis
Rice University
187 CITI/Fondren MS 41
Houston, TX 77251–1892 USA
713–527–4094
713–285–5318 (FAX)
dennis@caam.rice.edu

Craig C. Douglas
IBM, Watson Research Center
PO Box 218
Yorktown, NY 10598–0218 USA
914–945–1475
914–945–3434 (FAX)
douglas-craig@cs.yale.edu

Jonathan Eckstein
Rutgers University
School of Business and RUTCOR
PO Box 5062
New Brunswick, NJ 08903–5062 USA
908–445–3596
908–445–5472 (FAX)
jeckstei@rutcor.rutgers.edu

Alexandre Ern
CERMICS, ENPC
Centre d'Enseignement et de
Recherche en Mathématiques
Informatique et Calcul Scientifique
93167 Noisy-le-Grand cedex, France
+33–1–49–14–35–70
+33–1–49–14–35–86 (FAX)
ern@cmapx.polytechnique.fr

Klaas Esselink
KSLA, Shell Research
Dept. MCA/3
PO BOX 38000
1030 BN Amsterdam, The
Netherlands
+31-20–630–3254
+31-20–630–4041 (FAX)
esselin1@ksla.nl

Martin Feyereisen
Cray Research, Inc.
655 E Lone Oak Drive
Eagan, MN 55121 USA
612–683–3608
612–683–3099 (FAX)
feyer@cray.com

Robert J. Geller
Dept. of Earth and Planetary Physics
Tokyo University
Yayoi 2–11–16
Bunkyo-ku, Tokyo 113, Japan
+81–3–5900–6973
+81–3–3818–3247 (FAX)
bob@global.geoph.s.u-tokyo.ac.jp

Richard J. Hanson
Visual Numerics, Inc.
9990 Richmond Ave.
Houston, TX 77042 USA
713–954–6461
713–781–9260 (FAX)
hanson@vni.com

Liam M. Healy
Naval Research Laboratory
Code 8233
Washington, DC 20375–5355 USA
202–404–8338
202–404–7785 (FAX)
Liam.Healy@nrl.navy.mil

Maurice Herlihy
Brown University
Computer Science Dept.
Providence, RI 02912 USA
401–863–7646
401–863–7657 (FAX)
herlihy@cs.brown.edu

Peter A.J. Hilbers
Shell Research B.V.
Dept. MCA
PO Box 38000
1030 BN Amsterdam, The
Netherlands
+31–20–6303892
+31–20–6304041 (FAX)
hilbers1@ksla.nl

Michael Karasick
IBM T.J. Watson Research Center
PO Box 218
Yorktown Heights, NY 10532 USA
914–784–6627
914–784–7455 (FAX)
msk@watson.ibm.com

Craig Kolb
Stanford University
Room 374, M/C 9035
Stanford, CA 94305–9035 USA
415–725–3728
415–725–6949 (FAX)
cek@graphics.stanford.edu

Håkan Lennerstad
University of Karlskrona/Ronneby
Dept. of Telecommunications and
Mathematics
S-371 79 Karlskrona, Sweden
+46–455–78052
+46–455–78057 (FAX)
Hakan.Lennerstad@itm.hk-r.se

Peter Lomdahl
Los Alamos National Laboratory
T–11, MS B262
Los Alamos, NM 87545 USA
505–665–0461
505–665–4063 (FAX)
pxl@lanl.gov

Glenn R. Luecke
Iowa State University
291 Durham Center
Ames, IA 50011 USA
515–294–6659
515–294–1717 (FAX)
grl@iastate.edu

Lars Lundberg
University of Karlskrona/Ronneby
Department of Computer Science and
Economics
S-372 25 Ronneby, Sweden
+46–457–78730
+46–457–27125 (FAX)
Lars.Lundberg@ide.hk-r.se

Michael Mascagni
Center for Computing Sciences,
I.D.A.
17100 Science Drive
Bowie, MD 20715–4300 USA
301–805–7421
301–805–7604 (FAX)
mascagni@super.org

John M. Mulvey
Princeton University
School of Engineering and Applied
Science
Princeton, NJ 08544 USA
609–258–5423
609–258–3791 (FAX)
mulvey@macbeth.princeton.edu

Anna Nagurney
University of Massachusetts
Department of Finance & Operations
Management
Amherst, MA 01003 USA
413–545–5635
413–545–3858 (FAX)
nagurney@gbfin.umass.edu

Sandy Nguyen
Southwest Research Institute
6220 Culebra Road
PO Drawer 28510
San Antonio, TX 78228–0510 USA
210–522–3868
210–647–4325 (FAX)
snguyen@swri.edu

Charles H. Romine
Oak Ridge National Laboratory
PO Box 2008
Oak Ridge, TN 37831–6367 USA
423–574–3141
423–574–0680 (FAX)
rominech@ornl.gov

Andrew L. Sargent
East Carolina University
Department of Chemistry
405 Flanagan Hall
Greenville, NC 27858 USA
919–328–1637
919–328–6210 (FAX)
chsargen@ecuvax.cis.ecu.edu

Jeremy Schneider
Northwest Airlines
5101 Northwest Drive
St. Paul, MN 55111–3034 USA
612–726–2615
612–726–0677 (FAX)
schnei@opsys.nwa.com

Robert T. Schumacher
Carnegie Mellon University
Department of Physics
5000 Forbes Ave.
Pittsburgh, PA 15213 USA
412–268–3532
412–681–0648 (FAX)
rts@andrew.cmu.edu

Nir Shavit
Tel-Aviv University
Dept. of Computer Science
Tel-Aviv 69978, Israel
+011–972–3–640–9616
+011–972–3–640–9357 (FAX)
shanir@cs.tau.ac.il

Andrew H. Sherman
Scientific Computing Associates, Inc.
One Century Tower
265 Church Street
New Haven, CT 06510–7010 USA
203–777–7442
203–776–4074 (FAX)
sherman@sca.com

Mitchell D. Smooke
Yale University
Department of Mechanical
Engineering
Becton Laboratory Room 205
New Haven, CT 06520 USA
203–432–4344
203–432–6775 (FAX)
smooke%smooke@biomed.med.yale.edu

David Strip
Sandia National Laboratories
Mail Stop 0951
PO Box 5800
Albuquerque, NM 87185–0951 USA
505–844–3962
505–844–6161 (FAX)
drstrip@sandia.gov

Francis Sullivan
IDA Center for Computing Sciences
17100 Science Drive
Bowie, MD 20715–4300 USA
301–805–7534
301–805–7604 (FAX)
fran@super.org

El-Ghazali Talbi
Laboratoire d'Informatique
Fondamentale de Lille
Bat. M3 59655
Villeneuve d'Ascq Cedex, France
+33–20–43–45–13
+33–20–43–65–66 (FAX)
talbi@lifl.fr

David C. Torney
Los Alamos National Laboratory
Los Alamos, NM 87545 USA
505–667–9452
505–665–3493 (FAX)
dct@lanl.gov

David W. Walker
Oak Ridge National Laboratory
PO Box 2008
Oak Ridge, TN 37831–6367 USA
423–574–7401
423–574–0680 (FAX)
walker@msr.epm.ornl.gov

Steven W. White
IBM RISC System/6000 Division
MS 9221
11400 Burnet Road
Austin, TX 78758 USA
512–838–1849
512–838–6486 (FAX)
white@austin.ibm.com

Theresa Hull Wise
Northwest Airlines
5101 Northwest Drive
St. Paul, MN 55111–3034 USA
612–726–0820
612–727–6112 (FAX)
theresa@opsys.nwa.com

Jim Woodhouse
Cambridge University
Engineering Department
Trumpington Street
Cambridge CB2 1PZ, England
+44–1223–332642
+44–1223–332662 (FAX)
jw12@eng.cam.ac.uk

Zhijun Wu
Argonne National Laboratory
Mathematics and Computer Science
Division
Argonne, IL 60439 USA
708–252–3336
708–252–5986 (FAX)
zwu@mcs.anl.gov

Stavros A. Zenios
University of Cyprus
Department of Public and Business
Administration
Kallipoleos 75
Nicosia, Cyprus
+357–2–338764
+357–2–339063 (FAX)
zenioss@atlas.pba.ucy.ac.cy

Xiaodong Zhang
University of Texas at San Antonio
High Performance Computing and
Software Lab
San Antonio, TX 78249 USA
210–691–5541
210–691–4437 (FAX)
zhang@ringer.cs.utsa.edu

Jianping Zhu
Mississippi State University
Department of Mathematics and
Statistics
NSF Engineering Research Center
Mississippi State, MS 39762 USA
601–325–3414
601–325–0005 (FAX)
jzhu@math.msstate.edu

Brian Zook
Southwest Research Institute
PO Drawer 28510
San Antonio, TX 78228–0510 USA
210–522–3630
210–647–4325 (FAX)
bzook@swri.edu

Contents

Preface

Parallel processing is widely seen and accepted as the only approach that can make available the computing resources that are necessary to solve large-scale computing models. Because of advances in technology and industry, there is an increased need to solve more complex physical models, with finer resolution, more particles, more dimensions, and for larger timescales. Generally, simple models will no longer suffice.

Several different application areas are presented in this book, along with their solutions on (generally parallel) advanced architecture computers. Often the applications have been lost in the volumes of literature about advanced architectures, parallel algorithms, hot chips, and integrated circuit technology. The new computer technology presented here, however, affords solutions to "what it is all about." Applications drive the sale of machines and make the end users' organizations more competitive and productive. Stated simply, applications are the raison d'être of parallel processing.

At this juncture, making effective use of parallel processing is a nontrivial task. This is especially true when we consider genuine applications codes. As you read the chapters you will implicitly, and in some cases explicitly, sense the amount of human effort that goes into devising, developing, and implementing a parallelized application code.

This book grew from a regular column, "Applications on Advanced Architecture Computers," that appeared in *SIAM News*. This is the set of articles that appeared beginning with the column's inception in March 1990 until the "cut-off" date for putting together this book—June 1995. The focus of the column was to present *applications* that have been successfully treated on advanced architecture computers. Our working definition of what constitutes an advanced architecture has been, and remains, quite liberal. It includes everything that is even slightly more exotic than the simple SISD (single-instruction single-data stream) class. In some cases we have had articles about general algorithmic issues, which are algorithms that are commonly used in building parallel codes for solving applications.

The column that is the breeding ground for the material in this book has continued for nearly six years. In the high-performance computing arena this is two to three epochs. The Connection Machine, the most cited machine in the articles, is no longer with us. Several of the later articles have used workstations as the basis for the computing "engine." This may well be a mild confirmation of the trend that a number of computer industry watchers believe is the future limit point—clusters of workstations or servers as *the* parallel architecture. The fact that applications continue to be developed and implemented on these advanced types of architectures, as the architectures themselves evolve, is an indication of the importance placed on the worth of the applications.

The 30 articles in the column cover a broad spectrum. While trying to identify the few tidbits that might serve as a theme, I admit to being bewildered. Some of the computers used most often in the past articles are no longer in use. Some of the languages used to code the applications are now rarely used. What is consistent is the diversity of areas to which advanced computing has been, and will continue to be, applied.

There is significant overlap in applications. Four articles deal with molecular dynamics, four articles deal with what we can fairly call optimization or mathematical programming, three articles deal with financial or economic applications, two articles deal with geometry, and five articles are solely focused on partial differential equations. Given this roll call, it is hard to say that any one area has dominance over any other.

There have been several notable events since the column's inception. Michael Mascagni's article was really the inaugural article. Craig Douglas was our first repeat author; both of his contributions appear in this book. David Beazley and Peter Lomdahl offered the first two-part article. Eduardo D'Azevedo, Charles Romine, and David Walker offered another exciting first for the column, the initial appearance of a new research result: an article describing the first *billion*-atom molecular dynamics simulation to ever appear in print.

By agreement of the contributing authors all proceeds from this book are donated to the SIAM Student Travel Fund (see *SIAM News*, Vol. 28, No. 4, April 1995). It is my hope that the book sells well so that many more students will have the opportunity to attend SIAM conferences that they might otherwise not be able to attend.

I found it nearly impossible to find a content- or topic-based way to order the chapters. Since many chapters have more than one author a simple strict alphabetization seemed inappropriate. So, in an attempt to induce some fairness about the order of the chapters, I used the following algorithm. I took the lowercase ASCII values of the letters of the chapter's authors' last names, summed them, and normalized them by the number of characters. An ascending numerical sort of these values determined the order of appearance in the book. Perhaps it is not optimal, but it eliminates any personal bias or interpretation. The editorial preface for each chapter indicates the date of its appearance in *SIAM News*.

During the preparation of this book we learned of the untimely and sad news of Jan Almlöf's death in January 1996. As a tribute to Jan I chose to violate the chapter ordering just described. To honor Jan and his work his chapter appears first in this book.

There are a great many acknowledgments that I need to offer, exactly 50, to each contributing author. Having never been an editor of a regular column before this experience I can't offer comparisons. I can, however, offer absolute testament that these 50 people have been a distinct pleasure to work with; each was cooperative and helpful. My job has been easier because of their efforts, and this book would not exist without them.

The genesis for this column, with proportionate thanks, goes to Ed Block, former managing director of SIAM. He started the whole thing by making the suggestion that a column focused on the union of mathematics and computing would be a welcome addition to *SIAM News*. A five-year retrospective that I wrote on the column is included as an appendix to this book.

The editorial staff at SIAM has been most helpful and thorough in getting this book completed, notably Jean Anderson and Susan Ciambrano.

During the course of the column's tenure, which continues to this day, the editor of *SIAM News*, Gail Corbett, has been the quiet, "behind the scenes" force. Gail has taken every article and added that special treatment to make it more readable. She approached each article as a naive reader would, and each has prospered from her efforts.

This book was produced with LaTeX, version 3.1. All of the figures are in PostScript.

Finally, any responsibility for errors in this book rests with me.

Greg Astfalk
Richardson, TX

Massively Parallel Algorithms for Electronic Structure Calculations in Quantum Chemistry

Andrew L. Sargent
Jan Almlöf
Martin W. Feyereisen

Editorial preface

The authors develop the underlying mathematics of the self-consistent field (SCF) method for electronic structure calculations. They focus on the parallelization of the formation of the so-called Fock matrix, then develop a method to solve the entire problem, formation and solution, on a heterogeneous system. In this approach each machine is used for that portion of the algorithm for which it is best suited. Finally, a modification is offered that permits large-scale electronic structure calculations while using a relatively small amount of memory.

This article originally appeared in *SIAM News*, Vol. 26, No. 1, January 1993. It was updated during the summer/fall of 1995.

One of the most significant challenges facing contemporary computational chemists involves the restructuring of application software to allow full utilization of current computer hardware. Considering the wide variety of available computer architectures and the ephemeral nature of the cutting-edge technology on which they are based, this is not a one-time task but rather an ongoing development project.

Massively parallel processing, the latest trend for the supercomputing community, is universally hailed as the vehicle by which grand challenge, i.e., teraflops, computing will be achieved in the future. However, the concept of parallel computing is not well defined, and there are many different ways in which calculations can be carried out in parallel. Key decisions to be made in this context include the number of processors to be utilized, the accessibility of memory (shared versus local), and the granularity of the parallel algorithm. Another important issue is the extent to which load balancing is pursued, a problem that is considerably more difficult in a multiuser, time-shared environment than with dedicated hardware.

Our ab initio computational chemistry code [3] is being continuously updated to keep abreast of parallel technology. The code has been modified to run on clusters of workstations [12]; one or several loosely coupled, dissimilar supercomputers [10]; and massively parallel hardware [7]. In this article, we review our recent work in adapting electronic structure codes to parallel processing and report benchmark results for a variety of parallel architectures. We also discuss new ways to address certain bottlenecks encountered in several of the parallel approaches.

1.1. Calculating Electronic Structure

In virtually all calculations of molecular electronic structure, one-electron wavefunctions $\phi_i(\mathbf{r})$ are expanded in a basis set:

$$(1.1) \qquad \phi_i(\mathbf{r}) = \sum_{p=1}^{N} C_{pi}\chi_p(\mathbf{r}).$$

The basis functions $\chi_p(\mathbf{r})$ are chosen to be atomic orbitals, i.e., previously computed and tabulated one-electron wavefunctions for the atoms constituting the molecule. The probability distribution corresponding to a wavefunction $\phi(\mathbf{r})$ is given by its square amplitude $|\phi(\mathbf{r})|^2$ and the total electron density in a molecule $\rho(\mathbf{r})$ is therefore obtained as an expansion in products of these basis functions:

$$(1.2) \qquad \rho(\mathbf{r}) = \sum_{p,q} D_{pq}\chi_p^*(\mathbf{r})\chi_q(\mathbf{r}),$$

where (*) denotes a complex conjugate. The coefficients D_{pq} are elements of a density matrix $\mathbf{D}$ and are obviously related to the expansion coefficients C_{pi}, the exact relation depending on details of the electronic structure model, which is not our concern here. The electrostatic interaction between electrons is given by the six-dimensional integral

$$(1.3) \qquad V = \int\int \frac{\rho(\mathbf{r})\rho(\mathbf{r}')}{|\mathbf{r}-\mathbf{r}'|}d\mathbf{r}d\mathbf{r}'.$$

For the purpose of the present discussion, the only significant observation we need to make is that the evaluation of the energy involves two-electron integrals of the form

$$(1.4) \qquad I_{pqrs} = \int\int \chi_p^*(\mathbf{r})\chi_q(\mathbf{r})\frac{1}{|\mathbf{r}-\mathbf{r}'|}\chi_r^*(\mathbf{r}')\chi_s(\mathbf{r}')d\mathbf{r}d\mathbf{r}',$$

the number of which scales as the fourth power of the number of basis functions N. The quantum mechanical description of the system is obtained by solving the generalized matrix eigenvalue equation

$$(1.5) \qquad \mathbf{FC} = \mathbf{SC}\epsilon.$$

In this expression, $\mathbf{F}$, the Fock matrix, is a matrix representation of the effective one-electron Hamiltonian operator in the basis set $\{\chi(\mathbf{r})\}$, $\mathbf{C}$ is the matrix of expansion coefficients in (1.1), $\mathbf{S}$ is a metric matrix defined as

$$(1.6) \qquad S_{pq} = \int \chi_p^*(\mathbf{r})\chi_q(\mathbf{r})d\mathbf{r},$$

and ϵ is a diagonal matrix. These matrices are all $N \times N$ square matrices, N being the length of the basis set expansion in (1.1). Equation (1.5), known as the Roothaan equation [13], is nonlinear because the Fock matrix $\mathbf{F}$ depends on the density matrix $\mathbf{D}$ introduced in (1.2) (and therefore indirectly on $\mathbf{C}$) through the relations

$$(1.7) \qquad \begin{aligned} F_{pq} &= F_{pq} + 4D_{rs}I_{pqrs}, \\ F_{rs} &= F_{rs} + 4D_{pq}I_{pqrs}, \\ F_{pr} &= F_{pr} - D_{qs}I_{pqrs}, \\ F_{qs} &= F_{qs} - D_{pr}I_{pqrs}, \\ F_{ps} &= F_{ps} - D_{qr}I_{pqrs}, \\ F_{qr} &= F_{qr} - D_{ps}I_{pqrs}. \end{aligned}$$

Because of this nonlinearity, the Roothaan equation must be solved iteratively.

The traditional ab initio Hartree–Fock approach involves two general steps. In the first step, the integrals I_{pqrs} in (1.4) are evaluated and written to disk. In the second step, these integrals are read back and combined with density matrix elements to form the Fock matrix $\mathbf{F}$, and the Roothaan equation is solved for the new set of expansion coefficients, from which a new density matrix is constructed. The second step is repeated until the change in the density matrix between successive iterations is below a given threshold.

Evaluation of a large number of complicated two-electron integrals I_{pqrs} to high accuracy is undoubtedly a gargantuan challenge. Nevertheless, as advances in CPU technology have outpaced improvements in I/O and storage capacity for decades, storage of integrals has replaced integral evaluation as the true bottleneck in these calculations. With the N^4 dependence, even modest-sized calculations ($N \approx 200$ basis functions) require disk storage space on the order of gigabytes. Furthermore, the I/O subsystem is severely taxed as these integrals are read from disk at every iteration.

More than a decade ago, we suggested an alternative approach, in which the integrals are recalculated in each iteration as needed [1, 2, 4, 5]. While often more CPU-demanding than the traditional approach, this *direct* Hartree–Fock (or direct SCF) method allowed much larger calculations than previously possible. With the storage problem effectively eliminated through the direct approach, CPU power is again the bottleneck, and our focus returns to more efficient methods for evaluation of large numbers of integrals as parallel processing technology provides the tool for extending accurate quantum chemical calculations to new classes of molecules.

1.2. Parallel Hartree–Fock Calculations

More than 95% of the CPU time in a typical direct SCF application is spent constructing the Fock matrix, and any improvement must necessarily focus on that part of the algorithm. It is obvious that the evaluation of the integrals constitutes a very large number of independent tasks. In fact, applications utilizing multiple processors to distribute the work involved in calculating the two-electron integrals predate the early parallel processing machines [6].

Clementi linked several single-processor machines in a loosely coupled array of processors (LCAP) to achieve parallel processing [6]. Several machines, designated as "slaves," were used to calculate a subset of the two-electron integrals, and one machine, the "master," administered these tasks, collected and further processed the integrals, and solved the Roothaan equations. The main bottleneck of this and other early parallel Hartree–Fock approaches was in the communication overhead, due to the large number of integrals that had to be moved from the slaves to the master. However, this problem can now be circumvented by the direct techniques, in which not only the evaluation of the integrals but also their further processing can be seen as independent tasks. Accordingly, the direct SCF method should constitute a very promising candidate for parallel implementations.

In parallel approaches to direct Hartree–Fock calculations, an integral is never moved from the node on which it is evaluated. Instead, batches of integrals are assigned to various nodes for evaluation and further processing. In addition to evaluating the integrals, the node combines them with the appropriate elements of the density matrix and adds these contributions to its private copy of the Fock matrix. Short coded messages from the master instructing the node to evaluate and process a batch of integrals, and requests from the nodes for new instructions constitute the only information transferred during the iteration. At the end of each iteration, the partial Fock matrices from each node are added, combined with matrix elements of the one-electron operators, and finally diagonalized to yield the expansion coefficients from which a new density matrix is formed. The communication load on the system with this approach is on the order of N^2 (N being the number of basis functions, typically 10^2 to 10^3), whereas the computational work is on the order of N^4.

The parallel implementations of our direct Hartree–Fock procedure have involved a variety of parallel architectures, e.g., a 16-processor Cray C90, a 512-processor Intel Delta, a 544-processor Connection Machine-5 (CM-5), various clusters of IBM RS/6000 and Silicon Graphics workstations, as well as arbitrary combinations of the above platforms. Four molecules, ranging in size from small to moderate, were used to obtain benchmarks for the calculations on these machines. The results, shown in Table 1.1, include the times required to build the Fock matrix $\mathbf{F}$ (t_1) and to solve the Roothaan equation once $\mathbf{F}$ and $\mathbf{S}$ are available (t_2).

The Cray C90 is the top performer. Its impressive per-processor throughput is due largely to its vector hardware, of which our code makes explicit

TABLE 1.1

DISCO timings for one Hartree–Fock iteration on four benchmark molecules (common names: fullerene, imidazole, diquinone, and polycarbonate, respectively) with molecular formulas: C_{32}, $C_3N_2H_4$, $C_{22}H_{10}O_4$, and $C_{17}H_{18}O_3$. All times are wallclock seconds with the calculations run in dedicated mode. The basis sets and the resulting total number of basis functions (shown in parentheses) are 3-21G(288), CCDZ(235), STO-3G(140), and 6-31G(314).

Benchmark	Cray C90			Intel Delta			TMC CM-5			IBM 350			Cray!Cray-2/CM-5		
	N	t_1	t_2	N	t_1	t_2	N	t_1	t_2	N	t_1	t_2	N	t_1	t_2
1	1	124	0	-	-	-	32	654	386	1	2152	76	-	-	-
	2	-	-	-	-	-	256	103	365	2	1128	67	-	-	-
	4	-	-	-	-	-	512	95	371	3	752	80	-	-	-
	8	17	9	-	-	-	-	-	-	4	569	70	-	-	-
	16	10	0	-	-	-	-	-	-	5	450	71	-	-	-
	-	-	-	-	-	-	-	-	-	6	380	72	-	-	-
2	1	100	0	16	797	30	32	565	215	1	1996	46	-	-	-
	2	51	0	64	193	30	256	131	203	2	1034	32	256	131	2
	4	26	0	128	99	30	512	129	202	3	691	54	-	-	-
	8	14	0	256	55	29	-	-	-	4	522	54	-	-	-
	16	7	0	512	39	29	-	-	-	5	420	36	-	-	-
	-	-	-	-	-	-	-	-	-	6	359	33	-	-	-
3	1	81	0	-	-	-	32	383	43	1	1374	12	-	-	-
	2	41	0	-	-	-	256	102	39	2	730	8	-	-	-
	4	21	0	-	-	-	512	94	42	3	486	9	-	-	-
	8	11	0	-	-	-	-	-	-	4	386	10	-	-	-
	16	6	0	-	-	-	-	-	-	5	295	8	-	-	-
	-	-	-	-	-	-	-	-	-	6	247	11	-	-	-
4	1	1517	2	-	-	-	32	10962	618	-	-	-	-	-	-
	16	113	0	256	500	-	256	1490	601	-	-	-	256	1490	5
	-	-	-	512	356	-	512	843	587	-	-	-	-	-	-

use. The performance of the Intel Delta exceeded that of the CM-5, partly due to the fact that the latter machine was not fully equipped with vector-processing units during the period of our tests. The distributed processing on the RS/6000 workstations yields impressive results and per-processor performance far superior to that of either the Delta or the CM-5.

Evaluating and processing the integrals associated with a unique set of four atoms can be viewed as a "task," which can be performed independently of all other tasks. The number of tasks for benchmarks 1–4 are 23,474, 1035, 111,391 and 274,911, respectively. Very good estimates of these numbers can be obtained with the formula

$$(1.8) \qquad n_{task} = \frac{n^4}{8G},$$

where n is the number of atoms; the factor of 8 in the denominator is due to the equivalence among integrals under permutations of the indices

$$(1.9) \qquad I_{pqrs} = I_{qprs} = I_{pqsr} = I_{qpsr} = I_{rspq} = I_{srpq} = I_{rsqp} = I_{srqp}$$

(assuming real basis functions), and G is a symmetry index that accounts for the fact that integrals with different indices are often related because of the symmetry of the molecule.

The speedups for the Intel Delta and CM-5 machines are far from linear for all calculations except the largest, benchmark 4. Two factors have a strong influence on the speedup curve. One is the number of tasks issued, and the other is the variation in the sizes of individual tasks. With efficient algorithms for integral evaluation, the sizes of the tasks (i.e., the number of operations required to evaluate a batch of integrals) can fluctuate by several orders of magnitude, and it is seldom practical to estimate these sizes before the tasks are actually performed. Effectively, this prohibits the effective use of schemes based on a static load-balancing strategy, and even in dynamically load-balanced schemes, it is essential to have a very large number of tasks to compensate for this imbalance. On the other hand, the communication overhead associated with an approach that is too fine-grained will eventually become the bottleneck, at which point it will be beneficial to redefine the task size to include several batches of integrals.

If a large discrepancy exists in the relative sizes of the tasks, particularly among the tasks issued late in the calculation, then some processors will be idle while the longer tasks are completed. For runs using 256 nodes on the CM-5, these time delays were 33 seconds for benchmark 2 and less than 1 second for the remaining benchmarks. These results indicate that our code is well load balanced for moderate and large calculations. However, the nonlinear speedups for benchmarks 1 and 3, which have small and relatively simple basis sets, indicate that communication is the bottleneck and that better results would be obtained with fewer but larger tasks.

It is a relatively straightforward procedure to optimize the makeup of the tasks for a calculation, given the architecture and the number of processors on which it will run.

1.3. Heterogeneous Distributed Computing

On a conventional supercomputer the t_2 times are highly insignificant, as expected: the power dependence of any component of t_2 is at most N^3, compared with the N^4 time requirement for t_1.

This is not the case for several of the parallel architectures studied here, especially for modest-sized calculations. The t_2 times are particularly large for the calculations on the CM-5, where the parallel linear algebra libraries, which operate in the data parallel mode, cannot easily be incorporated into a code that uses a message-passing mode, which, unfortunately, is the only reasonable way to evaluate the integrals. While awaiting the development of system software that would enable users to link these two modes in the same program, an alternative approach was investigated. Because the Crays perform superbly on the t_2 part while the CM-5 evaluates and processes integrals very efficiently, we have combined the different machines to utilize the strengths of each.

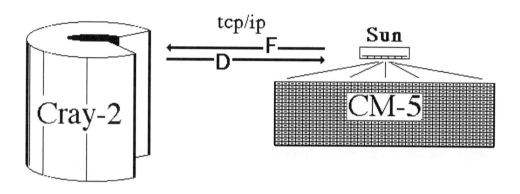

FIG. 1.1. *Schematic representation of the heterogeneous distributed computing approach.*

Figure 1.1 illustrates the components of such an "eclectic" approach. The Sun front end to the CM-5 is connected to the Cray-2 via a tcp/ip network, through which the two machines communicate and transfer data using standard UNIX sockets. The Cray-2, with superior single-processor power, does everything except build the Fock matrix. As an additional benefit, the CM-5 executes a much smaller program, and significant amounts of memory are also saved in data handling.

The approach is simple; only three components need to be passed across the network. The first, the nuclear coordinate and basis set information, is passed only once from the Cray-2 to the Sun at the beginning of the calculation. The second, the density matrix, is passed from the Cray to the Sun at the beginning of each SCF iteration. Finally, after the partial Fock matrices on the nodes have been reduced to one and returned to the Sun, it is shipped to the Cray-2. There it is combined with other matrix elements whose evaluation is totally insignificant on the Cray, and (1.5) is solved. Because the binary representations of numbers differ between the Cray-2 and the Sun, intermachine conversion of data is necessary; efficient system routines are used to perform the conversion on the Cray-2.

Since both the density and the Fock matrices are symmetric, triangular arrays can be used throughout the calculation to save memory. Utilizing full triangularity, this code, running on a CM-5 configuration of 16 megabytes of memory per node, can accommodate calculations with up to 1200 basis functions; each of the triangular matrices requires 5.8 megabytes, and 3 to 4 megabytes should be reserved for the executable and the arrays required by the integral algorithm. If the calculation was performed entirely on the CM-5, the required storage of the other $N \times N$ matrices in the parallel processing approach would reduce the functional limit to approximately 450 basis functions.

This "eclectic" approach was implemented to perform calculations on benchmarks 2 and 4, using 256 nodes on the CM-5. Timings for these calculations are shown in the last column of Table 1.1. As expected, the t_1 times, which are measured on the CM-5, remain the same, whereas the t_2 times become insignificant. It is essentially impossible to obtain reproducible numbers for the communication times without exclusive (i.e., dedicated) access to the Cray-2, the CM-5, and the tcp/ip network. This is not a realistic situation for production calculations, but it is of interest that the communication speed measured in the environment used for these calculations (the Supercomputer Institute/Supercomputer Center/Army High-Performance Computing Research Center at the University of Minnesota) is typically several hundred megabytes/second even during peak hours. This translates into fractions of a second per iteration for calculations with about 1000 basis functions, which typically spend several minutes per iteration in building the Fock matrix. The communication time is therefore a small (negligible) fraction of t_2, because they both scale in approximately the same way with the number of basis functions.

The benchmark calculations presented in Table 1.1 correspond to relatively small systems, where the sizes range from 140 to 314 basis functions. Such calculations are routinely performed on conventional nonparallel computers. Since the ultimate goal of parallel processing technology in quantum chemistry is to facilitate the computational analysis of new classes of molecules, the litmus test that determines the success of new parallel algorithms should be the extent to which they achieve this goal. To this end, we reference recent production calculations on cluster models for lithium intercalated graphite [9] in which calculations incorporating over 100 atoms and 1000 basis functions were performed on the dual architecture CM-5/Cray-2 hardware platform. The largest of these calculations, a 145-atom 1053 basis function calculation on a lithium intercalated bis-circumcoronene complex with D_{6h} point group symmetry, required approximately 370 seconds to complete the t_1 (Fock matrix build) portion of the SCF iteration on the 512-node partition of the CM-5, while the t_2 (linear algebra) portion required approximately 70 seconds on the Cray-2.

The replication of the full (or triangular) density and Fock matrices on each node rapidly depletes the amount of fast access memory available on that node. For example, in the 1053 basis function calculation cited above, nearly 18 megabytes of memory is required for the storage of both matrices. Even when the symmetry of these matrices is exploited, nine megabytes of memory is required for matrix storage in the canonical form. Considering that the nodes in most distributed-memory parallel environments have only 32 megabytes of memory, the storage of the private copies of these matrices inherently limits the size of the calculations to approximately 1400 or 2000 basis functions for the square and canonical matrices, respectively. In the absence of dedicated access to parallel processing hardware, the functional limits on the size of the calculations is much smaller for the simple reason that calculations that request a substantial portion of a precious system resource (such as nodal memory) incur a penalty in the form of lower job priority and reduced job turnaround. For a large-scale parallel SCF algorithm to be practical in a multiuser environment, the bottleneck in nodal memory must be alleviated through additional algorithmic modifications. Such modifications are described below.

1.4. Small-Memory Parallel SCF

As previously outlined in equation (1.8), during the direct computation of the SCF energy, a given integral makes six contributions to the Fock matrix, two of which describe Coulombic interactions and four of which describe exchange interactions. Notice that during this stage of the Fock matrix construction, only small subsets of the full $N \times N$ density and Fock matrices are addressed. Specifically, for the Coulomb contributions, only the elements in the rows corresponding to basis functions p and r are addressed, while for the exchange contributions, only those elements in the rows corresponding to functions p and q are addressed. For all processors, the distribution and storage of only these rows or "strips" of the density matrix, followed by the evaluation of the relevant two-electron integrals and concomitant construction of the corresponding rows of the Fock matrix, drastically reduces nodal memory demands.

If two passes are made through the integral algorithm to separately construct Coulomb and exchange contributions to the Fock matrix, global broadcast operations may be used outside of the two innermost loops in the four-index looping scheme to distribute the rows of the density matrix, as illustrated in Figure 1.2. Furthermore, the separate evaluation of Coulomb and exchange contributions allows for the use of one-center density expansions or fragment multipoles in constructing the Coulomb part, as well as enhanced density-based prescreening of the integrals in the exchange part, both of which can result in an overall decrease in SCF time compared with that of the traditional one-pass approach [11].

The generalized outline for the small-memory parallel strip algorithm in Figure 1.2 is easily adapted to different levels of granularity. For calculations with a large number of atoms and modest basis sets, the best load-balancing scheme is achieved at the level of granularity where the looping described in Figure 1.2 is over atom labels rather than basis functions. Instead of distributing individual rows of the density matrix, multiple rows, or strips, which correspond to the basis functions centered on atoms p and r (for the Coulomb part) or atoms p and q (for the exchange part), are distributed, and the corresponding strips of the Fock matrix are constructed. Aside from the additional pass through the integral algorithm, the total number of computational tasks remains the same as that in the traditional parallel

```
##Coulomb contributions: {ΔF_pq=4 D_rs(pq|rs)
                          {ΔF_rs=4 D_pq(pq|rs)
loop p ≤ N
  loop r ≤ p
                   {d1(*) = D(p,*)
   broadcast rows: {and                 to all processors
                   {d2(*) = D(r,*)
    loop q ≤ r
      loop s ≤ r
        evaluate (pq|rs) and process: } {distribute
        f1(q) = f1(q)+4 d2(s)(pq|rs)    } {across
        f2(s) = f2(s)+4 d1(q)(pq|rs)    } {processors
      end loop s
    end loop q
    return f1 and f2 to master;
    F(r,*) = F(r,*)+f2(*) }
    F(p,*) = F(p,*)+f1(*) } on master
  end loop r
end loop p
                          {ΔF_pr=-D_qs(pq|rs)
                          {ΔF_ps=-D_qr(pq|rs)
##Exchange contributions: {ΔF_qr=-D_ps(pq|rs)
                          {ΔF_qs=-D_pr(pq|rs)
loop p ≤ N
  loop q ≤ p
                   {d1(*) = D(p,*)
   broadcast rows: {and                 to all processors
                   {d2(*) = D(q,*)
    loop r ≤ p
      loop s ≤ r
        evaluate (pq|rs) and process:   }
        f1(r) = f1(r)-d2(s)(pq|rs)       } {distribute
        f1(s) = f1(s)-d2(r)(pq|rs)       } {across
        f2(r) = f2(r)-d1(s)(pq|rs)       } {processors
        f2(s) = f2(s)-d1(r)(pq|rs)       }
      end loop s
    end loop r
    return f1 and f2 to master;
    F(q,*) = F(q,*)+f2(*) }
    F(p,*) = F(p,*)+f1(*) } on master
  end loop q
end loop p
```

FIG. 1.2. *Pseudocode outline of the small-memory parallel strip algorithm for the construction of the Coulomb and exchange contributions to the Fock matrix.*

approach (1.4), while the communication scales as N^3 rather than N^2 but is split over smaller parcels. Interprocessor communication is facilitated by the parallel virtual machine (PVM) message-passing software [8] that utilizes a standard Ethernet network and UNIX UDP socket connectivity.

To illustrate the scalability of the new parallel algorithm, the results from benchmark calculations on two small complexes are reported in Table 1.2. Despite the increased communication overhead compared with the traditional parallel approach, the new algorithm scales nearly as well as the traditional algorithm: the parallel strip calculations with five slave processors ran between 4.4 and 4.0 times faster than the calculations using a single slave processor, whereas the traditional parallel algorithm experienced a 4.5-fold speed increase.

Since the current version of the algorithm exploits only the simplest form of enhanced density-based integral prescreening, we would expect that the timings for the traditional parallel approach, which evaluates the Coulomb and exchange contributions to the Fock matrix in a single integral pass, would be roughly half of that for the new algorithm. The results presented in Table 1.2 confirm this expectation.

TABLE 1.2

Benchmark calculations for one complete SCF iteration for two small molecular complexes in a nondedicated parallel environment of one to five SGI Indigo (150 MHz) workstation computers. One complex is the benzene dimer with a CCDZ basis set (240 basis functions) in the D_{2h} molecular point group symmetry. The other is the same planar diquinone complex as benchmark 3 from Table 1.1 but with a CCDZ basis set minus the polarization functions (254 basis functions) in C_{2v} symmetry. Timings reflect elapsed wallclock time and are reported in seconds.

Complex no. slaves	Traditional parallel algorithm	Parallel strip algorithm	Speedup factor
Benzene dimer			
1 Slave	1148	2377	1.0
2 Slaves	-	1315	1.8
5 Slaves	-	537	4.4
Diquinone			
1 Slave	10405	19096	1.0
2 Slaves	-	9768	2.0
5 Slaves	2321	4726	4.0

To illustrate the applicability of the new parallel algorithm to new classes of chemical compounds, the results of calculations for a relatively low-symmetry graphite intercalation compound in two different basis sets are reported in Table 1.3. These results highlight the difference in the memory requirements for the traditional parallel versus the new small-memory algorithms. Nearly 53 megabytes of memory are required for the storage of the square density and Fock matrices alone in the calculation with 1815 basis functions. In contrast, only 1.3 megabytes of memory are required for the storage of the strips of the density and Fock matrices residing on a node at any given time. The new algorithm's total nodal memory requirement, which includes space for integral batches, data, and all scratch space, is only 2.4 megabytes for the 1815 basis function calculation.

A separate calculation was performed that was identical to the run with 1053 basis functions but that used another SGI workstation as the master process, rather than the Cray C90. The serial portion of the code required an additional 40 minutes to execute over that used by the Cray C90, emphasizing the utility of the heterogeneous distributed approach to parallel processing.

TABLE 1.3

Time and memory requirements for parallel strip calculations on a lithium-intercalated bis-circumcoronene complex (145 atoms) with a 3-21G basis set (1053 basis functions) and a 3-21G* basis set (1815 basis functions). The calculations were performed in C_{2v} symmetry. The parallel environment consisted of 16 SGI Challenge (150 MHz) slave processors attached to a Cray C90 master process. Memory usage is listed in megabytes; times are listed in seconds.

Number of basis functions	Memory used by full D and F matrices	Memory used by D and F strips	Total memory used by strip code	Wallclock time for one SCF iteration
1053	17.8	0.8	1.8	650
1815	52.7	1.3	2.4	3550

1.5. Conclusions

We have shown that it is not only possible, but also practical, to use massively parallel hardware for the very complex computational tasks that occur in electronic structure calculations. Very encouraging speedups and scalability are achieved on different architectures. However, small fractions of the code remain serial, which, given the feeble front-end capacity available on, for example, the CM-5, constitutes a severe bottleneck. The most promising results are therefore obtained with a dual, heterogeneous architecture, where these small, serial parts of the calculation are performed on a conventional supercomputer.

In addition, a large-scale, small-memory algorithm for the parallel computation of SCF energies has been presented. The defining characteristic of this method is that only a small subset of the full $N \times N$ density and Fock matrices is addressed by a given task on a node. The amount of memory allocated by the nodes for the processing of a batch of integrals is therefore drastically reduced. Considering that the nodal memory in most parallel environments is small, the parallel strip algorithm opens up a whole new class of complexes to computational analysis using inexpensive hardware.

1.6. Acknowledgments

This work was supported in part by a contract between the Army Research Office for the Army High-Performance Computing Research Center at the University of Minnesota. The authors would like to thank the Concurrent

Supercomputing Consortium and the U.S. Department of Energy for access to the Intel Delta. Thanks are also due to Rick Kendall, Jeff Nichols, and Steve Fetherston for their help with the benchmarks, and Joel Parriott for his assistance with the socket code.

References

[1] J. ALMLÖF, K. FAEGRI, AND K. KORSELL, *Principles for a direct SCF approach to LCAO-MO ab initio calculations*, J. Comput. Chem., 3(1982), pp. 385–399.

[2] J. ALMLÖF AND P. TAYLOR, *Computational aspects of direct SCF and MC SCF methods*, in Advanced Theories and Computational Approaches to the Electronic Structure of Molecules, C. Dykstra, ed., NATO ASI Ser., Ser. C, 133, Reidel, Dordrecht, 1984, pp. 107–125.

[3] J. ALMLÖF, K. FAEGRI, M. FEYEREISEN, AND K. KORSELL, *DISCO: A direct Hartree-Fock code*. Newer versions of the code are better known as SUPERMOLECULE; M. Feyereisen and J. Almlöf, University of Minnesota, 1993.

[4] J. ALMLÖF, *Direct methods in electronic structure theory*, in Modern Electronic Structure Theory, D. Yarkony and C-Y. Ng, eds., Singapore Publishing, 1994, pp. 121–180.

[5] ———, *Methods of modern Hartree Fock theory*, in Lecture Notes in Chemistry, B. Roos, ed., Springer-Verlag, Berlin, New York, 1994, pp. 1–90.

[6] E. CLEMENTI, *Introduction to MOTECC*, in Modern Techniques in Computational Chemistry, E. Clementi, ed., Escom Science Publishers, 1990, pp. 1–54.

[7] M. FEYEREISEN AND R.A. KENDALL, *An efficient implementation of the direct-SCF algorithm on parallel computer architectures*, Theoret. Chim. Acta., 84(1993), pp. 289–298.

[8] A. GEIST, A. BEGUELIN, J. DONGARRA, W. JIANG, R. MANCHEK, AND V. SUNDERAM, *PVM 3 User's Guide and Reference Manual*, Oak Ridge National Laboratory, Oak Ridge, TN, the University of Tennessee, Knoxville, TN, and Emory University, Atlanta, GA, 1993.

[9] D. HANKINSON AND J. ALMLÖF, *Cluster models for lithium intercalated graphite: Electronic structures and energetics*, J. Mol. Struc. (THEOCEM), to appear.

[10] H.-P. LÜTHI, J.E. MERTZ, M. FEYEREISEN, AND J. ALMLÖF, *A coarse-grain parallel implementation of the direct SCF method*, J. Comput. Chem., 13(1992), pp. 160–164. See also Cray Research, Inc., 1990 *Gigaflop Performance Awards*, p. 25.

[11] I. PANAS, J. ALMLÖF, AND M.W. FEYEREISEN, *Ab initio methods for large systems*, Internat. J. Quant. Chem., 40(1991), pp. 797–807.

[12] L.G.M. PETTERSON AND T. FAXEN, *Massively parallel direct SCF calculations on large metal clusters*, Theoret. Chim. Acta., 85(1993), pp. 345–352.

[13] C.C.J. ROOTHAAN, *New developments in molecular orbital theory*, Rev. Modern Phys., 23(1951), pp. 69–89.

Chapter 2

Massively Parallel Lattice QCD Calculations

Clive F. Baillie

Editorial preface

Quantum chromodynamics (QCD) is, at realistic and meaningful problem sizes, a daunting computational task. This article describes the application of hybrid Monte Carlo to lattice QCD calculations. Implementations on both the CM-2 and the CM-5 are described and specific programming features of the CM-2 and CM-5 that were required to achieve over 6 Gflop and 40 Gflops, respectively, are discussed. This clearly illustrates the algorithmic and coding efforts that are a necessary part of wringing the best performance from large-scale parallel computers.

This article originally appeared in *SIAM News*, Vol. 24, No. 3, May 1991. It was updated during the summer/fall of 1995.

Quantum chromodynamics (QCD) is the best available theory of the strong nuclear force. To explain to nonspecialists what QCD calculations are, we describe exactly what state-of-the-art QCD calculations involve and thereby show why they consume so much supercomputer and parallel computer time. In fact, these calculations are becoming too big for traditional supercomputers like the Cray Y-MP so one has to use either commercial parallel machines or home-grown special-purpose machines. Here, we focus on the former and discuss in detail implementations on the Thinking Machine Corporation (TMC) Connection Machines 2 and 5 (CM-2 and CM-5). More information on the latter approach can be found in [1, 2].

There are four forces found in nature: strong, weak, electromagnetic, and gravitational. In the everyday world we all experience the gravitational force, and we see and hear pictures and sounds brought to us via electromagnetic radiation. On the other hand, the strong and weak forces operate only within the nucleus of the atom and are therefore less familiar. The weak force is responsible for things like beta-decay, which is part of the process of fusion going on inside the sun. The strong force keeps matter together; it is what binds the most elementary particles of matter—*quarks*—into the basic constituents of the world, including protons, neutrons, pions, etc.

Quarks come in three different varieties, denoted "red," "green," and "blue" for no particular reason other than they had to be called something (these quark "colors" have nothing to do with color in the "real" world, i.e., the world outside the nucleus). All particles have antiparticles and so antiquarks can be "antired," "antigreen," or "antiblue." Thus the theory describing the interaction of these quarks and antiquarks is called quantum chromodynamics (the quantum dynamics of "color"). It predicts that quarks bind together in only two possible configurations. In the first, the "qqq" configuration, three quarks, one of each color, bind together to form particles like protons and neutrons (or, equivalently, three antiquarks comprise antiprotons and antineutrons). In the second, the "$q\bar{q}$" configuration, a quark of one color and an antiquark of the corresponding anticolor join to make particles like pions. In other words, the strongly interacting particles found in nature are colorless composites of quarks.

One of the most important predictions of QCD is that quarks cannot exist outside nuclear matter; they are *confined* within the proton or pion. This prediction has not been made analytically; it has instead been demonstrated numerically using a computer. This is because the QCD theory can only be solved analytically in the high-energy limit; at the lower energies experienced by quarks inside protons the theory cannot be solved in this way and so we turn to computer simulations in order to obtain a numerical solution.

2.1. QCD

In order to explain the computer simulations of QCD, we shall have to look at the theory in a bit more detail. QCD is a four-dimensional (three space and one time) quantum field theory containing two basic fields and a symmetry: the quark field representing the "colored" quarks and the so-called *gluon* field representing fields which mediate the color force between quarks. Unlike the analogous field that mediates electromagnetic radiation—the photon—gluons also experience the force they mediate; i.e., they have color. This makes the theory very nonlinear and is the basic reason why it cannot be solved analytically (except at high energy where the nonlinearity is small enough to be considered as a perturbation). The symmetry in the theory is a *local gauge symmetry*, and since the gluon fields obey this symmetry they are called gauge fields. Gauge symmetry means that one can change the phases of all the fields leaving the theory unchanged; local refers to the fact that these phase changes can be different at each point in space. Gauge symmetries are very important in field theories: by insisting that the basic fields obey the gauge symmetry one can derive the dynamical equations governing the interaction of these fields.

As the quarks have three colors, the gauge fields mediating the force between them are 3×3 matrices. In fact they are 3×3 complex $SU(3)$ matrices; SU means that the matrices are unitary with unit determinant (i.e., *special unitary*). These matrices describe how the colors of two quarks change when they interact; for example, a red quark may change into a green quark emitting a red antigreen gluon, which on encountering a green quark will turn it red. We could at this point write down the *continuum* equations for QCD, but since in order to simulate the theory on a computer we will have to discretize these equations anyway we shall skip this unnecessary detail.

Field theories can be formulated in various ways; here we use Feynman's path integral formulation as it lends itself most easily to discretization. Any theory involving fields ϕ interacting according to an action (i.e., energy) S has a path integral representation

$$(2.1) \qquad Z = \int D\phi e^{-S(\phi)}.$$

This function, known in statistical mechanics as the partition function, describes the field theory completely. It represents the sum of all possible states and from it one can calculate any physical quantity (i.e., *observable*) $\mathcal{O}$ as the following *expectation value*:

$$(2.2) \qquad \langle \mathcal{O}(\phi) \rangle = \frac{1}{Z} \int D\phi \mathcal{O}(\phi) e^{-S(\phi)}.$$

In words, this equation says that the total observable is the sum of all the partial observables from each state of the system. For example, a typical observable would be a product of fields $\mathcal{O} = \phi(x)\phi(y)$, which says how the fluctuations in the field are correlated, and in turn tells us something about the particles that can propagate from point x to point y. The appropriate correlation functions give us, for example, the masses of the various particles in the theory. Thus to evaluate almost any quantity in field theories like QCD one must simply evaluate the corresponding path integral. Unfortunately, in the continuum, the integrals are over a space of infinite dimension.

2.2. Lattice QCD

Hence we study QCD numerically by discretizing space and time into a lattice of points. Then the functional integral is simply defined as the product of the integrals over the fields at every site of the lattice $\phi(n)$:

$$(2.3) \qquad \int D\phi \rightarrow \int \prod_n [d\phi(n)].$$

Restricting space and time to a finite box, we end up with a finite (but large) number of ordinary integrals. This is "lattice QCD," something we might imagine simulating directly on a computer. However, the high dimensionality of these integrals renders conventional mesh techniques impractical. For example, consider a 10^4 site lattice (which nowadays would be considered small for QCD). If we take the simplest possible field theory in which the fields only have two states (the Ising spin model), the partition function becomes an ordinary sum. But this sum has an enormous number of terms: 2^{10000}. Even if we could add one term in the time it takes light to pass by a proton and continue adding for the age of the universe, we would not make perceptible progress in evaluating it! Fortunately, the presence of the exponential e^{-S} means that the integrand is sharply peaked in one region of configuration space. Hence we resort to a statistical treatment and use Monte Carlo-type algorithms to sample the important parts of the integration region.

Monte Carlo algorithms typically begin with some initial configuration of fields and then make pseudorandom changes on the fields such that the ultimate probability P of generating a particular field configuration ϕ is proportional to the Boltzmann factor

$$(2.4) \qquad\qquad P(\phi) = e^{-S(\phi)},$$

where $S(\phi)$ is the action associated with the given configuration. There are several ways to implement such a scheme, but for many theories the simple Metropolis algorithm is effective. In this algorithm a new configuration ϕ' is generated by updating a single variable in the old configuration ϕ and calculating the change in energy (action)

$$(2.5) \qquad\qquad \Delta S = S(\phi') - S(\phi).$$

If $\Delta S \leq 0$, the change is accepted. If $\Delta S > 0$ the change is accepted with probability $\exp(-\Delta S)$. In practice this is done by generating a pseudorandom number r in the interval $[0, 1]$ with uniform probability distribution and accepting the change if $r < \exp(-\Delta S)$. This is guaranteed to generate the correct (Boltzmann) distribution of configurations, provided "detailed balance" is satisfied. This condition means that the probability of proposing the change $\phi \to \phi'$ is the same as that of proposing the reverse process $\phi' \to \phi$. In practice this is true if we never simultaneously update two fields which interact directly via the action.

Whichever method one chooses to generate field configurations, one updates the fields for some equilibration time of E steps and then calculates the expectation value of $\mathcal{O}$ in (2.2) from the next T configurations as

$$(2.6) \qquad\qquad \langle \mathcal{O} \rangle = \frac{1}{T} \sum_{i=E+1}^{E+T} \mathcal{O}(\phi_i).$$

The statistical error in Monte Carlo behaves as $1/\sqrt{N}$, where N is the number of effectively independent configurations. $N = T/2\tau$, where τ is the autocorrelation time. This autocorrelation time can easily be large, and most of the computer time is then spent in generating statistically independent configurations.

Now we can write down the basic discretized equations for lattice QCD. On the computer the four-dimensional space–time continuum is replaced by a four-dimensional hypercubic periodic lattice of size $N = N_s \times N_s \times N_s \times N_t$ with lattice spacing a. The quarks sit on the sites and the gluons live on the links

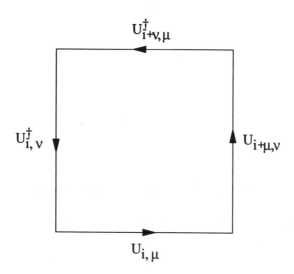

FIG. 2.1. *Illustration of plaquette calculation.*

of the lattice. $N_s a$ and $N_t a$ are the spatial and the temporal extents of the lattice, respectively. The action for the purely gluonic part of QCD is

$$(2.7) \qquad S_G = \beta \sum_P \left(1 - \frac{1}{3} ReTrU_P \right),$$

where

$$(2.8) \qquad U_P = U_{i,\mu} U_{i+\mu,\nu} U_{i+\nu,\mu}^{\dagger} U_{i,\nu}^{\dagger}$$

is the product of link matrices around an elementary square or plaquette on the lattice; see Figure 2.1. $\beta = 6/g^2$, where g is the coupling between the gauge fields. The action for the quarks is

$$(2.9) \qquad S_Q = \bar{\psi} M \psi,$$

where M is a large sparse matrix the size of the lattice squared. In particular, we use the Wilson quark representation, in which M is given by

$$(2.10) \quad M_{ij} = \delta_{ij} + \kappa \sum_{\mu} [\, (\gamma_{\mu} - r) U_{i,\mu} \delta_{i,j-\mu} - (\gamma_{\mu} + r) U_{i-\mu,\mu}^{\dagger} \delta_{i,j+\mu} \,],$$

where $\kappa = 1/2m$, with m being the quark mass, γ_{μ} are the Dirac gamma matrices, and one can choose $r = 1$. For state-of-the-art simulations the lattice size $N = 32^4$, and with ψ having four spin and three color complex components, M is a $(24 \times 1048576) \times (24 \times 1048576)$ matrix. Unfortunately, quarks are fermionic particles, which means that the quark fields ψ are anticommuting Grassmann numbers. This means that for two quark fields ψ_i and ψ_j, the anticommutator $\{\psi_i, \psi_j\} \equiv \psi_i \psi_j + \psi_j \psi_i = 0$. Therefore the order in which the quark fields are combined is important and as there is no simple representation of this on the computer these quark fields cannot be dealt with numerically. Fortunately, there is an analytical trick whereby the quark action can be rewritten in terms of *pseudofermion* fields ϕ, which are normal numbers,

$$(2.11) \qquad S_Q = \phi^\dagger (M^\dagger M)^{-1} \phi.$$

Thus the path integral we want to evaluate numerically is

$$(2.12) \qquad Z = \int DU D\phi D\phi^\dagger \exp(-S_G - S_Q).$$

Note that the lattice is a mathematical construct used to solve the theory— at the end of the day, the lattice spacing a must be taken to zero to get back to the continuum limit. The lattice spacing itself does not show up explicitly in the partition function Z above. Instead the lattice spacing is an implicit function of the parameter $\beta = 6/g^2$, which plays the role of an inverse temperature. To take the continuum limit $a \to 0$ of lattice QCD, one tunes $g \to 0$ or $\beta \to \infty$. Typical values used in simulations range from $\beta = 5.4$ to $\beta = 6.0$. This corresponds to $a \approx .1$ Fermi $= 10^{-16}$ meter. Thus at typical values of β a lattice with $N_s = 32$ will correspond to a physical box about 3 Fermi on an edge, which is thought to be large enough to hold one proton without crushing it too much in the finite volume. However, the spacing $a = .1$ Fermi is not fine enough to be close to the continuum limit. One can estimate that we still need to shrink the lattice spacing by something like a factor of 4, leading to an increase of a factor 4^4 in the number of points in the lattice in order to keep the box the same physical volume.

The biggest stumbling block preventing a large increase in the number of lattice points is the presence of the matrix inverse $(M^\dagger M)^{-1}$ in the partition function. There have been many proposals for dealing with this problem. The first algorithms tried to compute the change in the inverse when a single gauge link matrix was updated. This obviously scales as the square of the volume of the lattice and is therefore prohibitively expensive. Today, the preferred approach is the so-called *hybrid Monte Carlo* algorithm. The basic idea is to invent some dynamics for the variables in the system in order to evolve the whole system forward in (simulation) time and then do a Metropolis accept/reject for the entire trajectory on the basis of the total energy change. In this case the dynamics is provided by molecular dynamics (MD) evolution of the fields. The first great advantage is that the whole system is updated at one fell swoop, and the algorithm scales with the volume of the lattice. (In practice, it is actually a little worse, $N^{5/4}$.) The Hamiltonian for the MD evolution is

$$(2.13) \qquad H = \frac{1}{2} Tr \sum P_{i,\mu}^2 + \frac{\beta}{3} Tr \sum (1 - U_P) + \phi^\dagger (M^\dagger M)^{-1} \phi,$$

where $P_{i,\mu}$ are momenta conjugate to the gauge link matrices $U_{i,\mu}$. The $P_{i,\mu}$ come from the "invented dynamics" and are also included in the partition function; therefore, (2.12) becomes

$$(2.14) \qquad\qquad Z = \int DU D\phi D\phi^\dagger DP \exp(-H).$$

The P and U fields are leapfrogged through MD time with some finite step size, which introduces numerical errors. However, a global Metropolis accept/reject step at the end of the trajectory corrects for this and makes the algorithm *exact*; this is the second great advantage of the hybrid Monte Carlo algorithm. At each step of the MD evolution, $\phi(M^\dagger M)^{-1}\phi$ has to be recalculated—this is the most time-consuming part of the QCD calculation. In practice, we rewrite $\chi = (M^\dagger M)^{-1}\phi$ as $(M^\dagger M)\chi = \phi$, a linear system of equations, and use conjugate gradient or one of its cousins (since M is sparse) to solve for χ.

Before turning to actual implementations of such a QCD calculation on the TMC Connection Machines, we want to comment upon the amount of computer time required for a realistic simulation. In order to get close to the continuum limit one can estimate that a lattice of size about 128^4 will be required. On computers with a performance of a few Gflops, it is possible to do simulations on 16^4 lattices in about a year, i.e., a total of roughly, say, 3 Gflops-years. In going to 128^4 there is a factor of 4000 increase in lattice volume so the hybrid Monte Carlo algorithm will require a factor of $4000^{5/4} \approx 32000$ more computer time. Therefore we estimate that a realistic simulation of QCD on a 128^4 lattice will take about 100 Tflops-years (Tflops = teraflops = 1000 Gflops). This seems like rather a lot of computer time but bearing in mind that supercomputer and parallel computer power has increased at an exponential rate of a factor of 10 every two years since the early 1980s [2], we should have 100 Tflops around the end of the century!

2.3. Implementation on the CM-2

The hybrid Monte Carlo QCD code was originally implemented on the Cray Y-MP. In 1989 it was re-engineered for the CM-2 and ran on this machine for several years until 1993, when the code was ported to the new CM-5 where it has been running to this day. We describe the original development for the CM-2 in detail to give a flavor of what is involved in using state-of-the-art massively parallel processors. For completeness we also include a brief summary of the similar work done to port the code to the CM-5.

The TMC CM-2 is often described as a distributed-memory, single-instruction multiple-data (SIMD) massively parallel processor comprising up to 65,536 (64K) processors. However, these processors are simple bit-serial processors, which are not used in doing floating point calculations. Instead, the floating-point units (FPUs) are used for applications such as QCD, and so it is more useful to think of the CM-2 as an 11-dimensional hypercube of 2048 FPUs (known as "sprint nodes"). The original assembly language of the machine, Paris, embodies a computational model, which is essentially independent of the specific nature of the hardware, but its implementation on the CM-2 leans very much toward the view of the machine as consisting of the bit-serial processors (known as the "fieldwise" view of the machine) instead of consisting of the FPUs (known as the "slicewise" view of the machine). Since all the originally provided high-level languages—*LISP, C*, CM Fortran—compile into Paris, any code written in these languages is subject to the implementation inefficiencies inherent in Paris. The main inefficiency is due to the fact that since groups of 32 processors share a 32-bit Weitek floating-point chip, there is a transposer chip that changes 32 bits stored bit-serially within 32 processors into 32 32-bit words for the Weitek, and vice versa. This means that all Paris operations using the FPU must be done in chunks of 32 words. Because the internal architecture of the Weitek chips used in the CM-2 includes only 32 registers, there is no room for storing intermediate results in the Weitek. Therefore, operations such as a complex multiply, which could employ internal registers to store the intermediate results, are forced into storing and reloading these in memory. Moreover, an overhead of 32 cycles is added to every load and store, since that is the time required to fill the transposer. The net effect on performance is that Paris peaks at around two Gflops on the whole machine for any code sequence that compiles into a series of simple multiplications and addition/subtractions.

The solution to these problems is found in going to the "slicewise" model of the machine. In this model, one uses the transposers to convert data into the form required by the FPU once and for all and thereafter bypasses the transposers when loading to or storing from the FPU (using the bypass register). After conversion, a memory address references all 32 bits of a floating point word corresponding to a one-bit serial processor, with words belonging to consecutive processors arranged consecutively in memory. One is now free to program the FPU in the most efficient manner. For example, the complex matrix multiply—which is the basis for all QCD simulations—is done for each processor in turn, thereby making full use of the internal Weitek registers and the memory bandwidth. The price one pays for this freedom is that one must write virtually all the slicewise code one will need. TMC provides an assembly language-level programming system called CMIS (Connection Machine Instruction Set), which allows one to construct completely pipelined code for the FPU with a minimum of fuss.

When the hybrid Monte Carlo QCD code was initially re-engineered for the CM-2, only Paris was available, and the most efficient language (in terms of how well the compiler compiled it into Paris) was *LISP. Therefore, the initial implementation was approximately 6000 lines of *LISP [3]. The *LISP variables defined include four 3×3 complex $SU(3)$ matrices at each site to represent the gluons (four because the lattice is four dimensional). The quark variables require one 3×4 complex matrix per site. During the calculation several other 3×3, 3×4, and 3×2 complex matrices are required to hold intermediate data such as the MD momenta $P_{i,\mu}$. The net result is that on a CM-2 with 256 Kbit memory chips, we are limited to a virtual processor (VP) ratio of four. Nevertheless, this ratio is high enough to give good efficiency: 97–99% CM utilization for most of the functions in the code. The basic, most time-consuming operation in the code is the matrix–vector product $M\chi$ or $M^{\dagger}(M\chi)$, which is performed by a routine called DSLASH. By inspection of (2.10) we see that M connects only nearest-neighbor lattice sites. Therefore, communication of quark matrices is done with the "news" system on the CM-2 (rather than the more general, but slower, "router" mechanism). With the Wilson parameter $r = 1$ we can use the characteristics of the Dirac gamma matrices to project the 3×4 quark matrices down to 3×2, do the communications and calculations required, and expand back up to 3×4. This reduces the communication time by a factor of 2 and the calculation time by 40%. Hence, apart from news, the other main routines that DSLASH uses are M32, which does complex matrix multiplies of 3×3 times 3×2 matrices; PROJECT to project 3×4 matrices to 3×2; and EXPAND to expand 3×2 matrices to 3×4. The original code was benchmarked on a 64K CM-2 with Sun 4/260 front end at VP ratio of 4 to obtain the performances shown in column 2 of Table 2.1. We see that the matrix multiply routine M32 runs at close to the peak *LISP rate of two Gflops, but when we add in communication as well as projects and expands we get a *sustained* performance of about one Gflop (as most of the time is spent in DSLASH).

TABLE 2.1

Performances in Gflops for various routines in the three versions of the code.

Function	*LISP	CMIS	Multiwire
M32	1.9	10.2	10.2
PROJECT	1.7	1.7	3.1
EXPAND	1.2	1.2	4.4
DSLASH	0.9	1.6	6.0

As soon as CMIS became available the low-level functions were rewritten to take advantage of it and a dramatic increase in the performance of M32 was obtained: 10.2 Gflops [4]. However the sustained rate only increases to 1.6 Gflops (third column of Table 2.1). To determine why this is so, the various components of DSLASH were timed to find out where most of the time was being spent (columns 2 and 3 of Table 2.2). We see immediately that the reason for the poor overall speedup is that despite the calculation time being more than halved, the communication time remains the same (there is actually a slight decrease in going to CMIS that results from a more favorable arrangement of the slicewise data in memory). Furthermore, for the CMIS version the time spent in communication is greater than that spent in calculation. Therefore, the last thing to do in order to get the maximum performance out of the CM-2 for QCD is to speed up the communication code. In turns out that this is possible to do using the "multiwire-news" software. (For a detailed discussion of this see [5].)

TABLE 2.2

Time in seconds spent in communication and calculation.

Component	*LISP	CMIS	Multiwire
Communication	4.5	3.9	0.4
Calculation	8.7	3.3	1.6
Total	13.2	7.2	2.0

We recall that the QCD code employs a four-dimensional lattice of points and the matrix inversion algorithm involves only nearest-neighbor communication. This communication consists of sends forward then backward for each of the four dimensions. Restructuring the code allows the communication to proceed in both directions in all four directions *concurrently*. This multiwire-news version of the code does the communication part of DSLASH 9.3 times faster than the CMIS version (column 4 of Table 2.2). There is one other optimization we can do with the multiwire-news part of the code and that is to merge it with the expands and projects so that there are no intermediate stores to memory. This results in a factor of 2 speedup in the calculation part over the CMIS version (column 4 of Table 2.2). Hence overall the final multiwire-news version of the QCD code achieves a sustained rate of six Gflops (column 4 of Table 2.1).

This figure of six Gflops is rather interesting because it is roughly the actual performance (rather than peak performance which is 16 Gflops) of the fastest special purpose 256-node parallel computer built especially for QCD simulations by Norman Christ at Columbia University. Therefore, in the 1990s it is possible to achieve as good a performance for QCD on commercial parallel computers as on specially built machines, which was not the case in the last decade [1]. Of course, in order to do that one has to devote many person-months to low-level programming. In both approaches the performances obtained are about 30 times that from one processor of a traditional Cray Y-MP supercomputer.

2.4. Implementation on the CM-5

TMC's follow-on machine to the CM-2 was called the CM-5 and became available in 1992. It is a multiple-instruction multiple-data (MIMD) machine but is usually programmed in a SIMD fashion using TMC's parallel Fortran called CM Fortran. Each node in the CM-5 consists of a Sun SPARC processor plus four custom designed floating point accelerator chips, giving a peak speed of 128 Mflops. The nodes are interconnected by a "fat-tree" network, which is the major bottleneck in the CM-5, yielding a realistic communication bandwidth of only 10 Mbyte/s (other massively parallel processors are typically 10 times faster).

The hybrid Monte Carlo QCD code was rewritten as 50,000 lines of CM Fortran for the CM-5 in 1993. Then the computationally intensive routines were converted to an assembly language called CDPEAC for optimal performance (this is analogous to the use of CMIS for the CM-2). Initially the CDPEAC code was more than four times faster than the CM Fortran code, but with subsequent improvements in the Fortran compiler this was reduced to a factor of 2. The CDPEAC computational intensive routines (essentially M32) achieve up to 100 Mflops/node, but with communications the overall sustained performance of the code is around 40 Mflops/node. As the code is typically run on 1024 nodes this is a total of over 40 Gflops. Due to memory constraints the CM-2 could only run 16^4 lattices; the CM-5 can do 32^4.

2.5. Conclusions

We have implemented a QCD code on the TMC CM-2 and CM-5 that uses a state-of-the-art algorithm—the hybrid Monte Carlo algorithm—on some of the largest lattices simulated so far: 16^4 and 32^4. This code, written in *LISP and CMIS for the CM-2, and rewritten in CM Fortran and CDPEAC for the CM-5, has been running on several Connection Machines in production mode since early 1989. Most of the results from the CM-2, using 16^4 lattices at β values of 5.4, 5.5, and 5.6 with various values of the quark mass, appear in [6]. Some results from the CM-5, using 32^4 lattices at $\beta = 6.0$, are reported in [7]. The performance on the CM-2 was initially about one Gflop, then later six Gflops, and on the CM-5 over 40 Gflops.

2.6. Acknowledgments

Most of this work was done in collaboration with Rajan Gupta and Ralph Brickner of Los Alamos National Laboratory while the author was at Caltech.

References

[1] C.F. BAILLIE, *Lattice QCD: Commercial vs. home-grown parallel computers*, in Proc. Fifth Distributed Memory Computing Conference (DMCC5), D.W. Walker and Q.F. Stout, eds., IEEE Computer Society Press, Los Alamitos, CA, 1990, p. 397.

[2] C.F. BAILLIE, D.A. JOHNSTON, AND G.W. KILCUP, *Status and prospects of the computational approach to high-energy physics*, J. Supercomputing, 4(1990), p. 277.

[3] C.F. BAILLIE, R.G. BRICKNER, R. GUPTA, AND L. JOHNSSON, *QCD with dynamical fermions on the Connection Machine*, in Proc. Supercomputing '89, ACM Press, New York, 1989, p. 2.

[4] R.G. BRICKNER, C.F. BAILLIE, AND S.L. JOHNSSON, *QCD on the Connection*

Machine: Beyond LISP, Comput. Phys. Comm., 65(1991), p. 39.

[5] R.G. BRICKNER, *CMIS arithmetic and multiwire news for QCD on the CM*, Nucl. Phys. B Proc. Suppl., 20(1991), p. 145.

[6] R. GUPTA, C.F. BAILLIE, R.G. BRICKNER, G.W. KILCUP, A. PATEL, AND S.R. SHARPE, *QCD with dynamical wilson fermions* II, Phys. Rev. D, 44(1991), p. 3272.

[7] T. BHATTACHARYA AND R. GUPTA, *A potpourri of results in QCD from large lattice simulations on the* CM5, Nucl. Phys. B Proc. Suppl., 34(1994), p. 341.

Chapter 3

Parallel Weiner Integral Methods for Elliptic BVPs: A Tale of Two Architectures

Michael Mascagni

Editorial preface

The use of probabilistic methods as a solution technique for elliptic boundary values problems (BVPs) is not new, but the availability of parallel computers offers a new perspective on this approach. This chapter illustrates the large difference in the structure of algorithms for the MIMD and SIMD architectures and how probabilistic methods are applied to random walk algorithms for each machine class. The conceptual view is that of a "forward" walk on a MIMD architecture and a "backward" walk on a SIMD architecture.

This article originally appeared in *SIAM News*, Vol. 23, No. 4, July 1990. It was updated during the summer/fall of 1995.

One of the most intriguing aspects of linear elliptic boundary value problems (BVPs) is their relationship to probability. The discovery of this relationship dates back to the beginnings of rigorous measure theory, or more specifically to the time when measure theorists began considering how to place measures on spaces of continuous functions. The measures placed on these infinite-dimensional spaces are first defined for simple sets of continuous functions with the help of the fundamental solution of certain linear parabolic partial differential equations (PDEs). Once these simple "cylinder sets" of continuous functions can be measured, it is rather easy to extend the measure to the entire space of continuous functions with standard techniques from measure theory. As a consequence of this construction, certain integrals with respect to these measures (which can be thought of as mathematical expectations with the measure thought of as a probability density) are solutions to particular linear parabolic and elliptic problems.

A curiosity of looking at a space of continuous functions with this measure is that almost all (full measure) of the continuous functions are nowhere differentiable.[1] Because of this fact, continuous functions with this measure (called Wiener measure) are almost all extremely jagged and monstrously wiggly. Thus it is straightforward to associate these spaces of continuous functions under Wiener measure with spaces of continuous Brownian motion paths. This makes sense when one recalls that diffusion can be thought of as the macroscopic manifestation of microscopic Brownian motion, and that parabolic PDEs (like the diffusion equation) play a fundamental role in the construction of these Wiener measures. It turns out that this probabilistic theory for representing the solutions of linear elliptic and parabolic PDEs has many applications in analysis [2, 4] and, as we will see below, in numerical computation [1, 3].

As a simple example of the application of these ideas to computation let us consider the Dirichlet BVP for the Laplace equation:

$$(3.1) \qquad -\Delta u(x) = 0, \quad x \in \Omega, \quad u(x) = g(x), \quad x \in \partial\Omega.$$

The probabilistic representation of (3.1), often called the Wiener integral representation, is denoted by

$$(3.2) \qquad u(x) = E_x \left[g(\beta(\tau_{\partial\Omega})) \right].$$

The interpretation of (3.2) is that the solution of (3.1) at an interior point x is the expectation of the boundary value at the first hitting location of the sample path $\beta(\cdot)$ started from x. The Markov time $\tau_{\partial\Omega}$ is the time at which a sample path first encounters the boundary and is called the mean first passage time. This quantity is defined for the sample path $\beta(\cdot)$ by $\tau_{\partial\Omega} = \inf_{\beta(t)\in\Omega} t$.

An alternate interpretation of (3.2) is as a probabilistic version of the traditional Green function representation of the solution to (3.1). This is because (3.2) is an integral of the boundary values against a boundary mass $p(x, y)$, the probability of a sample path starting at x and first encountering the boundary at y. It is an elementary fact that $p(x, y)$ so defined is the Green function of (2.1) in [1]. If we think in terms of Brownian motion, which is intimately related to the Laplacian, then (3.2) states that the solution to (3.1) is the expected value of the first hitting boundary value of a Brownian motion started at x.

It is rather easy to see that the Wiener integral in (3.2) solves the Dirichlet problem for the Laplace equation. A function is a solution to (3.1) if (i) it has the mean value property and (ii) it has the correct boundary values. In two dimensions a function has the mean value property if its value at the center of a circle is the average of the function on the circle. Pick a point x in the interior of Ω and, using x as the center, draw a circle (call it C) totally within Ω. Equation (3.2) states that $u(x)$ is the expected boundary value at the point of first passage. By continuity, any path that started at x will encounter C

---

[1]Recall the fuss created by Weierstrass's construction of a single continuous, nowhere differential function by Fourier series.

before hitting the boundary. Thus $u(x)$ is the expected first passage boundary value of paths started on C conditional on where the path that started at x first hit C. Since Brownian motion is isotropic, a Brownian path started at x will first encounter any point on C with equal probability. Thus $u(x)$ is the average of the first passage boundary values from walks started on C (which are the values of $u(x)$ on C by (3.2)) and so $u(x)$ has the mean value property. In addition, if the boundary is smooth, we can see that $u(x)$ takes on the appropriate boundary values by letting x approach the boundary while using the above argument.

This functional integration approach can be used to solve discretizations of these continuous problems by utilizing random walks in place of Brownian motion. Through this formalism, and extensions of the probability to different elliptic PDEs and different BVPs, a large class of Monte Carlo methods for these problems emerge. Implementation of these random walk-based Monte Carlo methods on multiple-instruction multiple-data (MIMD) and single-instruction multiple-data (SIMD) machines will be considered in the subsequent section. It will be shown how different implementations lead to different algorithms, which in turn lead to different practical and analytic considerations.

3.1. Architecture and Implementation

Given that we wish to implement algorithms related to these probabilistic ideas on a parallel computer, it is incumbent on us to consider what aspects of these algorithms we wish to exploit in a parallel implementation. In random walk-based algorithms there are two natural ways to use parallelism based on certain replicated aspects of random walks. Since the discrete versions of these algorithms are all based on collecting statistics from random walks over some grid, it is natural to associate processing elements for a parallel implementation with either the walkers or the places they walk, i.e., the grid points.

In some sense, the mapping of groups of walkers onto parallel processors is the most natural parallel decomposition and is readily mapped onto a MIMD machine. The second mapping, that of grid points to processors, maps very naturally onto massively parallel SIMD computers. Below we discuss how the choice of one mapping over another influences the details of the algorithm and the performance aspects of the implementations. To make our discussion more concrete, let us think about MIMD implementations on either a shared-memory MIMD machine (like the Cray C90) or a distributed-memory machine like the IBM SP2. For the massively parallel SIMD implementation let us keep the rather old Thinking Machines CM-2 or the current MasPar MP-2 in mind.

The choice of mapping groups of random walkers onto parallel processors naturally leads us to the following algorithm for the evaluation of first passage time statistics like those required in (3.2). Each processor starts with random walkers with different starting locations on the grid. During each iteration all the walkers take a random step on the grid. Those that encounter the boundary are removed and their starting locations are scored with the boundary value of the first passage location, and new walkers are started somewhere on the grid to replace them. It is obvious that this algorithm faithfully implements the collection of statistics implied in (3.2) in an "embarrassingly" parallel fashion. In fact, this algorithm is such that it may be implemented asynchronously on independent processors until it becomes necessary to gather the statistics from each independent processor into centralized memory locations. It also makes little difference if we implement this algorithm on a shared- or distributed-memory machine (or a loosely coupled group of workstations) since there is no interprocessor communication until the statistics are centrally collected.

This idea of exploiting the parallel nature of independent statistical sampling arising from certain Monte Carlo calculations is an old one, and quite easy to implement in the case of the Dirichlet BVP for the Laplace equation. The nature of this algorithm also makes the choice of a stopping criterion rather simple. Since we are statistically sampling the solution to a problem that has a well-behaved underlying stochastic process, the computational error, which is only due to the statistical sampling error, should have law-of-large-number behavior. Thus if we have p processors, each of which samples the solution at every grid point, and we desire an overall variance in the sampling error of size ϵ^2, then we may run the p independent processes until all of them achieve at least an $\epsilon^2 p$ sampling variance. Then when we accumulate the p independent samples from the processors, we will be left with an overall variance of no greater than $(\epsilon^2 p)/p$, by the addition of variance. Finally, we consider the cost of generating each sample. Since our algorithm advances walkers from the interior until they hit the boundary, we average one sample every $\tau_{\partial\Omega}$ (the average length of a walk) iterations per walker. We will refer to this algorithm as the MIMD or "forward" random walk algorithm.

In contrast, the choice of mapping grid points onto the processor of a massively parallel SIMD machine, such as the CM-2 or MP-2, leads to a different set of considerations. In fact, the algorithm described above is an extremely poor choice on a SIMD machine. This is due to the fact that the CM-2 and MP-2 are physically an array of processors, each with local memory, upon which two types of interprocessor communication are implemented. General interprocessor communication is implemented via the "router," which is a large multiple slower than comparable communication over nearest-neighbor connections. This later is called "NEWS" communication. Thus the task of communicating the boundary value at the first hitting location back to the walker's starting point, which will generally require the router, seriously degrades the above forward random walk algorithm's performance. Because of this difference between NEWS and router performance, it is worthwhile to consider a variation on the above algorithm, which abolishes the need for router-based communication.

If we assume that the grid points in our calculation are such that they can be embedded into a regular d-dimensional grid, then all nearest-neighbor communication on this grid can be implemented via the fast NEWS communication on a CM-2. If $d = 2$, this is also true on an MP-2. A simple variation of the forward random walk algorithm allows us to get by with only nearest-neighbor communication. If, when generating the random walks from the starting interior points to the boundary, we save the path taken through the grid, this path can be retraced to bring the boundary values into the interior via only nearest-neighbor communication. In reality, this improves the situation very little, as random walks generate extremely suboptimal routes from the boundary to given interior points. It is, however, the case that while retracing walks from a given boundary point, every grid point along the retraced path may be considered as the starting point of a new forward random walk that first encountered the boundary at the given boundary location. In fact, it can be proven that scoring the boundary value at each point in the retraced path is probabilistically equivalent to the forward random walk algorithm discussed above [3]. In addition, retracing has the advantage that we obtain one sample per walker per step. Finally, it should be obvious that the notion of retracing is superfluous, as it is more efficient to start our walkers at the boundary.

Thus by trying to avoid an extremely inefficient aspect of a particular parallel computer's design, we have been led to a variation on our original algorithm. This "backward" random walk algorithm is an improvement over the original forward random walk algorithm in two obvious ways: (1) it requires only nearest-neighbor communication on the computational grid, and (2) it generates samples at the rate of one per walker per iteration instead of one per walker per complete random walk $(O(\tau_{\partial\Omega}))$. A not so obvious difference is based on the fact that on a SIMD machine, the MIMD rationale for the design of a stopping rule is not at all applicable. When we consider a more reasonable stopping rule for a SIMD implementation we will encounter yet another advantage of the backward over the forward random walk algorithm.

In a SIMD implementation, it makes much more sense with the backward random walk algorithm to start off a large cohort of walkers from the boundary with their boundary values, and then after some number of iterations terminate all of the walking and compute the statistics. This stopping rule makes more sense than waiting for an acceptable level of variance at each grid point, as was suggested for the MIMD implementation. This is because in the backward algorithm we are specifying the end not the beginning of random walks, and so starting new walkers will not necessarily reduce the sampling error at specified interior grid points. Since it is more natural in the SIMD case to consider the termination of all the walks uniformly and accumulate statistics at that point, one must be able to calculate the effect this has on the evaluation of the Wiener integral in (3.2). This effect is precisely due to the fact that by placing a limit on the number of iterations in the backward random walk algorithm, we are sampling the random walk expected value in (3.2) over only a portion of the entire space of random walks possible on our grid. We are evaluating (3.2) over only the space of random walks up to a given length equal to the number of iterations before termination. Fortunately, this truncated expected value can be explicitly computed. Surprisingly it is a nonlinear object.

It has been shown that this expected value over the space of random walks up to a given length can be computed as the *quotient* of two Jacobi method solutions of related discrete Dirichlet BVPs for the Laplace equation [3]. In the simple case of the discrete Laplacian on the two-dimensional square with a uniform grid, this statistic has an expected value that is the quotient of the Jacobi solution of the BVP with the given boundary values over the Jacobi solution with unit boundary values. Since unit boundary values asymptotically yield the constant unit function solution, the asymptotic behavior is that of the ordinary Jacobi method. However, for small iteration numbers, this quotient gives remarkably good empirical convergence results as demonstrated in the comparative figure below (Figure 3.1).

As Figure 3.1 shows, the backward random walk algorithm has initial convergence behavior comparable to the method of successive overrelaxation (SOR) with optimal relaxation parameter.

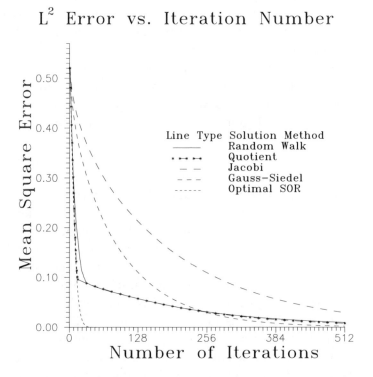

FIG. 3.1. *Empirical comparison of the Jacobi method, red-black Gauss–Seidel method, optimal red-black SOR method, the backward random walk method, and the nonlinear quotient method (which is the expected value of the backward method). For the random walk method, iteration number refers to maximal length of random walks.*

3.2. Concluding Comments

It is well known that these Monte Carlo methods are far inferior to many deterministic methods for these types of problems. However, in very high dimensions variants of these Monte Carlo methods are often used to solve problems in quantum mechanics. In addition, the Monte Carlo methods often serve to motivate the design and analysis of deterministic analogues, which may offer some unique advantages over more conventional algorithms. Another property of these Monte Carlo algorithms that may prove useful in real computations is the fact that with them one may sample the solution at as few as one grid point.

We have seen a rather simple example of how implementing a given mathematical formulation for a particular problem on different types of parallel computers not only leads to different implementations but also to different questions of the numerical analysis. The MIMD forward random walk implementation of these Wiener integral representations naturally motivates a stopping rule based on a sampling error tolerance. The SIMD backward implementation makes this type of stopping rule awkward and leads to the idea of numerically evaluating Wiener integrals over certain natural truncations of the space of all random walks. The analysis of these SIMD inspired truncations leads to a nonlinear iterative method that is based on Jacobi iterations (and hence can be implemented in parallel without grid point coloration) and gives initial behavior similar to optimal red-black SOR, without having to choose a relaxation parameter. Thus we have an object lesson on how parallel architectures can influence not only the design but the analysis of numerical algorithms.

References

[1] R. COURANT, K.O. FRIEDRICHS, AND H. LEWY, *Über die partiellen Differenzengleichungen der mathematischen Physik*, Math. Ann., 100(1928), pp. 32–74.

[2] M. FREIDLIN, *Functional Integration and Partial Differential Equations*, Princeton University Press, Princeton, NJ, 1985.

[3] M. MASCAGNI, *High dimensional numerical integration and massively parallel computing*, Contemp. Math., 115(1991), pp. 53–73.

[4] J. STROOCK AND R. VARADHAN, *Multidimensional Diffusion Processes*, Springer-Verlag, New York, Berlin, 1976.

Chapter 4

A Parallel Genetic Algorithm Applied to the Mapping Problem

El-Ghazali Talbi
Pierre Bessiere

Editorial preface

The graph partitioning problem has a number of possible applications. In this chapter the mapping problem—mapping processes to processors in a parallel architecture—is approached as a graph partitioning problem. The solution method is via a parallel genetic algorithm (GA), which naturally raises the interesting question: Can you use the mapping problem solution to solve, in parallel, the mapping problem?

This article originally appeared in *SIAM News*, Vol. 24, No. 4, July 1991. It was updated during the summer/fall of 1995.

The computing power of parallel processors makes their use appropriate and attractive for a variety of applications. We illustrate the application of an algorithm that, although more compute intensive relative to other similar algorithms, is well suited to a specific problem we are interested in solving. What makes this particularly interesting is that the problem to be solved relates to the efficient partitioning of processes on a parallel processor.

The problem we address consists of placing communicating processes on the processors of a distributed-memory parallel machine. This is referred to as the mapping problem. Indeed, to execute an application on a parallel machine, the mere translation step from a high-level language to binary code used for sequential computers is not sufficient. The code and data have to be split into loadable code objects (processes) and these objects have to be placed on the network of processors. There can exist an optimal placement of the processes. Most existing programming environments for parallel machines (Transputer development system, iPSC/2, CM-2) do not propose any solution to this problem. The burden is on the programmer, with the result that, in the worst case, a program's design may not be independent of hardware configuration.

A parallel *program* can be modeled by a graph, where the vertices represent the processes and the vertices' weights represent known, or estimated, computation costs of these processes. The edges represent communication links required between processes and the edges' weights estimate the relative amount of communication necessary along those links. A parallel *architecture* can also be modeled by an undirected, connected graph, where the vertices represent processors and the edges represent communication links between processors. When the number of processes exceeds the number of available processors, as is usually the case in massively parallel programming, the mapping problem encompasses the contraction problem, which is equivalent to the graph partitioning problem. This is illustrated in Figure 4.1.

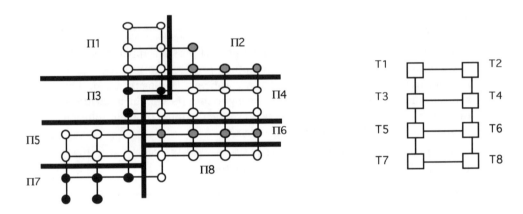

FIG. 4.1. *Mapping a parallel program on a parallel architecture.*

Given a graph, the "graph partitioning problem" searches for a partition of the graph's nodes that optimizes a given cost function. In addition to the mapping problem we discuss, there are numerous practical applications of this problem. Consider, for instance,

- the design of very large scale integration (VLSI) circuits, where, given a set of components and a set of modules, one wants to place the components in order to minimize the number of connections between modules, yet preserve some balance concerning the number of components on each module;
- routing in distributed systems, where the considered problem is to subdivide the computer network into smaller clusters so that the control overhead for routing is minimized;
- image segmentation in the field of computer vision, where segmented images are represented as graphs in which each vertex represents a segment and each weighted edge between two vertices represents a topological relationship between two segments of the image;
- virtual memory paging systems, where one wants to distribute the

different objects on memory pages in order to minimize the number of references between objects stored on different pages.

The graph partitioning problem is NP-complete. Consequently, heuristic methods are used to deal with it. Heuristics may find solutions that only approximate the optimum, but they do so in a reasonable amount of time. The many different approaches that have been proposed for this problem may be divided in two main classes: on one hand, the general purpose optimization algorithms independent of the given application and, on the other hand, the heuristic approaches especially designed for a specific problem. As we want to avoid the intrinsic disadvantage of the algorithms of this second class (their limited applicability due to the problem dependence) our concern in this paper is only for the first class of algorithms.

Two widely used general purpose optimization techniques are the hill-climbing algorithm and simulated annealing. The hill-climbing algorithm is certain to find the global minimum only in convex spaces. Otherwise, it is generally a local rather than a global minimum that is found. Simulated annealing offers a way to overcome this major drawback of hill climbing but at the price of longer computation time. A more serious drawback is the sequential nature of the simulated annealing algorithm. Its parallelization is extremely difficult.

Other distributed optimization techniques matrix–vector intrinsically parallel may also be considered. Some of them are closely related to neural network-based algorithms. Other specific examples are the genetic algorithms (GAs), which are considered in this paper. These are stochastic search techniques introduced by Holland twenty years ago [2].

GAs are inspired by the biological evolution of species. Development of massively parallel architectures made them very popular in the last few years. GAs have recently been applied to combinatorial optimization problems in various fields, such as the traveling salesman problem, the optimization of connections and connectivity of neural networks and classification systems. GAs are theoretically and empirically proven to provide a robust search in complex spaces, but they take an extremely long time to execute. We therefore propose a GA, implemented on a parallel processor, to reduce the execution time.

4.1. The Graph Partitioning Problem

Given

- an undirected graph $G = (V, E)$;
- an application Ω_1 from V into Z^+, such that $\Omega_1(v_i) = w_{1_i}$ is the weight of vertex v_i;
- an application Ω_2 from E into Z^+, such that $\Omega_2(e_j) = w_{2_j}$ is the weight of edge e_j; and
- a set of numerical constraints $\Phi = \{\phi_1, \phi_2, \ldots, \phi_m\}$ on these weights,

the graph partitioning problem has to find a partition Π of V ($\Pi = \pi_1$, π_2, $\ldots$, π_n) satisfying the constraints Φ.

Most applications (VLSI design, image segmentation) correspond to the following set of constraints, Φ_1, with the weights of all nodes set to 1:

— for each subset π_i of V belonging to the partition Φ, the number of nodes in π_i is equal to a given value B_i:

$$\sum_{v \in \pi_i} \Omega_1(v) = B_i \qquad \forall \pi_i \in \Pi$$

with

$$\Omega_1(v) = 1 \qquad \forall v \in V;$$

— the total cost of the edges going from π_i to π_j should be minimum

$$\min \left(\sum_{e \in \varepsilon} \Omega_2(e) \right)$$

with

$$\varepsilon = \{(x,y) \mid (x,y) \in E \ \wedge\ x \in \pi_i \ \wedge\ y \in \pi_j \ \wedge\ i \neq j\}.$$

The graph partitioning problem under constraints Φ_1 has been proven to be NP-complete [1]. For our application, the mapping of parallel programs on parallel architectures, we have to consider the following set of constraints, Φ_2: minimize the sum of communication costs between processors (total cost of the edges going from π_i to π_j) and the variance of the loads of the different processors (variance of cost of vertices belonging to a given π_i):

$$\min \left(\sum_{e \in \varepsilon} \Omega_2(e) + \left\{ K \left[\frac{\sum_{\pi_j \in \Pi} \left(\sum_{v \in \pi_i} \Omega_1(v) \right)^2}{|\Pi|} - \left(\frac{\sum_{v \in V} \Omega_1(v)}{|\Pi|} \right)^2 \right] \right\} \right).$$

With $K = 0$ the set of constraints Φ_2 reduces to Φ_1. This proves that the mapping problem under constraints Φ_2 is NP-complete. K is the weight of the contribution of the communication cost relative to the computational load balance across the system. Choosing a suitable value for K depends on knowledge of the characteristics of the parallel architecture involved in the mapping problem. Very small values of K would suggest a uniprocessor solution, and very large values of K would reduce the problem to one of multiprocessor scheduling without communication costs. The parallel architecture used was a network of transputers and $K = 2$ has been estimated by experiment.

4.2. The GA Solution to the Mapping Problem

GAs compose a very interesting family of optimization algorithms. Their basic principle is quite simple. Given a search space Σ of size M^N, and an alphabet of M symbols, then any point of this space may be represented by a string of N of these M symbols. The strings are analogous to chromosomes in biological systems. In natural systems, chromosomes combine to form the total genetic prescription for the construction of some organism.

In a phase of the process called "reproduction," some genetic operators are used to generate new points of Σ from existing points. During this phase, some points of Σ are replaced, keeping the size of the population fixed. We will inevitably have a competition for survival of the chromosomes in the next generation. The fundamental principle of the GA is "the fitter a string, the most probable its reproduction." We can assume that we start from some given initial population and that there exists a fitness function F from Σ into $\mathbb{R}$ that associates a real value to any point in Σ. Then in mathematical terms the fundamental principle of the GA means that the probability P of reproduction is increasing as F increases:

$$F(\sigma_1) > F(\sigma_2) \quad \Rightarrow \quad P(\sigma_1) > P(\sigma_2) \qquad \forall \sigma_1, \sigma_2 \in \Sigma.$$

Thus, over many generations, the average fitness of the population increases.

The standard GA, in a pseudolanguage, is shown in Figure 4.2.

```
Generate a population of random strings.
While number-of-generations <= max-number-of-generations
    Do
            Evaluation   - assign a fitness value to each string.
            Selection    - make a list of pairs of strings likely
                           to mate, with fitter strings listed
                           more frequently.
            Reproduction - apply genetic operators to the selected
                           pairs. New strings produced constitute
                           the new population.
```

FIG. 4.2. *A standard genetic algorithm.*

The most common genetic operators used during reproduction are crossover and mutation. Both of these operators are derived by analogy from the biological process of evolution. Given two strings, crossover involves cutting both strings at the same randomly selected point and exchanging the two portions. Mutation is simply a random state exchange for a single bit in the string. Two parameters need to be defined: P_c and P_m. They represent, respectively, the probability of application of the crossover and mutations operators. Other genetic operators may be found in the literature. For instance, there is the inversion operator and many variants of the crossover operator designed for specific problem domains.

To apply the GA to the mapping problem, the following formalism is used. Let us suppose that we have a graph of N processes to place on a parallel architecture of M processors. A symbol (for instance an integer between 0 and $M - 1$) is assigned to each processor. A given mapping is represented by an N string of those symbols, where symbol p in position q means that process q of the graph is in the subset p. We use the usual version of crossover, but mutation is a random trial of one of the M possible symbols.

Standard GAs with large populations suffer from lack of efficiency (long execution times). Two approaches to parallel GAs have been considered so far: the standard parallel approach and the decomposition approach. In the first approach, the evaluation and the reproduction are done in parallel. However, the selection is still done sequentially, because parallel selection would require a fully connected graph of strings, as any two strings in the population may be mated. The decomposition approach consists of dividing the population into equally sized subpopulations. Each processor runs the GA on its own subpopulation, periodically selecting good strings to send to its neighbors and periodically receiving copies of its neighbors' good strings to replace bad ones in its own subpopulation. The processor neighborhood, the frequency of exchange, and the number of strings exchanged are adjustable parameters.

The standard parallel model is not flexible in the sense that the communication overhead grows in proportion to the square of the population's size. Therefore, this approach is not well suited to distributed-memory architectures, where the cost of communication has a significant impact on the performance of parallel programs. In the decomposition model, the inherent parallelism is not fully exploited since the treatment of subpopulations may be further decomposed. This approach should be considered only when the number of available processors is less than the required size of the population.

For implementation on massively parallel architectures with numerous processors, we chose a fine-grained model, where the population is mapped on a connected processor graph such as a grid, with one string per processor. We have a bijection between the string set and the processor set. The selection is done locally in a neighborhood of each string. The choice of the neighborhood is an adjustable parameter. To avoid the overhead and complexity of general routing algorithms in parallel distributed machines a good choice is to restrict the neighborhood to only directly connected strings (i.e., processors). It is important to notice that these modifications to the standard model do not cause a degradation in the search efficiency of the standard GA.

4.3. Supernode Implementation

The Supernode is a loosely coupled, highly parallel machine based on the Inmos T800 Transputer. One of its most important characteristics is its ability to dynamically reconfigure the network topology by using a programmable VLSI switch device. This architecture offers a range of 16 to 1024 processors, delivering from 24 to 1500 Mflops of peak performance. To achieve these performance levels, a hierarchical structure has been adopted [3]. The programming environment used in our experiments is Parallel C 3L. The dynamic configurator of the physical network, developed in our laboratory, has been used to obtain the desired topology of the architecture.

We assume that each string in the population resides on a processor, and communication is carried out via message-passing. Figure 4.3 shows the pseudo-Occam description of the process executed by each processor. The Occam language provides a very interesting framework for the parallel or sequential execution of processes. The parallel construct PAR combines a number of processes that are performed concurrently. Values are passed between concurrent processes by communication on channels using input and output. Each channel provides unbuffered unidirectional point-to-point communication between two concurrent processes.

```
SEQ
    Generate (local-string)
    Evaluate (local-string)
    While number-of-generations <= max-number-of-generations
        SEQ
            -- communication phase --
            PAR i=0 FOR number-of-neighbors
                PAR
                    neighbor-in[i] ? neighbor-string[i]
                    neighbor-out[i] ! local-string
            -- computation phase --
            PAR i=0 FOR number-of-neighbors
                Reproduction (local-string, neighbor-string[i])
            Replacement
```

FIG. 4.3. *Pseudo-Occam program executed on each Supernode.*

Each reproduction produces two offsprings. Our strategy is to randomly choose one of the offsprings. The replacement phase consists of replacing the current local string with the best local offspring produced in the reproduction phase. The population is placed on the processors. The processors are configured such that the topology of the machine is a torus. Given the four links of the transputer, each string will have four neighbors. No routing is needed in the processor network because only directly connected processors (i.e., nearest neighbors) need to exchange information. We do not attempt to find the best solution globally because the communication involved in determining this solution would be considerable. Instead we pick up the best solution via a "spy process" placed on the "root processor." The root processor is the processor that connects the network to the host computer (e.g., a PC for PC-hosted systems or the 68000 or SPARC processor for Sun workstation systems).

4.4. Performance Evaluation

A performance evaluation of the algorithm has been performed. Speedups were studied to determine the quality of a given solution when the parallel GA was run on different size tori and with populations of different sizes (the two being equal given that there is one string per processor). Figure 4.4 shows the influence of the number of processors, and population size, on the time needed to reach a solution scoring 8. The specific mapping problem studied was a pipeline of 32 processes to be mapped on a pipeline of 8 processors. For this problem the optimal solution scores 7. The execution cost and the communication cost between processes are set to 1.

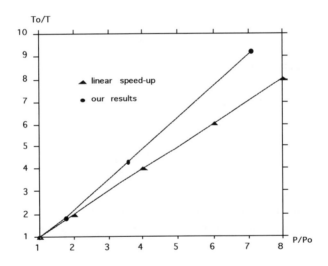

FIG. 4.4. *Execution times on different sizes of parallel architectures.*

We have a "superlinear" speedup of the parallel GA, in the sense that when multiplying the number of processors by p the execution time is divided by kp $(k > 1)$. Premature convergence may occur for a population that is too small, and the desired solution will never be reached. A comparative study of our approach with hill-climbing algorithms and simulated annealing has been done on different benchmarks [4]. The experimental measures show that our algorithm gives better results concerning both the quality of the solution and the time needed to reach it.

4.5. Conclusions and Future Directions

A massively parallel GA was used to solve the mapping problem, resulting in a superlinear speedup. An important characteristic of GAs is that they may be used to solve a great variety of combinatorial optimization problems. We are using them to solve such optimization problems in the field of robot control, computer vision, and neural networks.

We are also studying an important improvement of the algorithm, which includes the dynamic variation of its parameters and particularly the mutation probability. The crossover operator becomes less effective over time as the strings in the population become more similar. One way to avoid the premature convergence and to sustain genetic diversity is by using adaptive mutation. During the first generations, when there is ample diversity in the population, mutation must occur at very low rates. However, as diversity decreases in the population, the mutation rate must increase.

More theoretical work is planned: a cellular automata-based model will be used to study the influence of the algorithm's parameters on its convergence.

References

[1] M.R. GAREY AND D.S. JOHNSON, *Computers and Intractability: A Guide to the Theory of NP-Completeness*, Freeman, San Francisco, 1979.

[2] J.H. HOLLAND, *Adaptation in Natural and Artificial Systems*, University of Michigan Press, Ann Arbor, MI, 1975.

[3] C.R. JESSHOPE, T. MUNTEAN, C. WHITBY-STREVENS, AND J.G. HARP, *Supernode project P1085: Development and application of a low cost high performance multiprocessor machine*, in Proc. ESPRIT '86, Brussels, 1986.

[4] E-G. TALBI AND P. BESSIERE, *A parallel genetic algorithm for the graph partitioning problem*, in Proc. ACM Int. Conf. on Supercomputing, Cologne, Germany, June 1991.

Chapter 5

Supercomputers in Seismology: Determining 3-D Earth Structure

Robert J. Geller

Editorial preface

A comprehensive understanding of the internal structure of the earth would be useful in a number of contexts. One such application is in the study of earthquakes. This chapter addresses the use of seismographic data to solve the inverse problem of determining the earth's structure, based on how the seismic waves have propagated through the interior. The direct solution method, developed by the author and his colleagues with the use of classic supercomputers, allows for more detailed results than were previously possible.

This article originally appeared in *SIAM News*, Vol. 24, No. 5, September 1991. It was updated during the summer/fall of 1995.

Supercomputers are transforming the way seismologists analyze seismograms to determine the global-scale three-dimensional (3-D) structure of the earth's interior. Seismograms are recordings of the ground motion caused by earthquakes. Traditionally seismologists used only a few selected data from each seismogram, such as the arrival time of the initial elastic waves, and "threw away" the rest of the data. However, supercomputers are now making it possible to extract all of the information that is contained in the recorded seismograms.

5.1. Introduction

Plate tectonics is the basic theory that explains the large-scale geological features of the earth's surface. The earth's surface is divided into a small number of rigid tectonic plates whose motion is the visible part of a thermal convection cycle in the earth's interior. The transport of heat energy from the earth's interior to the earth's surface drives this convection cycle.

Seismology is the study of how elastic waves propagate through the earth and of the earthquakes that generate them. There are two types of elastic waves: P-waves (longitudinally polarized compressional waves) and S-waves (transversely polarized shear waves). The former propagate in both solids and fluids, but the latter do not propagate in fluid regions, such as in the earth's oceans or outer core (a region deep in the earth's interior). P-waves are always faster than S-waves. Letting α denote the velocity of P-waves and β denote the velocity of S-waves, we seek to determine the 3-D distribution $\alpha(r, \theta, \phi)$ and $\beta(r, \theta, \phi)$, where (r, θ, ϕ) are spherical polar coordinates.

Based primarily on analyses of seismological data, it is known that the earth's interior is divided into several major regions: the crust, the upper mantle, the lower mantle, the outer core, and the inner core. The elastic wave velocities tend to increase with depth within each of these layers, due to the effect of increasing pressure. Lateral heterogeneity of the elastic wave velocities, which is primarily due to lateral temperature variations, is on the order of several percent of the spherically averaged velocities. Higher temperature regions, which correspond to rising material in the convection cycle, have slower than average elastic wave velocities; the opposite is true of regions with lower than average temperatures. Three-dimensional images of the distribution of elastic wave velocities in the earth's interior can thus provide important, albeit indirect, information on the convection process.

5.2. Seismic Data Analysis

Seismograms are recorded by sensitive instruments called seismographs. By analyzing seismograms, seismologists obtain information on the source of the earthquake and the structure of the earth. Data from newly deployed broad-band, high dynamic range, seismographs have provided the impetus for the development of the new methods presented in this paper.

The most basic seismological datum for a given earthquake is the arrival time of the first elastic energy—the initial P-wave—at each observatory. If P-wave arrival times from at least four observatories are available, the location of the earthquake and its origin time can be determined, provided that the P-wave velocity distribution in the earth is known. In practice a kind of bootstrapping procedure is required. A provisional P-wave velocity model is assumed, earthquake locations are determined, and the velocity model is then refined. Although 3-D earth structure can be studied by analyzing arrival time data, we concentrate on waveform analysis techniques in this paper.

5.3. Surface waves

As P-waves and S-waves propagate away from the earthquake source, some of their energy is trapped in the form of surface waves whose amplitudes decay roughly exponentially with depth. The depth to which significant energy penetrates is roughly proportional to the horizontal wavelength of the surface waves. The waves that are not trapped, and continue to propagate through the earth's deep interior in the form of P-waves or S-waves, are called body waves.

There are two distinct types of surface waves. If we construct a vertical plane that contains both the source and the receiver, Love waves involve only horizontal motion perpendicular to the plane, whereas Rayleigh waves involve only ground motion (both horizontal and vertical) in the vertical plane. Love and Rayleigh waves exhibit dispersion—their phase velocity is a function of their period. With some exceptions, the phase velocity tends to increase with increasing period, as the surface waves travel through progressively deeper regions of the earth. The forward problem of computing the Rayleigh and Love wave dispersion $c_R(\omega)$ and $c_L(\omega)$, where ω is the frequency, can be easily solved if $\alpha(r)$, $\beta(r)$, and the density $\rho(r)$ are known. However, the inverse problem of determining $\alpha(r)$, $\beta(r)$, and $\rho(r)$ from $c_R(\omega)$ and $c_L(\omega)$ would be underdetermined even if there were no measurement errors. Auxiliary constraints can be imposed to obtain a well-posed problem, but subjective decisions are required.

Surface wave phase velocity data do not have sufficient resolving power to warrant inverting for P-wave velocities or density. However, if we could measure the Rayleigh and Love wave phase velocities as a function of frequency at every point on the earth's surface, $c_R(\theta, \phi, \omega)$ and $c_L(\theta, \phi, \omega)$, respectively, we could invert for the 3-D S-wave velocity distribution $\beta(r, \theta, \phi)$. Such research has been extensively carried out, but there are several major problems.

The quantity actually measured by surface wave studies is the phase difference between a surface wave of a given frequency and the same surface wave after it has made one complete revolution around the earth. The actual phase velocities can be expressed in terms of a spherical harmonic expansion. When we try to recover the coefficients for this expansion we find that all information on odd-order heterogeneity is lost, and that information on even-order heterogeneities is low-pass filtered. Furthermore, the great circle analyses are based on treating the lateral heterogeneity as though it were infinitesimal. Unfortunately, only fundamental mode surface wave phase velocities, which afford the poorest depth resolution, can be easily and reliably measured.

5.4. Free Oscillations

For a laterally homogeneous earth the surface-dependent (θ- and ϕ-dependent) part of the eigenfunctions of the modes of free oscillation is given by spherical harmonic functions Y_ℓ^m, but the eigenfrequencies of the modes depend only on the angular order ℓ and not on the azimuthal order m. The modes of a spherically symmetric model are thus degenerate—all $2\ell + 1$ modes that constitute the multiplet with angular order ℓ have exactly the same eigenfrequency. Lateral heterogeneity removes this degeneracy and splits the multiplet of $2\ell + 1$ degenerate modes into $2\ell + 1$ singlets, each having a unique natural frequency.

An obvious strategy for studying the earth's lateral heterogeneity is the "spectroscopic" approach—measuring each of the split eigenfrequencies, and then inverting them for 3-D earth structure. However, this approach fails for several reasons, the most important of which is related to anelastic attenuation of elastic waves, i.e., friction. Seismic waves in the earth decay as $\exp(-\omega t/(2Q))$; the quality factor Q is not a constant, but, in general, is on the order of several hundred. Since the broadening of the spectral peaks due to anelastic attenuation is almost always much wider than the separation between neighboring split modes, measuring the frequencies of the individual split modes is essentially impossible.

5.5. Waveform Inversion for 3-D Earth Structure

In order to extract all possible information on 3-D earth structure from recorded seismograms it is necessary to analyze the seismic waveforms themselves, rather than intermediate parameters such as surface wave phase velocities. By doing so, we gain two major advantages: we can avoid inaccuracies due to approximations that treat the entire laterally heterogeneous part of the structure as an "infinitesimal" perturbation, and we can fully use all the information contained in the recorded seismograms in our analysis.

Let us consider the quantities that must be calculated in order to determine 3-D earth structure by directly inverting recorded seismograms. First, we require the ability to compute the theoretical waveforms, which seismologists call synthetic seismograms. Second, we must be able to compute the change in the synthetic seismograms due to any given infinitesimal change in the earth model, i.e., the partial derivatives of the synthetic seismograms. Finally, we must formulate a systematic inversion procedure to find the earth model whose synthetic seismograms best fit the observed data.

The first successful application of waveform inversion was the landmark work of Woodhouse and Dziewonski [9], who inverted waveform data from surface waves to determine the 3-D variation of elastic wave velocities to depths of about 650 km. Like great circle phase velocity studies, the method used by these authors also assumed that all of the energy traveled around the great circle containing the source and the receiver. However, because they computed separate synthetic seismograms for the major and minor arcs of the great circle, they were able to resolve odd-order lateral heterogeneity to some extent.

The starting model used by Woodhouse and Dziewonski was laterally homogeneous. Considering the S-wave velocity as an example, the laterally homogeneous starting model, which depends only on the depth r and not on θ or ϕ, can be expressed as $\beta^{(0)}(r, \theta, \phi) = \beta^{(0)}(r)$. The laterally heterogeneous (3-D) perturbation to the laterally homogeneous (1-D) initial model, which is determined by inverting the observed waveforms, is $\delta\beta(r, \theta, \phi)$. The final model is the sum of the initial model and the perturbation determined by the inversion

$$(5.1) \qquad\qquad \beta(r, \theta, \phi) = \beta^{(0)}(r) + \delta\beta(r, \theta, \phi).$$

Almost all work on waveform inversion expands the unknowns in spherical harmonics and truncates the expansion at low order. We represent $\delta\beta(r, \theta, \phi)$

as follows:

$$\text{(5.2)} \qquad \delta\beta(r,\theta,\phi) = \sum_{s=0}^{SMAX} \sum_{t=-s}^{s} \delta B_s^t(r)\, Y_s^t(\theta,\phi),$$

where $Y_s^t(\theta,\phi)$ are spherical harmonics. A spherical harmonic of angular order s corresponds (approximately) to features with a horizontal wavelength $2\pi a/(s+1/2)$, where $a = 6371$ km is the radius of the earth. Woodhouse and Dziewonski use $SMAX = 8$; their model thus includes laterally heterogeneous structure with wavelengths of 5000 km and longer. $\delta B_s^t(r)$ in (5.2) must be replaced by a discrete representation to obtain a well-posed problem. Two popular choices are to divide the earth into layers, and treat $\delta B_s^t(r)$ as a constant within each layer, or to expand $\delta B_s^t(r)$ in terms of splines.

Figure 5.1 shows a laterally heterogeneous model (PUM0: Preliminary Upper Mantle model 0) of S-wave velocity obtained by this author and his colleagues [7] using the techniques explained below.[2] As only surface wave data were used, we could obtain accurate information only on S-wave velocities in the upper 600 km or so. Some of the geological implications of this model are discussed in [6]. Perhaps the most important conclusion is that the lateral heterogeneity of S-wave velocities in the depth range around 400 km can be quantitatively accounted for by the deep roots of continents (which have high velocities), mid-ocean spreading ridges (which have low velocities), and the presence of subducting slabs. Some authors had previously asserted that there had to be additional large-scale features with no surface manifestation in this depth range, but we showed this was unnecessary.

In order to further improve our knowledge of laterally heterogeneous earth structure several steps are needed. First, body waves, as well as surface waves, should be included in the waveform dataset. This in turn requires that our techniques be extended to permit efficient and accurate calculations of body waves and their partial derivatives with respect to the parameters of the 3-D earth structure model. Our efforts in this area are briefly summarized in the final section. Second, the correction for extremely shallow crustal structure (analogous to "static corrections" in seismic prospecting for petroleum) must be improved. Third, the volume of data must be increased greatly as compared with the dataset that was used to obtain the model in Figure 5.1.

5.6. Iterative Linearized Inversion

Synthetic seismograms are computed by solving a linear partial differential equation (PDE), but the change in the synthetic seismograms is a nonlinear functional of the perturbation to the earth model. Because of this nonlinearity, the change in the synthetic seismograms for a finite change in the earth model will not be equal to that predicted by linear extrapolation using the partial

---

[2]A color version of this figure is available from the SIAM WWW server at `http://www.siam.org/` `books/astfalk/`.

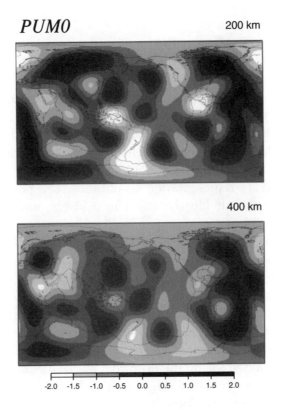

FIG. 5.1. *Lateral heterogeneity of S-wave velocity as a percentage of the spherically averaged S-velocity at depths of* 200 *km (top) and* 400 *km (bottom) as determined in* [7]. *Darker areas show high velocity regions and lighter areas show low velocity regions. The models shown in this figure include angular orders* $1 \leq s \leq 8$ *and azimuthal orders* $-s \leq t \leq s$. *Thus, roughly speaking, the model includes lateral heterogeneity with spatial wavelengths greater than* 5000 *km. The vertical resolution is roughly* 100 *km, so that the upper figure, for example, can be regarded as a spatial average of lateral heterogeneity at depths from* 150 *km to* 250 *km.*

derivatives. This error becomes worse as the perturbation to the starting model increases.

Woodhouse and Dziewonski [9] and many later workers used a spherically symmetric starting model and treated the entire laterally heterogeneous part of the earth model as an "infinitesimal" perturbation. The 3-D earth models obtained by such a linearized inversion will be inaccurate, perhaps seriously so.

If we start with a spherically symmetric model, we obtain a 3-D model from the linearized inversion, as shown by (5.1). Previous workers stopped at this point, but we can use this 3-D model as the starting model for the next iteration of an inversion and iterate until convergence is obtained. Since the model perturbations will become progressively smaller, the errors due to nonlinearity will decrease at each successive iteration.

We illustrate iterative linearized inversion using the S-wave velocity as an example. The unknowns are still the coefficients $\delta B_s^t(r)$ defined in (5.2), but the starting model is now laterally heterogeneous:

$$(5.3) \qquad \beta^{(0)}(r,\theta,\phi) = \sum_{s=0}^{SMAX} \sum_{t=-s}^{s} B_s^t(r)\, Y_s^t(\theta,\phi).$$

The 3-D perturbation to the earth model determined by the inversion (5.2) is added to the 3-D starting model (5.3) to obtain the new model:

$$(5.4) \qquad \beta(r,\theta,\phi) = \beta^{(0)}(r,\theta,\phi) + \delta\beta(r,\theta,\phi).$$

The new model defined in (5.4) becomes the starting model for the next iteration; we iterate until convergence is obtained.

Previous workers thought that the above iterative linearized inversion required an impractically large volume of computations. Here is what Woodhouse and Dziewonski said in 1984.

> "The [inverse] problem requires the calculation of not only synthetic seismograms but also their partial derivatives.... In this study we use...2000 records...and we construct models in terms of 324 parameters. A naive calculation would lead to the conclusion that more than 600,000 [partial derivatives] must be calculated. Even for a spherically symmetric model this would be a formidable undertaking, but for an aspherical model it is totally unfeasible. In addition, since the inverse problem is non-linear, the procedure must be reiterated a number of times."

5.7. The Direct Solution Method

The calculations described above as "totally unfeasible" have now, as the result of two new algorithms developed by our group at Tokyo University [3, 4], combined with the use of supercomputers, become routine. It thus has become possible to carry out iterative linearized inversion for laterally heterogeneous earth structure. We expect that this algorithm will rapidly be adopted as a standard method in seismology. We used these new algorithms to invert for the 3-D earth structure shown in Figure 5.1.

Our new algorithm, the direct solution method, is based on solving the "weak form" (Galerkin form) of the elastic equation of motion. It can thus be considered a generalization of the finite-element method or variational method.

Several previous workers calculated synthetic seismograms for a 3-D starting model following a two-step approach. They first calculated the modes of free oscillation of the 3-D model, and then calculated the synthetic seismograms by summing these modes. In contrast, we directly solve for the synthetic seismograms of the laterally heterogeneous model and their partial derivatives, without the unnecessary intermediate step of computing the modes of the 3-D model. This approach has a number of advantages, and no obvious disadvantages.

Our actual work uses spherical polar coordinates, but we simplify the following explanation by using cartesian coordinates $\boldsymbol{x} = (x, y, z)$. The $i = x, y,$ or z components of the vector trial functions are denoted by $\phi_i^{(m)}(\boldsymbol{x})$. The superscript (m) denotes the mth trial function $m = 1, \ldots, M$. We compute the synthetic seismograms and their partial derivatives in the frequency (Fourier transform) domain, where ω is the Fourier transform variable.

We expand the displacement in terms of the trial functions:

$$(5.5) \qquad \boldsymbol{u}(\boldsymbol{x}) = \sum_m D_m \, \phi^{(m)}(\boldsymbol{x}) ,$$

where the expansion coefficients D_m are the unknowns. Equation (5.5) gives the displacement everywhere in the earth, i.e., for all values of $\boldsymbol{x}$. Thus we only have to solve for the expansion coefficients D_m once for each earthquake. We are using the surface wave portion of the seismogram as the data in our present research. Therefore, we use the fundamental mode and the first several overtones, which correspond to surface waves, as trial functions.

The following system of linear equations is derived by substituting (5.5) into the elastic equation of motion:

$$(5.6) \qquad (\boldsymbol{H} - \omega^2 \boldsymbol{T})\boldsymbol{D} = \boldsymbol{f},$$

where $\boldsymbol{f}$ is the force that generates a particular earthquake. In actual practice the observed data must be inverted to determine both the 3-D earth model and the vector $\boldsymbol{f}$ for each earthquake.

The elastic moduli $\lambda(r, \theta, \phi)$ and $\mu(r, \theta, \phi)$ are defined in terms of the P-wave velocity α, the S-wave velocity β, and the density ρ.

$$(5.7) \qquad \lambda = \rho(\alpha^2 - 2\beta^2),$$
$$(5.8) \qquad \mu = \rho\beta^2.$$

The elastic tensor $C_{ijkl}(r, \theta, \phi)$ is defined in terms of λ and μ:

$$(5.9) \qquad C_{ijkl} = \lambda\delta_{ij}\delta_{kl} + \mu(\delta_{ik}\delta_{jl} + \delta_{il}\delta_{jk}),$$

where the δ's are Kronecker deltas.

The elements of the matrices $\boldsymbol{H}$ and $\boldsymbol{T}$ are defined as follows, where summation over repeated dummy subscripts is implied, $*$ denotes complex conjugation, the subscript i, j denotes spatial differentiation of the i-component with respect to the j-coordinate, and the volume integral is evaluated over the whole earth:

$$(5.10) \qquad H_{mn} = \int dV \, [\phi_{i,j}^{(m)}]^* C_{ijkl} \, \phi_{k,l}^{(n)},$$

$$(5.11) \qquad T_{mn} \;=\; \int dV\, [\phi_i^{(m)}]^* \rho \phi_i^{(n)}.$$

H and T are both diagonally dominant, but this is not necessarily true for $(H - \omega^2 T)$.

A perturbation to the elastic wave velocities causes perturbations in the elastic moduli and density, which in turn perturb the matrix elements. The perturbation to the expansion coefficients of the trial functions δD is found by solving the perturbed (5.6):

$$(5.12) \qquad ((H + \delta H) - \omega^2(T + \delta T))(D + \delta D) = f.$$

If the perturbation is infinitesimal, we can drop all terms higher than first order in (5.12) and use (5.6) to eliminate zero-order terms. We obtain the key equation for computing the partial derivatives of the expansion coefficients:

$$(5.13) \qquad (H - \omega^2 T)\delta D = -(\delta H - \omega^2 \delta T)D.$$

For each earthquake, we only have to solve (5.6) once, but for each earthquake we have to solve (5.13) once for each model perturbation. The latter is therefore the time-consuming part of the calculation. We substitute (5.13) into (5.5) to obtain the partial derivatives of the synthetic seismograms:

$$(5.14) \qquad \delta u(x) \;=\; \sum_m \delta D_m\, \phi^{(m)}(x) \;.$$

Note that x is the location of the observatory at which the seismograms are recorded.

The above was contained in our 1991 *SIAM News* article, but we subsequently realized that by rearranging the above equations we could further reduce the computational requirements [3]. Let us define the vector y to be the vector of the values of the trial functions at the observatory:

$$(5.15) \qquad y \;=\; \begin{pmatrix} \phi^{(1)}(x) \\ \phi^{(2)}(x) \\ \vdots \\ \phi^{(N)}(x) \end{pmatrix}.$$

Now we can use (5.13) and (5.15) to rewrite (5.13) as

$$\begin{aligned} \delta u(x) \;&=\; \left[y^T\, (H - \omega^2 T)^{-1} \right] (\delta H - \omega^2 \delta T)D \\ (5.16) \qquad &=\; z^T\, (\delta H - \omega^2 \delta T)D, \end{aligned}$$

where the superscript T denotes the transverse, and the superscript -1 denotes the inverse of the matrix. We never actually have to compute the inverse of the matrix. We instead compute z by solving the following equation:

$$(5.17) \qquad (H - \omega^2 T)^T z = y.$$

Thus we have to solve (5.6) once for each earthquake, and (5.17) once for each receiver. The number of times that (5.16) has to be evaluated is given by the product of the number of sources, receivers, and model parameters. However, the model parametrization is chosen so that the matrix $(\delta H - \omega^2 \delta T)$ is highly sparse. The right-hand side of (5.16) can therefore be evaluated efficiently.

When our partial derivatives are expressed in the form of (5.16) it is possible to show an extremely close connection between our approach and an approach proposed by Tarantola [8]. Space does not permit a detailed discussion; see reference [3] for details.

5.8. Recent and Future Research

One of the most important future topics is applying our methods to inversion of the body wave portion of the seismogram, as well as the surface waves. The key step is the calculation of the synthetic seismograms and their partial derivatives, which must be both computationally efficient and accurate. Some workers have used approaches in which the earth is approximated by a series of flat layers, but this leads to errors that are difficult to estimate. We therefore solve the exact problem in spherical coordinates. We recently presented results for calculations in a 1-D (spherically symmetric) model, using an approach that can easily be generalized to the laterally heterogeneous problem [1, 2].

The work described in previous sections of this paper used the vertically dependent part of the degenerate modes of the laterally homogeneous model as trial functions, but in our work on body waves we initially used linear spline functions. This choice, however, led to relatively large errors due to grid dispersion. We recently succeeded in deriving modified numerical operators that greatly reduce the error of the solutions (by a factor of 25–50) without increasing the CPU time [5]. The key to our derivation was to use (5.13) to make formal estimates of the errors of the numerical solution using an eigenfunction expansion. We then deliberately modified the error of the matrix operators (without changing their bandwidth) to minimize the error of the numerical solutions. We have extended this approach to the 3-D problem and are now preparing a paper on this work. We also have used this approach to derive optimally accurate time domain finite-difference operators for purely local bases in the time domain.

Iterative linearized waveform inversion for 3-D earth structure is becoming an increasingly important research topic. The direct solution method makes the necessary computations practical on a routine basis. We are optimistic that the direct solution method will become an essential tool of seismologists, playing much the same role as seismic tomography in the 1970s and 1980s. We are also optimistic that the direct solution method can be applied to analyses of shorter wavelength seismic waves recorded by regional or local seismic networks.

5.9. Acknowledgments

This paper discusses work done in collaboration with P. Cummins, T. Hara, T. Hatori, T. Ohminato, N. Takeuchi, and S. Tsuboi. The author thanks T. Hara for assistance with the figure and for critically reading the final manuscript.

References

[1] P.R. CUMMINS, R.J. GELLER, T. HATORI, AND N. TAKEUCHI, *DSM complete synthetic seismograms: SH, spherically symmetric, case*, Geophys. Res. Lett., 21(1994), pp. 533–536.

[2] P.R. CUMMINS, R.J. GELLER, AND N. TAKEUCHI, *DSM complete synthetic seismograms: P-SV, spherically symmetric, case*, Geophys. Res. Lett., 21(1994), pp. 1663–1666.

[3] R.J. GELLER AND T. HARA, *Two efficient algorithms for iterative linearized inversion of seismic waveform data*, Geophys. J. Internat., 115(1993), pp. 699–710.

[4] R.J. GELLER AND T. OHMINATO, *Computation of synthetic seismograms and their partial derivatives for heterogeneous media with arbitrary natural boundary conditions using the Direct Solution Method (DSM)*, Geophys. J. Internat., 116(1994), pp. 421–446.

[5] R.J. GELLER AND N. TAKEUCHI, *A new method for computing highly accurate DSM synthetic seismograms*, Geophys. J. Internat., 123(1995), pp. 449–470.

[6] T. HARA AND R.J. GELLER, *The geological origin of long wavelength lateral heterogeneity at depths of* 300–400 *km*, Geophys. Res. Lett., 21(1994), pp. 907–910.

[7] T. HARA, S. TSUBOI, AND R.J. GELLER, *Laterally heterogeneous upper mantle S-wave velocity structure obtained by iterative linearized waveform inversion*, Geophys. J. Internat., 115(1993), pp. 667–698.

[8] A. TARANTOLA, *A strategy for non-linear elastic inversion of seismic reflection data*, Geophysics, 51(1986), pp. 1893–1903.

[9] J.H. WOODHOUSE AND A.M. DZIEWONSKI, *Mapping the upper mantle: Three dimensional modeling of earth structure by inversion of seismic waveforms*, J. Geophys. Res., 89(1984), pp. 5953–5986.

Chapter 6

Advanced Architectures: Current and Future

Greg Astfalk

Editorial preface
_____

The author describes where high-performance computing (HPC) is today
as an adjunct to much of the architecture-related material in the chapters
in this book. This discussion leads into the projection(s) of technologies
that will dictate future computer performance and architectures. The
important issues of memory organization and memory latency are covered
in the context of current and future machines. Some discussion of
programming models and languages is also offered, with some closure
on the prospects for automatically generated parallelism.

The original article, to which this one bears no resemblance, appeared in
SIAM News, Vol. 23, No. 2, March 1990.

High-performance computing (HPC) has always been ill defined. In the
past we might have been tempted to define HPC as the computations being
done on the classic supercomputers of the day. This is clearly not the case
today. With the increase in computational power available on the desktop, or
perhaps at deskside, the notion of what constitutes HPC has changed.

In this paper we look at the current state of HPC from both a hardware
and a software perspective. We have a bias toward hardware since it is our
expectation that the HPC user who desires, or requires, good performance must
be cognizant of the hardware. This is, in itself, a distinct change from the past.
The software that an individual uses is relatively static compared with the
changes that have occurred with the hardware. Additionally, software is more
of a "religious" and personal issue than is the hardware. We go "on the record"
here as stating that software *is* more important to the user's productivity than
the hardware. It is only in this article that we show this hardware bias.

With the rapid rate of change in HPC, it is almost impossible to look
too far into the future with any accuracy. In spite of this we attempt in the
following sections to describe the setting for HPC users a few years from now,

weaving in past trends to bring us to the current state and then projecting forward from there. We attempt to bias our perspective and comments toward the industrial user rather than the user in a research laboratory or academic setting.

6.1. Processor Type

The processor provides the computing power for HPC calculations, so it is logical that we discuss it first. We assert that the most common computing engine for doing HPC is now, and will continue to be, the RISC processor. A gut reaction might be that it is "supercomputers" that have this honor. Yet if we consider all the computations done on workstations and servers and, to some extent, parallel processors, the number of cycles exceeds those of the classic supercomputers. Three main factors are pushing the RISC processor to be the dominant force that it is now. Only one is technological; the other two, as discussed in the following paragraphs, are economic.

The rapid rate of change in HPC forces computer vendors to provide newer and faster processors on about a two to three-year cycle. It would be very difficult for a small computer vendor to design its own full custom processor every two or three years due to limited resources (human and financial). Additionally, the absolute level of performance of today's processors dictates that sophisticated designs be used. This is an expensive proposition, and the obvious way to avoid it is to use a processor that someone else has designed and built.

To quantify this by way of two examples (with no malice meant toward anyone) let's consider Cray Computer Corp. and Hewlett-Packard Company. Cray Computer was working to build a full custom processor that, with a clock cycle of approximately one nanosecond, was leading edge. The development effort cost approximately 200 million dollars. Simple economics dictate that development money spent must be recovered in sales of the developed product. Because of the specialized nature of the Cray Computer processor and its attendant high price tag, however, the number of customers is quite limited and the revenue generated has not been sufficient to fund the development. Even worse, this processor development cycle must be repeated every two or three years.

Hewlett-Packard, on the other hand, designs a product that sells for a price that is widely affordable. To substantiate this we simply need to look at the number of Hewlett-Packard workstations and servers that are sold each year. Hewlett-Packard is still faced with the two- to three-year development cycle, but the volume of sales gives the company a healthier business plan. Still open is the question of whether the computations that were formerly the province of the supercomputer can be performed on a desktop engine, such as Hewlett-Packard's. We address this central issue throughout this paper.

Let's assume that, to a first approximation, the computing power of a processor is proportional to its clock cycle (i.e., reciprocal of frequency).

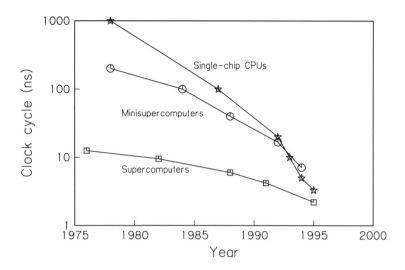

FIG. 6.1. *Historical behavior of the processor cycle time for three distinct classes of computers.*

Figure 6.1 shows the progress of the clock cycle for three different classes of processors. Four things are evident from this figure. First, the power of the individual processors in the supercomputer class has not changed significantly in 19 years. The Cray-1 of 1976 had a cycle time of 12.5 nanoseconds, and the soon-to-be-released Cray T90 has a cycle time of 2.2 nanoseconds. This is "only"[3] a factor of 6 in 19 years. Second, in the same period, performance increases have been more substantial for RISC processors than for other classes of machines. Third, the cycle times of processors are approaching a limit of close to one nanosecond. The ability of the computer industry to provide increases in single-processor performance by factors of 2 to 3 every one to two years is ending. Fourth, the delta between the cycle times of the RISC processor and the classic supercomputer has been reduced to nearly zero.

As of 1995, it is true that for some codes the high-end RISC processor outperforms the high-end classic supercomputer. In the late 1970s, using a supercomputer was the only way to achieve good performance. The faster cycle times made even nonvector code substantially faster when run on supercomputers. This is no longer necessarily true given the narrowing of the gap between supercomputers and RISC processors. This is *not* to say that classic supercomputers are bad, only that their overall dominance on high-end computing is being eroded by RISC processors.

An industrial perspective of the RISC processor offers another compelling reason for the dominance and future usage—third-party software packages.

---

[3]We don't mean any offense to Cray Research, Inc. Pushing the technology at that level of performance is a very demanding challenge.

Because the availability of this software is so important, especially to industrial users, we defer a detailed discussion of it to a later section.

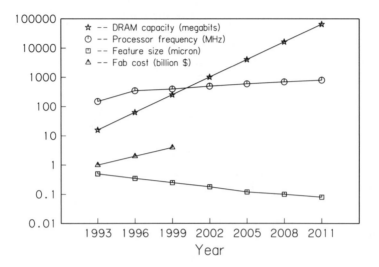

FIG. 6.2.   *Projected future trends for several key metrics in semiconductor technology.*

In Figure 6.2 a number of trends in the semiconductor industry are shown. The data come from the Semiconductor Industry Association, an organization that tracks the entire industry. Its data ought to be the best available and at least should be quite accurate. The curve for processor frequency shows that we can expect that by the year 2005 (less than 10 years from now) the frequency will be 600 MHz. We can make an assumption that the chip will deliver approximately two to four floating-point operations per Hz. The result will be a 2+-Gflop chip in 10 years or less, which means that a single-chip processor will have more than twice the peak performance of a single processor in today's most powerful supercomputer. When the cost differences are factored in, it is a very impressive and compelling point in favor of the RISC processor.

No magic technology is going to change the rate of single-processor performance growth. It is possible, however, that for an end-user's application, performance gains which are greater than the historical rate of single-processor performance increase can be realized through parallelism.

6.2. Parallel Computing

End users need to enlarge their problem sizes for better accuracy, to model more complex physics, to perform increased numbers of simulations for parametric analysis, and the list goes on. Some observers have indicated that the demand for computing power is growing "exponentially." We prefer not to attempt to quantify the growth rate but rather to appeal to the universal impression that the rate is increasing dramatically. From the previous section, we understand that the performance gain from single processors will be limited relative to the demands for computing power. It suffices to say that the demand for computing power far exceeds what can be offered via single processors. This logically leads to the notion of parallel processing.

Conceptually, parallel processing is so simple and appealing that it is hard to be convinced that it isn't a trivial solution to the problem of meeting the increased demand for computing resources. Parallel processing simply involves dividing the problem into several pieces and giving each piece to an individual processor to work on concurrently with other processors. In the limit we could reduce the execution time by a factor of $1/p$, where p is the number of processors we apply to the problem. The simplicity ends there. It is beyond the scope of this paper to cover all the details and intricacies of parallel processing, which would require a tome of epic proportions. What we hope to convey in this section is an overall sense of parallelism and the architectures of parallel machines.

The onus is always on the users to actually "map" their codes and algorithms onto parallel machines. As we discuss later, this can be accomplished at a variety of levels. What is invariant is that the code designer must give thought to the layout of data, the programming model, the synchronization of the processors, and many other issues. Getting parallel performance is unlike getting sequential performance.

In practice, the application of parallel processing is generally difficult. The primary issues in parallel processing are the partitioning of the problem into "separate" pieces, the required communication between the concurrently executing processes, and the balancing of the load on all the processors involved in the computation. We have found that in a majority of cases the first two issues are the more problematic. Despite the message of the preceding paragraphs, there are indications that parallel processing, in some form, will eventually mature into the common methodology for high-end computing.

In the recent past it seemed that in the absence of a large number of processors, the parallel machine was uninteresting. Experience with really large systems has been generally unfavorable. Some applications, after significant efforts by knowledgeable people, do yield very impressive results on $\mathcal{O}(10^3)$ processors, but this is definitely the exception rather than the rule. With the increased power of the individual processors and the inherent difficulty of dealing with a large number of processors, the pendulum is swinging in favor of systems that have 32–128 processors. In this regime the machines can be made useful on a larger spectrum of applications with less human effort.

There has been recent acceptance of the notion that the use of low processor count nodes— symmetric multiprocessors (SMP)—is an attractive approach to building large parallel machines. SMP nodes are a viable product for a broader market and thus allow computer vendors and independent software companies to leverage the economies of scale of the larger volume of sales. We say more about this in a subsequent section.

A final point we want to bring to light in this section is that the parallel computer architecture of choice is the MIMD (multiple-instruction multiple-data) class. A MIMD machine has multiple processors, each of which can concurrently execute a different set of instructions on a different set of data. Today, most MIMD machines use a commodity RISC processor as the fundamental engine. The actual synchronization of the algorithms executing on the asynchronous processors in a MIMD machine is the province of the programming language; it is done explicitly by the user in the coding or, in the case of automatically generated parallel code, by the compiler.

The machines in the other major class of parallel machines, the SIMD (single-instruction multiple-data) class, have proven to be less useful than those in the MIMD class. At this time there is only a single manufacturer of SIMD machines. In contrast, there are approximately 10 vendors of MIMD machines. However, it is true that for certain algorithms SIMD is the most appropriate architecture. Virtually no third-party software exists for SIMD machines, and this is quite detrimental to their acceptance and proliferation.

6.3. Memory Organization

Within the realm of parallel architectures in the MIMD classification, we can make a further division based on the memory architecture, or organization. The two most prevalent memory organizations are shared-memory and distributed-memory. Each has strengths and weaknesses. The type of memory organization has a profound effect on the usability, especially the programming, of the system.

In the shared-memory case, the multiple, individual processors in the "system" have equal access to a single pool of physical memory. Stated another way, all the virtual address spaces for all the processors lie within a single physical address space, i.e., memory. The interconnection of the processors to the memory is generally a bus or a crossbar. The important distinction is that all the physical memory is shared by the processors. The latency to memory, if second-order effects are neglected, is generally uniform for all processors. This type of machine is also described as having a uniform memory access (UMA) characteristic.

The shared-memory architecture offers the user a familiar and easy-to-program environment in which compiler-generated automatic parallelism can also be achieved. Unfortunately, the shared-memory architecture does not permit scaling the system to a large number of processors. Of the two most common interconnections of the processors and the memory, the bus presents a potential bottleneck in that all the processors must share this single resource. It is problematic to build a bus that has sufficient bandwidth to support a "large" number of today's high-performance processors. The crossbar avoids the bottleneck of a bus, but the complexity of the crossbar scales as $\mathcal{O}(n^2)$ (where n is the number of ports, usually one port per processor, in the crossbar).

The structure of distributed-memory is obvious. Each processor has its own physical memory. Therefore, no processor can directly address the memory of another processor. The address space for any process executing on a given processor is confined to the physical memory associated with that processor. If data are to be exchanged between processors, an explicit transaction, commonly referred to as a message, must take place between the processors. In contrast to shared-memory machines' uniform memory access (UMA), distributed-memory machines have a very large difference in latency between accessing the processor's local memory and passing a message to send data to or receive data from another processor's local memory.

A distributed-memory machine has the advantage of scalability. Individual processors are connected by some path, or paths, over which the processors communicate to exchange data. The connection scheme between the processors constitutes the "topology" of the system. Some of the more common topologies are the ring, the mesh, and the torus. All messages are passed over the interconnection from the sending processor to the receiving processor. Often the message will need to pass through other nodes en route to its final destination node. On any contemporary distributed-memory machine, the processors on intervening nodes are not involved in passing the message along. The generalization, of course, is that the size of the machine can be increased by adding additional processors to the topology.

The drawback of distributed-memory machines is the message-passing, which is generally accepted as a tedious and error-prone programming method. This isn't a global condemnation; as a parallel programming methodology, message-passing is widely used and accepted. Because it is so topical, a subsequent section is devoted to message-passing alone.

6.4. Distributed–shared-memory

As a continuation of the previous section, we point out that there is yet another variant of memory organization—distributed–shared-memory (DSM)—that, while not (yet) commonplace, is considered by some to be *the* memory organization of choice for the future. In the case of DSM the memory is physically distributed among the nodes, or processors, within the machine, much as in a distributed-memory machine. However, *all* the memory is globally addressable by any processor. If we view the physical addresses as an ordered doublet of node number and memory offset within the node, then the description of virtual address spaces for the shared-memory machine applies here. DSM is a true hybrid of shared-memory and distributed-memory; from a user perspective, a DSM machine is logically the same to program as a shared-memory machine. DSM architectures also permit message-passing programming, if desired. It may therefore be fair to view DSM architectures as a proper superset of the shared-memory and distributed-memory architectures.

An important distinction between DSM and shared-memory is that the memory latency for a DSM machine is not uniform as in the case of a true shared-memory machine. Access is faster to local memory than to remote memory. The difference is much smaller than for a distributed-memory machine. The memory of a DSM machine is generally classified as being NUMA, which stands for nonuniform memory access. We are more specific about the respective latencies of all three memory organizations in the next section.

Why is this notion of shared-memory versus distributed-memory so important? Of the entire community of users whose work could benefit from large-scale machines, a majority will find that writing new code, or modifying existing code, to utilize the message-passing required for distributed-memory machines is too difficult and time consuming. For this reason, and for the ability to efficiently deal with the massive amounts of existing code, the shared-memory or DSM approach is much more viable.

The number of address bits of current and future RISC chips will easily allow all the memory, even on a DSM machine, to be directly addressed. This should be taken to mean that the address bits can be used to address the nodes of the machine as well as the offset in the memory in each node. Of course this means that additional logic must be designed and built to enable the global addressing.

There is one additional issue associated with the DSM architecture—cache coherency. With a shared address space and multiple cache-based processors, it is crucial that the memory be cache coherent. This simply means that each processor knows about the current value of data contained in the caches of other processors. It is our assertion that a DSM architecture without cache coherence is ultimately programmed in a fashion that is reminiscent of message-passing.

Their ease of programming and low remote memory latency lend support to the notion that DSM machines are the trend of the future. To close this discussion, we note that future machines *will* have physically distributed-memory, and they will *probably* support global addressing (DSM). When programming these machines it will still be essential for top performance to pay attention to remote memory latency.

6.5. Memory Latency

While substantial progress has been made in processor speeds, as shown in Figure 6.1, the same is *not* true for memory speeds. The memory of most computers is composed of DRAM chips. In Figure 6.3a we show the increase in bit-per-chip density of DRAMs over roughly the past decade. Many people view memory chip density as the barometer of the integrated circuit industry (we won't debate whether this is correct or not). The figure shows amazing progress—a factor of 1000 increase in DRAM density in 13 years. The increases in memory chip density are projected to continue unabated for the foreseeable future (see Figure 6.2).

Of greater concern to the end user of the computer is the speed of the

DRAMs used in the computer's memory.[4] It is the access time of the memory chips that controls the latency and bandwidth for providing the data to the processor. In Figure 6.3b we show, for the same period as in Figure 6.3a, the decrease in access time for DRAM. Here we see a decrease of only a factor of 4 in the same 13-year period. Compared with the advances in processor frequencies, DRAM access times are lagging. This has widened the gap between the processor's demands on the memory and the memory's ability to meet them.[5]

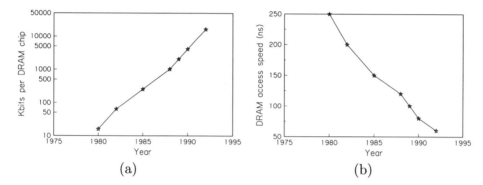

(a)   (b)

FIG. 6.3. *DRAM density and access time as a function of chronological time.*

A distinct and important hierarchy exists in the memory of contemporary machines. If it weren't for this hierarchy, the performance of RISC machines would be severely compromised. However, by not paying attention to the hierarchy, performance can be severely compromised. Table 6.1 is an approximate representation of the latency of the memory hierarchy in contemporary RISC-based machines. At each subsequent level of the hierarchy, the latency is larger and the amount of data that the level can contain is increased. The result of the technological improvements in processor and memory speeds is that the "distance" (here distance is taken to mean the number of processor clock cycles) from processor to memory is getting larger. Couple this with the rising prominence of parallel processing and we find a deepening of the memory hierarchy that users need to deal with in coding and algorithm design. Obviously, the goal is to make the most efficient use of the fastest possible level of the hierarchy that the algorithm and data set will permit.

One simplistic way to view the impact of this hierarchy is to consider how many instructions (potentially useful arithmetic operations) can be performed while accessing the various levels of the hierarchy. Latency-hiding

---

[4]Admittedly this is not completely correct since low latency memories can be built from DRAMs; however, continue reading.

[5]There are methods to construct high bandwidth memory subsystems, but these are in the province of the supercomputer and the cost leads us to the same economic argument that was asserted in the earlier section that discussed the types of processors.

TABLE 6.1

Approximate memory latency of the memory hierarchy in contemporary computers. Not all computers contain all levels of the hierarchy shown.

Level	Clocks
register	1
primary (L1) cache	2–3
secondary (L2) cache	6–12
tertiary (L3) cache	14–20
local memory	15–90
remote shared-memory	$\mathcal{O}(10^2)$
message-passing	$\mathcal{O}(10^3)$–$\mathcal{O}(10^4)$

techniques, an example of which is prefetching of data from memory, are useful in mitigating, or at least reducing, the negative effect of the latency on performance.

As a specific example of the effect of this memory hierarchy on performance, let's consider only the interaction of local memory and the primary cache. In Figure 6.4 we show the performance of a code that is executing an LU factorization of a dense unsymmetric matrix. In the case of the matrix–matrix formulation, the algorithm is making better use of the cache than in the case of the other two formulations. However, this still is not enough. In order to avoid (or at the least forestall) the drop-off in performance, we need to go even further and "block" the algorithm to make optimal use of the cache (i.e., memory hierarchy). By doing so, we increase the problem size at the point at which the severe drop in performance occurs.

The memory latency and hierarchy are issues that will not be quickly remedied by some new technology. Since these problems are going to stay with us, we need to address their presence and design algorithms and code to make the most effective use of existing architectures. Naturally, the deeper the memory hierarchy or, stated another way, the larger the latencies, the greater the algorithmic and coding effort required to achieve good performance. The hierarchy in memory latency is one of the reasons that the performance-sensitive user needs to be cognizant of the architecture on which the algorithm and code are executing.

6.6. Languages

The choice of language isn't that much more complicated for a parallel processor than for a sequential processor. Notice that this applies only to the *choice* of language—programming in that language to achieve good parallel performance is another discussion. Unfortunately, we have insufficient space to

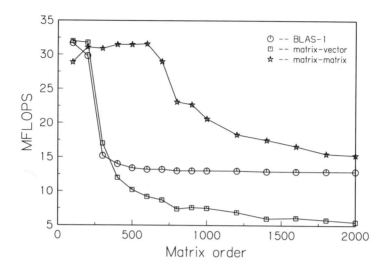

FIG. 6.4. *The behavior of three variants of LU factorization on a RISC processor as the problem (i.e., matrix) size is increased. Each of the formulations has the same computational complexity; they only differ in how they treat the memory hierarchy.*

do justice to this topic. We try, however, to make a few of the most important points about languages and expand on the future direction in programming language(s) in this section.

The most prevalent programming languages in use in HPC today are Fortran-77 and C. In parallel programming, message-passing, combined with Fortran or C, is the greatest common denominator. Message-passing has some strong points and some distinct disadvantages, both of which we discuss in a later section.

Fortran-90 is the current standard for Fortran. As of this writing, the general lack of availability of compilers has limited the use of Fortran-90. Many of today's Fortran-77 compilers support a subset of the features of the Fortran-90 definition. Fortunately, this subset is what most users want from the Fortran-90 language. Naturally, this reduces the need to have a full-featured Fortran-90 compiler. Another frequent and publicly stated complaint against Fortran-90 is that it is much more complex than its predecessor, Fortran-77. For these reasons, Fortran-90 is not a major presence in HPC.

There are two possible emerging languages within the HPC community. One, C++, is being used more and more frequently. The other, high-performance Fortran (HPF), has not yet made a practical "grand entrance," although there exists a tremendous body of literature related to it.

An increasing number of new software projects is being done in C++. Most often a hybrid language approach is used. In these instances C++ is used for all the high-level operations, such as pre- and postprocessing, data manipulation, and others tasks of this type. Fortran or C is called from C++ to do the computationally intensive operations. It is seldom efficient to use C++ for the computationally intensive kernels of application codes. The features of C++, such as encapsulation, result in a large amount of address taking, which is inefficient. Also, C++ compilers have not yet matured to the same degree as Fortran or C. We expect to see a continued increase in usage and importance of C++ in the future of HPC.

Similar in spirit, although definitely not in programming semantics, to message-passing, HPF offers the possibility of a truly portable programming language for parallel machines. The portability could exist across different machines and, perhaps, machine types. At issue are the acceptance of HPF by the user community and the development of robust compilers that produce efficient code. As of the writing of this paper, HPF compilers are generally *not* "ready for prime time," and only one vendor has a native HPF compiler. Translators exist for converting HPF into Fortran or C with message-passing. Compiler availability must occur before there will be widespread usage of HPF. Yet we could argue that serious efforts to develop compilers won't be undertaken until HPF is widely used, which is a Catch-22. A question of timing is whether HPF can establish itself in competition with the increasingly dominant message-passing and C++ over time. A final point is that HPF has a number of shortcomings, such as a lack of process management and capabilities for irregular problems, and a strong bias toward SIMD or data-parallel programming. HPF-2 (or is that HPF-II?) will supposedly address these points, but the open question is, again, "Is this enough, and will it be done soon enough?"

In addition to the languages mentioned in this section, there are a large number of what we'll call "few-user languages." These are generally modifications of either C or Fortran that offer some useful new constructs for parallel programming. While the implementations might be very well done and the features useful, these languages stand little chance of becoming widely used. Other than the few users who are in general proximity, either geographically or professionally, to the developers, the language does not propagate widely or quickly. These languages do serve a useful purpose as test-beds for new features that may eventually find their way into the mainstream big-three languages. They serve the purpose of pushing and developing parallel programming technology. Direct current usefulness to industry is minimal.

6.7. Message-Passing Programming

We stated earlier that message-passing is a tedious programming style. While we believe this to be a *generally* correct statement, there are many advocates of message-passing. Message-passing has also been used in many application codes that get impressive parallel performance and scale to large numbers of processors. These may seem like contradictory statements, but they really aren't.

Several people have been quoted as saying that message-passing is the "assembly language" of parallel processors. This is stretching the point a bit too far, but it remains true that for most users message-passing is difficult to master. It is also easy to get wrong, and when coupled with the lack of effective tools for debugging and profiling, this drawback makes for less productive programming than its shared-memory counterpart. Retrofitting an existing code to message-passing can be quite time consuming.

To its credit, message-passing offers at least two distinct advantages. First, it is the most portable parallel programming model that exists today. Message-passing is firmly established in the community, and it is well understood. Second, message-passing code, if it runs well at all, has forced the user to "do the right thing" with regard to data decomposition.

Message-passing programming has benefited from two efforts that have produced "standard" message-passing libraries. PVM (Parallel Virtual Machine) is the first, and still the most popular, message-passing library. It offers the user a syntax and library that enable the code to run on almost any existing computer. The strong points of PVM are that it supports heterogeneous collections of machines and allows explicit task management.

An alternative message-passing library, MPI (Message-Passing Interface), is also growing in popularity. MPI is considered a higher performance message-passing library than PVM. Additionally, MPI has more features than PVM, but it neither currently allows for heterogeneous processors nor has explicit task management.

Both libraries are freely available from various repositories on the Internet. Computer vendors generally offer a version of either, or both, of these libraries on their equipment. The vendors' versions are often highly tuned to give the end user the best possible performance, i.e., lower latency and higher bandwidth. We hope that vendors won't take this too far and modify the syntax, thereby destroying the raison d'être for using the libraries in the first place—portability.

As time passes we believe that neither PVM nor MPI will put the other out of business. We do expect the feature sets of both libraries to coincide to a larger degree than they currently do. Pragmatically, it is generally true that an application code written with one or the other of these two libraries can be translated relatively painlessly into the other.[6]

What message-passing doesn't offer is an effective way to deal with all the existing code that is needed for the software "infrastructure." The subject of legacy code(s) is not easily addressed by any current approach in parallel programming. Unfortunately, this is an important problem that awaits a solution that does not involve rewriting or re-architecting all the existing software for use on distributed-memory machines.

---

[6]Naturally this assumes that you have used only those features that are contained in the intersection of the two libraries.

6.8. Coarse-Grained Parallelism

The term "grain size" refers to the size of an independent set of calculations within a parallel algorithm. For example, if the two threads within a two-way parallel application can perform 100 operations before they need to interact in some fashion (either exchange data or synchronize), this would be considered a fine-grained application. While no specific numbers of operations make up fine-, medium-, and coarse-grained parallelism, intuition should be the guide. To be a bit more specific, fine grained might be taken to mean that the computation time is only slightly larger than the overhead required to invoke the parallelism. Medium-grained might be some multiple of the overhead, say $\mathcal{O}(10)$. Coarse grained would naturally be that much larger again.

It is well established and should be an accepted axiom to those working in parallel processing that the coarser (i.e., larger) the parallel grain size, the better. It is the "getting" of this coarse-grained parallelism that is the difficult issue. The contemporary compilers of today, whether for Fortran or C, cannot *automatically* find coarse-grained parallelism. The scope of automatic parallelization is, with few exceptions, constrained to the loop level.

In Figure 6.5a we show the granularity, i.e., CPU time, for each loop in a suite of 127 real application codes. There are nearly 21,000 data points showing the average CPU time per occurrence of each loop. The per-occurrence CPU time is not the time per iteration of the loop. The data include loops that contain I/O and procedure calls, so a great many of them cannot be parallelized automatically or manually.

The approach to automatic parallelization of loops would be "strip-mining." Strip-mining takes the iteration space of the loop and divides it among the processors, thereby giving each processor a subset of the induction variable's values. This imposes two constraints on the efficiency of this approach. First, it takes the initial, sequential granularity of the loop (as shown in Figure 6.5a) and reduces it by a factor of p, where p is the number of processors involved in the strip-mining. Second, there will be an implicit, and required, synchronization after the loop. Synchronizations impose overhead.

In Figure 6.5b we show the subset of the loops that are in fact amenable to automatic parallelism, as found by a contemporary auto-parallelizing compiler. The issues are apparent: virtually none of the very time consuming loops is automatically parallelizable, and of those loops that are parallelizable the loop span is quite short. The short loop span will possibly be too fine grained after strip-mining to be profitably parallelized. From the example illustrated by the data in Figure 6.5, it is clear that automatic parallelism via loop-based parallelism is not likely to be a big win in achieving coarse-grained parallelism. The inadequacy of the automatic approach leads us to perform the parallelization manually. Most users do not have the time or knowledge to accomplish this, however. There are no really bright spots on the immediate horizon in this area.

It is a simple concept to prove that the most efficient parallelism is the

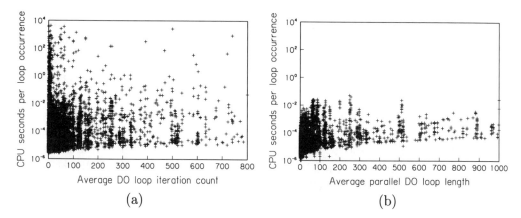

FIG. 6.5. (a) *The processor time used in all* 21,000 *of the loops in* 127 *real applications versus the loop span.* (b) *The same data for only those loops that can be automatically parallelized by a contemporary production parallelizing compiler.*

least amount of and the coarsest-grained parallelism. One of the primary goals of a parallel algorithm is to achieve coarse granularity and minimal communication. At this point, coarse-grained parallelism is achievable by human intervention only in either the algorithm design or in the alteration of an existing sequential code. For parallel processing to flourish and become a mainstream HPC vehicle, there needs to be a way to achieve coarse-grained parallelism for legacy code and solution-oriented users.

6.9. Hierarchical Parallelism

We previously developed the projection that the "nodes" of HPC machines will be multiprocessor based. Architectures of this type offer the possibility of what we call hierarchical parallelism. Hierarchical parallelism simply means that we concurrently have multiple levels of parallel execution. Consider a loop nest. It may be possible to execute the inner loop in parallel across the processors within a multiprocessor node. Iterations of the outer loop might then be strip-mined across the nodes of a multinode machine. Obviously, this is an attractive possibility, since it can lead to substantial reductions in time to solution.

As a more practical example, one that we have dealt with repeatedly in our work, consider a finite-element code. Typically, some form of mesh partitioning is applied to produce a number of parallel tasks that are assigned to the individual processors within the system. These subdomains, each a subset of the original complete mesh, are fairly coarse grained since we will need to solve for all the interior unknowns within each subdomain. Each of these subdomain problems looks almost exactly like the complete finite-element codes that we have parallelized on classic supercomputers. This parallelization was restricted to fine-grained parallelism within the solver (either direct or iterative). So we assign the subdomains to nodes and then parallelize the solver itself across the processors within the node. We have used message-passing as the programming method for the coarse-grained subdomain problem. Within each subdomain problem on a node, we use the compiler-generated loop-based parallelism across the individual processors.

Hierarchical parallelism has been used to advantage for several real application codes. We can go even further with this notion. With a hierarchical architecture and the above example, we could use fine-grained parallelism within the node, medium-grained across nodes with DSM, and, finally, coarse grained across SMP clusters with message-passing.

6.10. Third-Party Software

While it may be ignored by some people, and by the press that popularizes parallel processing, the fact remains that third-party software is a *crucial* issue for the success of any parallel system. Third-party software is also commonly referred to as ISV (independent software vendor) code. It must be understood that a majority of the machine cycles used on the union of all HPC machines are spent within third-party codes. This may seem incorrect, but you need only to realize the relative numbers of machines sold and used in universities and research institutes versus the number in Fortune-1000 companies. In the latter case, third-party codes consume most of the cycles used. It is a common experience that a supercomputer is purchased largely to run a small number of ISV codes. Industry tends to treat large problems that consume significant amounts of the capability of high-end computers.

Let's do a thought experiment to develop what we'll call the ISV "Catch-22." ISVs are businesses that are obligated to make a profit from developing and selling their application software. Almost always, the code that ISVs develop is large and complex and requires a significant porting and quality-assurance procedure. This requires significant resources on the part of the ISV. Before a port and subsequent quality-assurance testing are undertaken for a new machine, there must be a business case that indicates a reasonable expectation of a return on the investment. This is simple economics. Now assume the ISV sees this new HPC machine as a niche architecture that, with its relatively complicated architecture and programming, will require a significant porting effort. The first questions asked are, How many such machines are there in existence? How many will be sold? Who will they be sold to? If the projected number of sales is not enough to justify the porting expense, then the port won't be done. Of course, for a new architecture there is seldom a clear business case that shows a large number of sales.

So here we are; special-purpose HPC machines often don't get ISV codes ported to them, and large numbers of machines can't be sold within this market segment without the ISV software—a distinct Catch-22!

The ISVs focus on the high-end deskside machines since they are so prevalent and competent in application areas. Additionally, the large number of systems sold offers the ISVs the opportunity to sell large numbers of licenses. Almost every important ISV code exists on every major workstation.

Simple induction leads us to the notion that a new machine should either be a simple architecture or be compatible with the deskside systems. This allows a leveraging of the ISV porting and support for the deskside architecture to be projected onto the new architecture, which lends support to the notion developed earlier of the parallel machines of the future being clustered SMPs.

With tongue in cheek, we could say that the parallel machine with the largest number of ISV codes ported to it will be the most successful. There is more than anecdotal evidence to support this conjecture.

6.11. Epilogue

It is probably a good idea to wrap up this paper with a concise list of our projections. We are mindful that extrapolation is not a good technique in numerical analysis, and projections in the HPC arena are equally unstable. However, we believe that there is no magic bullet that is going to change the historical trends in any *substantial* way. This gives us some confidence that the projections won't be completely wrong.

We offer the following snapshot of HPC for the next four years:

- SMP nodes with 128 or fewer processors in a MIMD architecture;
- 1+-Gflop commodity RISC processors;
- large physical memory, $\mathcal{O}(10)$ Gbyte, within each node;
- hardware-based DSM within the node and possibly across a small number of clustered nodes;
- memory latency that is reduced from current levels but is still large enough to be a factor in performance;
- low-degree interconnect such as ring, mesh, or torus;
- inadequate I/O bandwidth;
- Fortran-77, C, and C++ as the dominant languages;
- increased usage of message-passing, both PVM and MPI, for coding parallel algorithms;
- nearly automatic fine-grained parallelism.

Clearly, the only definitive proof of the correctness of these assertions will be time. We obviously face an interesting four years in HPC.

Chapter 7

Large-Scale Molecular Dynamics on MPPs: Part I

David M. Beazley
Peter S. Lomdahl

Editorial preface

This chapter is the first of two by the same authors on the subject of "large-scale" (several hundred million atoms) molecular dynamics (MD). The use of large-scale MD enables us to approach the notion of macroscopic material properties from a molecular basis. The authors develop the multicell algorithm to enable the decomposition for parallel processing. In this chapter they also use explicit message-passing in a code that is quite portable between the Cray T3D and the Thinking Machine CM-5.

This article originally appeared in *SIAM News*, Vol. 28, No. 2, February 1995. It was updated during the summer/fall of 1995.

How does a piece of metal break? How do cracks propagate? How do impurities and grain boundaries affect the strength properties of a material? These are a few of the many interesting questions that arise in many areas of materials science and engineering research.

Because of the rapidly increasing capabilities of high-performance super-computers, computer simulation is playing a greater role than ever before in investigations of many of these questions. For several decades, molecular dynamics (MD) simulations have been used to study material properties [1]. The idea behind an MD simulation is very simple: a large collection of atoms is represented (in a crystal lattice, for example), and Newton's equations of motion, $F = ma$, are solved directly. While conceptually simple, this task presents a formidable computing problem. If the atoms interact according to a pair potential (e.g., gravity, Coulomb, Van der Waals), the direct solution of this general N-body problem will require the calculation of nearly $N(N-1)/2$ forces. To complicate matters, the atoms in many materials simulations may interact via more complicated embedded-atom or many-body potentials. Since direct methods quickly overwhelm the computing capabilities of even the fastest supercomputers, many schemes have been developed to solve both long- and short-range problems.

Despite the development of clever algorithms for reducing the complexity of MD simulations, until recently most simulations have been limited to a few hundred thousand atoms and a relatively small number of timesteps. To apply the method to realistic simulations of material properties in three dimensions, however, may require the inclusion of tens of millions to billions of atoms. Even a billion-atom MD simulation, if it could be performed, would be a "small" simulation, considering that a speck of dust can contain more than a billion atoms. A single timestep in most MD simulations could be on the order of ten femtoseconds, yet for a realistic experiment it may be desirable to follow a simulation for several microseconds. It should also be noted that the goal of large-scale MD simulations is to provide a means for studying material properties on a macroscopic level, and the classical approximations used would be inappropriate for studying such things as the basic electronic structure of a material. In many cases the "atoms" in an MD simulation may actually be large molecules or grains.

With the availability of massively parallel supercomputers, researchers have shown considerable interest in the development of fast parallel MD algorithms [2, 6, 7]. As a result of their work, simulation sizes have jumped to more than 100 million atoms and the time required to perform a simulation has been significantly reduced [5, 6]. In this two-part article, we describe our efforts at Los Alamos National Laboratory to develop a fast code for performing large-scale MD simulations (more than 100 million atoms) on two massively parallel supercomputers, a Connection Machine 5 (CM-5) and a Cray T3D. In Part I we discuss the molecular dynamics problem in general. Part II (which also appears in this book) will present some of the programming difficulties we have encountered in seeking high performance on state-of-the-art massively parallel supercomputers.

7.1. Short-Range MD

In many materials simulations, it is possible to assume that the atoms interact only with other nearby atoms (as a result of screening effects that mask out the long-range forces). In this case, we say that the atoms have "short-range" interactions. A cut-off distance r_c is specified, and any two atoms separated by more than this distance do not interact. The short-range MD problem has two critical aspects: (1) development of a scheme for determining which atoms interact with each other and (2) implementation of an efficient method for calculating the forces between those atoms.

Researchers have approached the short-range MD problem in two main ways. One approach, called the Verlet list method, searches the space around each atom and constructs a list of all of nearby atoms found [1]. When forces are calculated, each atom simply checks its list to find out which set of atoms

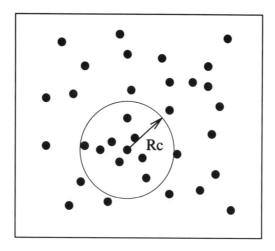

FIG. 7.1. *Short-range molecular dynamics.*

contributes to its total force. Since the list-building process is extremely expensive, the search space is usually given a radius slightly larger than the cutoff r_c so that the lists can be used again for several timesteps.

The second popular approach to the short-range MD problem is based on a cell method. Space is subdivided into a large collection of cubical cells [2, 4], and each atom, according to its coordinates, is assigned to a particular cell. Calculation of the total force on each atom includes the forces from other atoms in the same cell and from all the atoms in neighboring cells. Cell methods allow larger simulation sizes because no neighbor lists have to be stored; more work must be performed, however, because the search for neighbors is done in the cubical region around each atom rather than in the spherical region used by the neighbor-list method. Since our primary interest has been large simulation sizes, our approach is based on a cell method, as described later in this article. More information about high-performance list methods can be found in [6] and [7].

7.2. The Multicell Algorithm

Although in this article we highlight the main features of our algorithm in two dimensions, it also extends naturally to three dimensions. The algorithm is described in detail in [2].

The algorithm begins by dividing space into large regions that are assigned to the different processing nodes available. Each node further subdivides its region into small cells, each with dimensions slightly larger than the cut-off distance r_c, as shown in Figure 7.1. An atom is placed in the proper node and subcell according to its coordinates. This structure organizes the atoms in a way that makes it easy to calculate forces. Since each cell is slightly larger

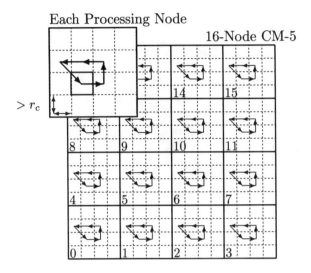

FIG. 7.2. *Processor layout and force calculation.*

than the cut-off distance r_c, only the atoms in the same cell or in the nearest neighboring cells will contribute to the total force on an atom.

To calculate the forces on the atoms in each cell, we introduce the idea of an "interaction path." The path is shown in two dimensions in Figure 7.2 and in three dimensions in Figure 7.3. The path serves two purposes. First, it specifies the order in which neighboring cells are processed in the force calculation. At each step, forces between the atoms in the starting cell and those in the neighboring cell are calculated. By Newton's third law, forces are accumulated by both cells; this cuts the number of force calculations in half and allows us to consider only half of a cell's neighboring cells (forces from a cell's lower neighbors will be calculated when those cells follow the path).

When all of a cell's neighbors reside on the same processor, the path simply specifies the order in which forces are calculated between those cells. When neighboring cells reside on different processing nodes, however, the path serves to coordinate the message-passing between nodes. When the path crosses a processor boundary, atom coordinates and accumulated forces are sent to the first neighboring processor on the path. This processor calculates the forces between the atoms whose coordinates it has received and the atoms in its own cells. The processor then passes the atom coordinates and accumulated forces to the next processor on the path. Eventually, the atom coordinates, along with all the forces that have been calculated, are sent back to the original processor. The carrying of both positions and forces along the path is an important aspect of the algorithm. For corner cells, the path may pass through as many as six different processing nodes in three dimensions, each of which requires atom positions and each of which contributes to the total force calculations before this information is returned to the original processor. This process of calculating forces proceeds serially on each node. The nodes run asynchronously except for boundary cells, where all the nodes participate in synchronous "send and receive"-type message-passing operations.

Once all the forces have been calculated, a second-order explicit finite-difference scheme is used to integrate the equations of motion. The atoms are then moved to new positions, and the data structures are updated to reflect the new positions. This is done by checking all the atom coordinates and, if necessary, moving atoms to new cells. Originally we used asynchronous message-passing to send particles to new processors, but this scheme proved to be undesirable under normal operating conditions. We have since switched to a new synchronous communication scheme based on a six-way data exchange described in [6] and [7].

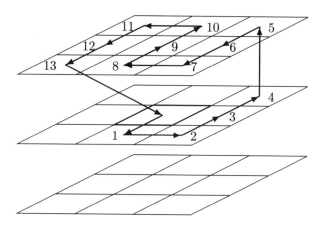

FIG. 7.3. *A three-dimensional interaction path.*

7.3. Implementation

The algorithm has been implemented in the code SPaSM (scalable parallel short-range molecular dynamics), which is written almost entirely in ANSI C and uses explicit message-passing for communications. For increased portability, SPaSM uses a custom message-passing library that we have developed. This library is implemented in whatever native message-passing environment is available.

We have used the CMMD message-passing library on the CM-5, and we have two versions that use PVM and the Cray shared-memory library on the TD3. Recently, we have run the SPaSM code on a Fujitsu VPP-500 using the p4 message-passing library. In addition, we have been able to port the code to single processor Sun, HP, IBM, and SGI workstations (in this case we only emulate message-passing). While workstations are limited in performance, we have found that this is an excellent way to develop and debug new code. It is important to emphasize that by writing our own message-passing library, the same source code can be used on all of the above machines.

The code allows the use of a variety of short-range potentials. A typical short-range potential is given by the truncated Lennard–Jones 6–12 potential:

$$(7.1) \qquad V(r) = \begin{cases} 4\epsilon \left(\left(\frac{\sigma}{r} \right)^{12} - \left(\frac{\sigma}{r} \right)^{6} \right), & 0 < r \le r_c , \\ 0, & r_c < r. \end{cases}$$

Because the potential quickly drops to zero, we truncate it at a distance equal to r_c. No atoms will interact beyond this point. While the Lennard–Jones potential is one of the most common short-range potentials for many MD studies, our code allows any short-range pair potential to be used through a table lookup and linear interpolation scheme. It is also possible to use more complicated potentials, such as the embedded-atom potentials that are useful for simulating metals.

7.4. Timing and Scaling Properties

Good scaling properties are a principal goal for developers of parallel machines and algorithms. Algorithms should scale well in two ways. First, algorithms should have good scaling properties as the problem size is increased (an $O(n \log n)$ solution will be better than an $O(n^2)$ solution). Second, these algorithms should scale well as the number of processors is increased. That is, a simulation run on eight processors will ideally run twice as fast on a machine with 16 processors. This latter goal is often the more difficult to achieve: as the number of processors is increased, it becomes more and more difficult to manage communications traffic between processors.

TABLE 7.1

Time for one timestep in seconds on the CM-5. Cutoff: $r_{max} = 2.5\sigma$.

Particles	Processors					
	32	64	128	256	512	1024
1024000	8.90	4.51	2.32	1.26	0.72	0.44
2048000	-	8.96	4.44	2.46	1.36	0.74
4096000	-	-	8.79	4.81	2.67	1.36
8192000	-	-	16.83	8.81	4.80	2.47
16384000	-	-	-	16.95	8.74	4.49
32768000	-	-	-	-	16.90	8.54
65536000	-	-	-	-	-	16.55
131072000	-	-	-	-	-	34.26

The scaling properties of SPaSM are summarized for a variety of problem sizes on the CM-5 in Table 7.1. The table shows the time required for a single timestep for a variety of simulation sizes; we see that time scales nearly linearly as the problem size is increased and as the number of processors is increased. This is shown graphically in Figure 7.4. Figure 7.4a shows nearly linear scaling as the problem size is increased from 1 million to 131 million atoms. Figure 7.4b shows the scaling properties as more processors are added. Again, we see a nearly linear relationship (doubling the number of processors roughly cuts the time per timestep in half). We get similar scaling properties on the T3D, but the timings are approximately 0.36 of that on the CM-5 (the code runs approximately 180% faster). More detailed performance results can be found in [5].

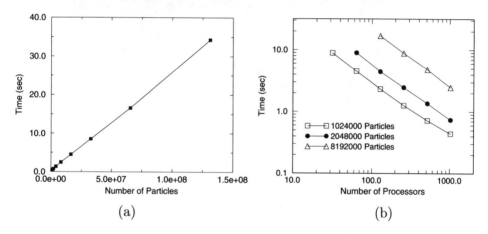

(a)   (b)

FIG. 7.4. *Time per timestep versus problem size and number of processors.*

7.5. Results and Applications

In tests, SPaSM has performed 3-D simulations with as many as 600 million atoms on the CM-5. For real simulations, we have been using SPaSM to perform large-scale fracture experiments in three dimensions with over 100 million atoms. While work along these lines is in progress, Figure 7.5 shows a snapshot of a test 3-D fracture experiment with 38 million ($1000 \times 2000 \times 19$) atoms in which a thin plate is pulled apart after a defect has been introduced.

7.6. The Need for Speed

The 38 million-atom fracture experiment exposes a fundamental problem of large MD simulations. The entire simulation, although it ran for only 6500 timesteps, required more than 50 hours of CPU time on a 512-processor CM-5. Because CPU time on large parallel machines is a scarce resource, code performance becomes a critical issue—achieving the highest performance possible is not only desirable but absolutely necessary if realistic large-scale simulations are to be performed.

In the second and concluding part of this article, we will discuss some of

FIG. 7.5. *Fracture experiment with* 38 *million atoms.*

the performance tactics and pitfalls we have used on the CM-5 and T3D.

7.7. Acknowledgments

We would like to acknowledge the valuable assistance of Adam Greenberg, Denny Dahl, Marc Bromley, and Mike Drumheller at Thinking Machines Corporation. We would also like to thank the Advanced Computing Laboratory for its generous support. In particular, Dave Rich and Mike Krogh have provided significant technical assistance. Last, we'd like to acknowledge our collaborators, Niels Grønbech-Jensen, Pablo Tamayo, Brad Holian, and Shujia Zhou. This work was supported by the U.S. Department of Energy.

References

[1] M.P. ALLEN AND D.J. TILDESLEY, *Computer Simulations of Liquids*, Clarendon Press, Oxford, 1987.

[2] D.M. BEAZLEY AND P.S. LOMDAHL, *Message-passing multi-cell molecular dynamics on the Connection Machine* 5, Parallel Comput., 20(1994), pp. 173–195.

[3] D.M. BEAZLEY, P.S. LOMDAHL, P. TAMAYO, AND N. GRØNBECH-JENSEN, *A high performance communications and memory caching scheme for molecular dynamics on the CM-5*, in Proc. 8th International Parallel Processing Symposium (IPPS '94), IEEE Computer Society, 1994, pp. 800–809.

[4] S. CHYNOWETH, Y. MICHOPOULOS, AND U.C. KLOMP, *A Fast Algorithm for the Computation of Interparticle Forces in Molecular Dynamics Simulations*, Parallel Computing and Transputer Applications, IOS Press/CIMNE, 1992, pp. 128–137.

[5] P.S. LOMDAHL, P. TAMAYO, N. GRØNBECH-JENSEN, AND D.M. BEAZLEY, *Proceedings of Supercomputing '93*, IEEE Computer Society, 1993, pp. 520–527.

[6] S. PLIMPTON, *Fast Parallel Algorithms for Short-Range Molecular Dynamics*, Report SAND91-1144, UC-705, Sandia National Laboratory, Albuquerque, NM, 1993.

[7] P. TAMAYO, J.P. MESIROV, AND B.M. BOGHOSIAN, *Proceedings of Supercomputing '91*, IEEE Computer Society, 1991, p. 462.

Chapter 8

Large-Scale Molecular Dynamics on MPPs: Part II

David M. Beazley
Peter S. Lomdahl

Editorial preface

This chapter looks at parallel processing from the bottom up. Specifically, the focus is on how to get the best possible performance from the individual processors in a parallel architecture. Although the CM-5 and the T3D are the machines targeted in this chapter, the issues are generic enough to apply to a broader range of machines. The conclusion is that in order to get the best possible performance, the code writers need to be aware of architectural and processor issues, some which are quite subtle. This attention to detail results not only in small gains but also in very large performance improvements.

This article originally appeared in *SIAM News*, Vol. 28, No. 3, March 1995. It was updated during the summer/fall of 1995.

With the development of sufficiently powerful supercomputers, molecular dynamics (MD) has become a useful tool for studying the dynamical properties of materials. In the first part of this article [1], we discussed the short-range MD algorithm we developed at Los Alamos National Laboratory for performing large-scale MD simulations with more than 100 million atoms. In this concluding part of the article, we explore some of the computational obstacles to high performance levels on state-of-the-art massively parallel machines.

Why is high-level performance necessary? Consider the computing requirements for a 100 million-atom MD simulation. Each atom is represented by a position, velocity, accumulated force, and type; storage of these data in double precision will require a minimum of 7.6 Gbytes of memory. For each step of the simulation, all the forces from each atom's neighbors (an average of 30 per atom) must be calculated; each force calculation will require approximately 30 floating-point operations, for a total of 90 billion floating-point operations per timestep. Since a simulation may be run for tens to hundreds of thousands of timesteps, the need for speed becomes clear.

8.1. Performance in the Real World

In a perfect world, an applications programmer would be able to write code in an abstract high-level language without worrying about the underlying machine architecture. Unfortunately, this cannot be done with current parallel, or even serial, machines. The complexity of modern parallel computers makes an understanding of the machine architecture critical: code performance can be severely affected by such factors as coding methodology, data layout, and choice of programming languages. Moreover, it is never as easy to achieve high performance levels as computer vendors would like you to believe! More often than not, it is much more difficult with real codes than with highly publicized benchmarks, such as LINPACK. Since our work has focused on massively parallel supercomputers, the CM-5 and the Cray T3D, we begin with a general overview of these two machines.

8.2. The CM-5 and T3D

The CM-5 consists of a large collection of processing nodes connected by two communications networks (data and control) arranged in a fat-tree topology. Each node consists of a 33-MHz SPARC microprocessor, 32 Mbytes of memory, and four vector units (VUs) capable of a combined speed of 128 Mflops (32 Mflops each). On the more recently introduced CM-5E, each node consists of a 40-MHz SuperSPARC microprocessor and 128 Mbytes of memory, and VU performance is upgraded to a peak of 160 Mflops. Figure 8.1 shows the layout of each CM-5 processing node.

The T3D also consists of a large collection of processing nodes, but in this case they are connected by a network in a 3-D torus topology. Each node consists of a 150-MHz DEC Alpha microprocessor that has 64 Mbytes of memory and is capable of 150 Mflops. Unlike the CM-5, the T3D does not have special-purpose vector processors for fast floating-point performance. Instead, it relies on the significantly faster clock rate of the Alpha microprocessors.

On both machines, the nodes operate independently, but they can communicate with each other using message-passing. We have used the message-passing programming model to implement our algorithm. Each node runs an independent copy of the same program and works on a small piece of the larger problem being solved. When communication is necessary, the nodes explicitly send "messages" to each other through the network. Periodically, the nodes may synchronize and perform global operations, such as I/O. Message-passing is supported on the CM-5 by the CMMD message-passing library and on the T3D by PVM and Cray's shared-memory library.

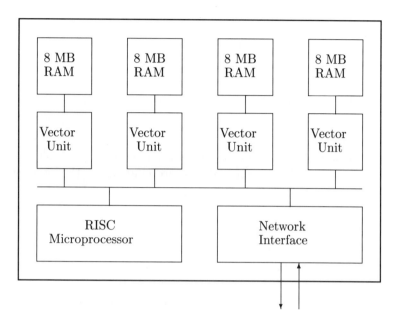

Fig. 8.1. *A* CM-5 *processing node.*

8.3. Writing Parallel Code

The complexity of modern massively parallel machines has led to considerable discussion in the parallel processing community of such issues as performance, code portability, and the need for new tools and programming paradigms that will allow better use of the parallel programming environment. While efforts to develop these new tools are well intentioned, it is our view that, with the field being in a time of rapid transition, tools often become available only several years after a machine is introduced; program performance is often sacrificed to "portability" or to an effort to keep Fortran on life support for a few more years. Ironically, it seems that many efforts to develop portable parallel programming languages have exactly the opposite result—a "portable" language on one machine may not be supported at all on another machine. As applications programmers, we don't have time to wait for software developers to sort out the mess and decide on, and then develop, the latest and greatest tool to make parallel programming easy. We have written our code in ANSI C with explicit message-passing, which provides us with a great deal of programming flexibility: ANSI C is highly portable, and virtually all massively parallel machines now support some form of message-passing. In addition, we have made every attempt to write code according to the POSIX.1 standard. This has allowed us to easily port the code to a wide range of workstations and parallel machines.

8.4. Why Message-Passing?

Since MD is an inherently unstructured problem, data-parallel languages (such as CM Fortran, C∗, and HPF) are not well suited to our problem (although data-parallel MD codes have been developed for the CM-2 [5, 8]). Our algorithm maps cleanly onto the message-passing programming model, which gives us explicit control over our data layout and communications. We organize the data into packets; this results in fast computation and minimizes communication between processors, which accounts for only approximately 5–20% of the overall processing time in a typical simulation [3]. Since almost all parallel machines support message-passing, our code is portable in the sense that moving to another machine will require changes in the message-passing function calls but not in the general algorithm.

8.5. Performance on the CM-5

The architecture of the CM-5 is oriented primarily toward high-level performance with the data-parallel languages CM Fortran and C∗, which were available on its predecessor—the CM-2. These languages make use of the performance of the VUs at a compiler level (which is usually hidden from the user). Unfortunately, codes written in ANSI C do not use the VUs, and maximum performance is thus limited to approximately 5 Mflops/node.[7] To make use of the power of the VUs, the C programmer must use CDPEAC, a set of C macros for programming in the VU assembler language DPEAC. To complicate matters further, the VUs operate in SIMD mode with a complicated memory-management scheme developed for the data-parallel programming model. As a result, the C programmer is given the sometimes daunting task of not only vectorizing a calculation but doing so in SIMD mode with special data layouts.

We decided to write our force calculation entirely in CDPEAC [3]. Because the programming of VUs is beyond the scope of this article, the development of our code is described here only very briefly. To solve data layout problems and improve memory bandwidth to the VUs, we implemented a memory-caching and memory-management scheme that doubled our VU performance. As a result, the code sustains calculation rates of 25–50 Mflops per node and runs approximately five times faster than the original C code that did not use the VUs. Furthermore, floating-point performance is better than that seen in most data-parallel applications. In November 1993 our MD code, SPaSM, was one of the winners of the IEEE Gordon Bell Prize; the code sustained a rate of 50 Gflops on the 1024-processor CM-5 at Los Alamos [6]. It is interesting that performance problems on the CM-5 are not limited to codes written in C; the overall winner in the 1993 Gordon Bell competition was a CM Fortran code that also used CDPEAC assembler kernels to achieve high performance levels [7].

While many people cringe at the thought of writing assembler code, we

---

[7]On the CM-5E the C performance is increased to a maximum of 40 Mflops/node.

improved the performance of our code substantially by doing so. Fortunately, it was possible to isolate all the assembler coding into a single code module. As a result, we were able to preserve most of the portability of ANSI C while improving performance by a factor of five. The benefits became clear when we were able to port the code to the T3D in only a few days (most of the time was spent changing the message-passing library to PVM). In fact, with the use of a small number of compiler directives, the code now compiles without modification on both the CM-5 and the T3D.

8.6. Performance on the T3D

One of the difficulties associated with programming the CM-5 is making efficient use of the vector units. This task is so difficult that the VU code performs nearly four times as many calculations as the original C code (because of the difficulty of coding an unstructured calculation efficiently in SIMD mode). The T3D has no VUs, which makes programming conceptually easier, but the performance problems are no less complicated.

One of the biggest performance problems on modern machines can be attributed to the use of superscalar RISC processors, such as Sun's SuperSPARC, IBM's RS/6000, IBM/Motorola's Power PC, Hewlett-Packard's Precision Architecture, and the DEC Alpha used in the T3D. These processors offer the potential for extremely good performance, but it is the compiler writers and programmers who are responsible for attaining this performance. Virtually all RISC microprocessors rely heavily on pipelining, which allows overlapping execution of several instructions (each will be at a different stage of completion). More advanced superscalar RISC microprocessors can actually begin to execute more than one instruction (typically from two to six) simultaneously. For example, the SuperSPARC can begin to execute a floating-point operation, an integer operation, and a memory operation all in the same clock cycle.

Unfortunately, making the most of this instruction-level parallelism requires careful arrangement of the instructions by either the compiler or the programmer. For example, a compiler that tries to interleave floating-point and memory instructions will generate faster code than a compiler that generates a series of memory instructions followed by a sequence of floating-point operations. A poorly arranged sequence of instructions can cause huge performance penalties by stalling the pipeline or failing to make effective use of the multiple-instruction-issue capability of superscalar processors. Further performance bottlenecks arise from the rapidly increasing performance gap between memory systems and microprocessors. Virtually all RISC microprocessors utilize caches and clever schemes for improving memory bandwidth, but all these efforts can be defeated by poorly written code. While a full discussion of RISC architectures is not possible here, it is important for programmers to realize that improper instruction scheduling and excessive memory accesses can substantially reduce performance. These are the kinds of problems that face applications programmers (and compiler writers) on machines such as the T3D.

The two code fragments shown in Figure 8.2, which are extracted from the force calculation in two short-range MD codes [2, 4], illustrate the problem. Each code fragment calculates r^2, the square of the distance between two atoms. If this distance is less than the cutoff r_c^2, a force is calculated; otherwise, the force is set to zero.

Code 1.

```
dx = x2 - x1;
dy = y2 - y1;
dz = z2 - z1;
r2 = dx*dx + dy*dx + dz*dz;
if (r2 < cutoff2) {

    /* Calculate forces */

}
```

Code 2.

```
dx = x2 - x1;
r2 = dx * dx;
if (r2 < cutoff2) {
    dy = y2 - y1;
    r2 += dy*dy;
    if (r2 < cutoff2) {
        dz = z2 - z1;
        r2 += dz*dz;
        if (r2 < cutoff2) {

            /* Calculate forces */

        }
    }
}
```

FIG. 8.2. _Two code fragments extracted from the force calculation of two separate MD codes._

In Code 1 r2 is calculated only after the values of dx, dy, and dz have all been calculated. In Code 2 r2 is calculated in stages, with the value checked at each stage to see whether it is greater than the cutoff; if it is, the calculation is aborted. On a test problem, Code 2 performed approximately 13% fewer floating-point operations and 20% fewer memory operations than Code 1. Yet Code 2 runs 16% slower on the T3D, 26% slower on the CM-5E, and approximately 25% slower on a wide range of RISC workstations.

This phenomenon is a direct result of the difficulty of programming modern RISC processors effectively. RISC processors are most efficient when instructions are overlapped as much as possible, which works best on sequential code with a minimal number of data dependencies. In Code 2, the extra compare instructions cause the processor to stall: several clock cycles may be required to complete the calculation of r^2, and the processor is unable to determine the proper outcome of the conditional until the value of r^2 is known. In Code 1, more work must be done to calculate r^2, but the additional instructions can easily be overlapped and executed in parallel. Thus, the processor will perform more work, but it will also avoid stalling. In the case of the T3D, the number of stall cycles created by Code 2 causes such a performance penalty that even with the 20% saving in floating-point operations, there is a 16% increase in the overall execution time (all of which is spent stalling the processor).

8.7. Performance Strategies

Before porting our code to the T3D, we looked into the question of how well our C code utilizes the SPARC processor on the CM-5. When we analyzed the assembler output created by the compiler, we found that we spent more than 65% of the time stalling the processor while waiting for results or memory accesses. We then experimented with various tactics for improving code performance, among them restructuring code to reduce memory accesses, using registers more effectively, inlining short subroutines by hand, and paying careful attention to potential stall situations. In fact, we made all of these modifications to the C code (none to the assembler code), and they produced a huge performance improvement. Code performance on the CM-5 improved by 120%, and we achieved nearly 80% of the peak floating-point performance of the SPARC (5.1 Mflops/node peak). On the CM-5E, code performance improved by 92%, and we obtained the remarkable result that the C code without VUs ran slightly faster than the code with VUs (the difficulty of programming VUs cannot be overstated). On the T3D these modifications resulted in a 132% speedup. Using additional optimizations, we were able to increase this to a 170% speedup. Despite this substantial speedup, our T3D code only achieves between 30–45 Mflops/node, or about 20–30% of the peak performance rate for the DEC Alpha.

While we have encountered many programming difficulties, we have found that most of the performance strategies work on a wide variety of RISC architectures. We have performed tests on a wide range of architectures and have measured speedups of 189% on the HP-PA7100, 155% on the IBM RIOS-2, and a staggering 285% on the MIPS R8000. Performance will surely improve as optimizing compilers mature, but no optimizing compiler will be able to make every code run fast. For this reason, we feel that an understanding of machine architecture is a very effective tool that allows us to work with the compiler to develop the fastest code possible.

8.8. Conclusions

Massively parallel machines hold a great deal of promise for large-scale molecular dynamics simulations, and future advances in computer architecture may soon allow billion-atom MD simulations. The complexity of modern microprocessors and parallel architectures has made an understanding of computer architecture an extremely important part of our research. Oftentimes efficient programming tools and languages are unavailable or unable to deliver the highest possible performance levels on a machine. In our experience with the T3D and CM-5E, we have found that substantial gains can be made by simply being a little more careful with our C coding and by remaining aware of how microprocessors and compilers interact. Certainly, such tactics will help greatly on next-generation machines as hardware complexity increases.

8.9. Acknowledgments

We would like to acknowledge the valuable assistance of Adam Greenberg, Denny Dahl, Marc Bromley, and Mike Drumheller at Thinking Machines. We would also like to acknowledge the Advanced Computing Laboratory for its generous support; in particular, Dave Rich and Mike Krogh have provided significant technical assistance. Last, we'd like to acknowledge our collaborators, Niels Grønbech-Jensen, Pablo Tamayo, Brad Holian, and Shujia Zhou. This work was supported by the U.S. Department of Energy.

References

[1] D.M. BEAZLEY AND P.S. LOMDAHL, *Large-scale molecular dynamics on MPPs: Part 1*, SIAM News, Vol. 28, No. 2, February 1995.

[2] ——, *Message-passing multi-cell molecular dynamics on the Connection Machine 5*, Parallel Comput., 20(1994), pp. 173–195.

[3] D.M. BEAZLEY, P.S. LOMDAHL, P. TAMAYO, AND N. GRØNBECH-JENSEN, *A high performance communications and memory caching scheme for molecular dynamics on the CM-5*, in Proc. 8th International Parallel Processing Symposium (IPPS '94), IEEE Computer Society, 1994, pp. 800–809.

[4] S. CHYNOWETH, Y. MICHOPOULOS, AND U.C. KLOMP, *A fast algorithm for the computation of interparticle forces in molecular dynamics simulations*, Parallel Computing and Transputer Applications, IOS Press/CIMNE, 1992, pp. 128–137.

[5] R.C. GILES AND P. TAMAYO, *Proceedings of SHPCC '92*, IEEE Computer Society, 1992, p. 240.

[6] P.S. LOMDAHL, P. TAMAYO, N. GRØNBECH-JENSEN, AND D.M. BEAZLEY, *Proc. of Supercomputing '93*, IEEE Computer Society, 1993, pp. 520–527.

[7] L.N. LONG AND J. MYCZKOWSKI, *Proc. of Supercomputing '93*, IEEE Computer Society, 1993, pp. 528–534.

[8] P. TAMAYO, J.P. MESIROV, AND B.M. BOGHOSIAN, *Proc. Supercomputing '91*, IEEE Computer Society, 1991, p. 462.

Chapter 9

Symbolic and Parallel Computation in Celestial Mechanics

Liam M. Healy

Editorial preface

One aspect of celestial mechanics is the computation of the long-term orbits of celestial bodies. This type of computation is complicted by the interaction of the many bodies that need to be considered to derive accurate long-term behavior. For reasons explained in this chapter, it is necessary to do this symbolically rather than "numerically." Symbolic computations performed on a LISP machine are described. The visualization of the solution is accomplished on a massively parallel SIMD machine.

This article originally appeared in *SIAM News*, Vol. 23, No. 5, September 1990. It was updated during the summer/fall of 1995.

The recent availability of advanced architecture computers has revolutionized the approach to the study of long-term behavior in celestial mechanics and allowed us to see things not understood previously. Although celestial mechanics is one of the oldest fields of research in the physical sciences, computers have been the stimulus for a reawakening of the field. The principle computational environments we use for these investigations are the LISP machine for object-oriented symbolic processing and the Connection Machine for mapping out the phase space in color. The LISP machine, while not a novel architecture in the sense of parallelism (being a serial machine) nor even particularly new (LISP machines date back to the mid-1970s), is not, for scientists, a conventional platform for working problems. The Connection Machine, a data-parallel architecture computer, is able to quickly map out the level curves of the integral in a one-degree-of-freedom system.

9.1. Understanding Long-Term Behavior

If there were only two point masses in an otherwise empty universe, their relative motion would be simple: as Kepler knew, the bodies would trace out ellipses about their mutual center of mass, with that center of mass at one focus. In the real universe, of course, there are many bodies. Moreover, each body, such as the earth, has a nonuniform distribution of mass. Both of these facts affect the motion of orbiting bodies: for example, a satellite orbiting the earth, while approximately tracing out a Keplerian ellipse, is in fact slightly perturbed by the nonuniform distribution of mass in the earth and, if the orbit is high enough, by the mass of the moon.

Many of these systems are still not fully understood. Among the open problems, in addition to the artificial satellite problem, are the restricted three-body (earth-moon-sun) and planetary problems. The reasons for wanting to know the long-term behavior of these systems are manifold. Increasingly, engineers and mission planners want to find a stable orbit, one where the ellipse remains fixed, to reduce the need for on-board fuel and constant orbit monitoring and correction. In addition, some orbits have become so popular that precise knowledge of long-term behavior is mandatory to avoid potential collisions. In the lunar theory, there is a need to predict the precise position of the moon for laser ranging studies. Initially, to gain insight into the dynamics, we explore where the equilibria are, what their stability is, and where the bifurcations occur. Further along in the study, it is useful to have more quantitative understanding, so that, e.g., satellites may be launched or the moon's position may be precisely determined.

Consider the zonal problem of the earth-orbiting satellite theory. In this case, we describe the earth's gravitational potential as a Keplerian term plus Legendre polynomial perturbations in the latitude, so the Hamiltonian is

$$(9.1) \qquad \mathcal{H} = \frac{1}{2}\left(R^2 + \frac{\Theta^2}{r^2}\right) - \frac{\mu}{r}\left(1 - \sum_{n\geq 2}\left(\frac{\alpha}{r}\right)^n J_n P_n(\sin B)\right),$$

where B is the latitude, $\sin B = \sin\theta\sqrt{1 - N^2/\Theta^2}$, the J_n are constants that describe the earth's mass distribution, α is the earth's radius, and μ is the Keplerian constant. The phase space coordinates r, θ, ν are, respectively, the distance from the center of the earth to the satellite, the angle in the plane of the orbit between the equatorial plane and the radius vector to the satellite, and the right ascension of the ascending node (the angle to the intersection of the orbital and equatorial planes). Their conjugate momenta are designated by the capital letters R, Θ, and N. This is a good model for satellites in low-earth orbit, as the satellite will not be high enough to be significantly influenced by the moon, and because of the nature of the orbit, the effects of the longitudinal variation in the earth's mass distribution will average out.

In the context of Hamiltonian systems, the normal form method, presented by Poincaré (and later by von Zeipel) as one of his *méthodes nouvelles*, is a common means of extracting the long-term behavior. The effect of the normalization is to give, from the original problem of the dynamics of, say, the satellite, the dynamics of the Keplerian ellipse: that is, the motion of the satellite *on* the ellipse is discarded, leaving us with the much slower motion *of* the ellipse. One obtains a canonical transformation from the original set of variables to a new set by creating a generating function of mixed old and new variables, such that the Hamiltonian in the new variables does not have the fast-phase behavior. Because of the implicit nature of the solution, however, it is hard to obtain the explicit averaging transformation and, thus, to find the correct generator and, once obtained, to solve for the actual transformation.

The Lie methods of Hori and Deprit for normalizing Hamiltonians, developed in the 1960s, have the same ultimate goal as Poincaré's, but overcome its drawbacks: with the generating function explicitly in terms of one set of variables, the effect of the transformation on an arbitrary function is easy to determine; moreover, the solution is easy to express as a recursive algorithmic procedure for a symbolic algebra processor on a computer.

In the normalized problem, we prefer to work with Delaunay coordinates ℓ, the mean anomaly (2π times the area swept out by the radius vector as a fraction of the total area of the ellipse), g, the argument of perigee (angle from the node to the point of closest approach), and h, the right ascension of the ascending node $\nu = h$, with their conjugate momenta L, G (the total angular momentum) and H (the polar component of the angular momentum). These variables have the advantage that all fast-phase behavior is contained in one variable (ℓ) rather than spread through two variables (r, θ). In the Kepler (unperturbed) problem, the Hamiltonian is a function only of L; in the full perturbed problem it is a function of all variables but h, and in the normalized problem, it is a function of all but ℓ and h. H is an integral in all cases, and L is an integral also in the normalized system.

Thus, we have extended a symmetry (in ℓ) of the unperturbed problem to the perturbed problem. All the long-term dynamics is now in one degree of freedom, g and G; together with the integrals H and L, we may describe the dynamics of the ellipse: the sine of the inclination of the orbital plane is $s = \sin I = \sqrt{1 - H^2/G^2}$, the eccentricity of the ellipse is $e = \sqrt{1 - G^2/L^2}$, and g itself gives the orientation of the ellipse in its plane.

The normalized Hamiltonian has one degree of freedom, yet the phase space is not a plane. Because g is an angular variable we must identify the points $g = 0$ and $g = 2\pi$ for all G, thus making phase space a cylinder. Further, for circular orbits ($G = L$) perigee has no meaning, and for equatorial orbits ($G = H$) the ascending node has no meaning, so we identify all values of g to one point at each of these two values of G; therefore, topologically, we have a sphere. We may choose to view the radius of the sphere to be dependent on the integral H, the longitude to be g, and the latitude to be related to G.

9.2. Symbolic Computation

The difficulty of normalizing a real system is the size and scope of the problem. There could be tens of thousands or even hundreds of thousands of terms, depending on how complex the physics and to what order we carry normalization. One complication is the implicit dependence of some variables on others; for example, r depends on ℓ implicitly through a transcendental equation, Kepler's equation. Another is the rather mundane task of algebraic simplification: Given multiple ways of writing an expression, which do we choose and why?

All this leads one to symbolic algebra manipulation on the computer. Commercial general-purpose packages provide a multitude of mathematical capabilities for all manner of complicated expressions. Ironically, they are poorly equipped to efficiently process huge expressions that belong to a restricted class of algebraic formulas, the Poisson series, that characterize these problems. In response to this, many people have developed specialized processors, dating back to the 1960s. Today, we use the code MAO (mechanized algebraic operations) of Miller and Deprit written in an object-oriented style of LISP for a Symbolics LISP machine.

MAO allows us to structure the problem in an algebra, either polynomial or Fourier (Poisson), built over a domain of coefficients that is itself an algebra, and so on, all the way down to a domain of numbers, e.g., the rationals (Figure 9.1). With this hierarchy and the object-oriented philosophy, we are able to isolate algebraic operations to a particular algebra. In addition to thinking through the problem algebraically and structuring it appropriately for the computer, there are different mathematical strategies that can significantly reduce the computational load. The canonical transformation involved in the normalization may be broken into two or more transformations, which we consider as successive simplifications. Although it seems more complicated, the total number of terms computed can be greatly reduced, thereby removing much of the burden on the computer.

As an example of the end result of a normalization, the so-called main problem (J_2 perturbation only) of the satellite theory yields, to second order, with $\eta = G/L$, $\beta = 1/(1+\eta)$, and $p = G^2/\mu$:

$$\mathcal{H}^* = -\frac{1}{2} + \frac{\alpha^2}{p^2} J_2\eta\left(\frac{3}{4}s^2 - \frac{1}{2}\right)$$

$$+ \frac{\alpha^4}{p^4} J_2^2\eta\left(\eta^2\left(-\frac{15}{128}s^4 - \frac{3}{16}s^2 + \frac{3}{16}\right) + \eta\left(-\frac{27}{32}s^4 + \frac{9}{8}s^2 - \frac{3}{8}\right)\right.$$

$$-\frac{105}{128}s^4 + \frac{15}{8}s^2 - \frac{15}{16}$$

$$+ \left(e^2\beta^2\left(\frac{15}{16}s^4 - \frac{3}{4}s^2\right) + e^2\beta\left(-\frac{15}{8}s^4 + \frac{3}{2}s^2\right)\right.$$

$$\left.\left. + e^2\left(-\frac{45}{64}s^4 + \frac{21}{32}s^2\right)\right)\cos 2g\right) + \mathcal{O}(J_2^2)$$

as the final Hamiltonian in units of μ^2/L^2. Note that there is no dependence on ℓ. This result, assuming knowledge of the methods, is not very difficult to

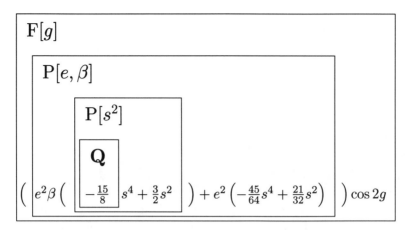

FIG. 9.1. *Hierarchy of algebras with a sample expression. The Fourier algebra in g has coefficients that are polynomials in e and β, which in turn has coefficients that are polynomials in s^2, which finally has rational coefficients.*

compute by hand. However, at higher orders, or with other zonal coefficients J_n, the task would be formidable.

For numerical evaluation, we convert the expressions from MAO to LISP or code for the Connection Machine using a simple pattern matcher, itself written in LISP, that also does constant folding for efficient numerical evaluation. For publication, expressions in MAO may be inserted as LaTeX directly into an editor buffer on the Symbolics via a built-in function, thereby eliminating the possibility of transcription error.

Our task is not complete when the normal form for the original Hamiltonian has been obtained: we still don't know the behavior qualitatively. By looking at the equations of motion, we may to lowest order extract the equilibria equations analytically. For example, we can solve the equations of motion ($\dot{G} = \partial\mathcal{H}/\partial g = 0$ and $\dot{g} = -\partial\mathcal{H}/\partial G = 0$) for the Hamiltonian above to first order. In that case, we find the so-called *critical inclination*: if $s^2 = 4/5$ ($I \approx 63.435°$), regardless of the value of G, there will be an equilibrium. At higher order, the degeneracy in G is broken. For a range of values of the parameter H near the bifurcation of the critical inclination equilibria, we can solve the second-order equilibria equations with an analytic Newton–Raphson method implemented in a general-purpose symbolic manipulation package such as Macsyma or Mathematica. As we decrease H, or at higher order, the analytic solutions become impractical because of the complexity of the expressions, and we must resort to numerical solutions to find the equilibria. The stability may be determined analytically for small equations, but we will again need numerical methods as they grow more complicated. Ultimately, though, we would like to know not just the stability, but to have a snapshot of the flows in some region of phase space.

9.3. Painting the Reduced Phase Space and Parallel Computation

One obvious way to find out what is happening is to do a numerical integration. While this is certainly a useful tool, it has a couple of problems. The first is speed: integration can take quite a long time, especially on a serial machine. Second, and perhaps more important, one must have an understanding of the answer in order to pick the right initial conditions. But an understanding is precisely what we are trying to obtain. How will we discover equilibria that are unknown to us initially?

Luckily, there is a solution. Recall that our system has one degree of freedom after the normalization; the variables are g and G. Because we have an integral, the Hamiltonian, the system is integrable. In such a case, we may exploit the fact that the flows are the level curves of the Hamiltonian. We need only identify bands of constant energy to find the flows; by evaluating the Hamiltonian at many different points and mapping the values into a color, we will see these bands. The mapping, however, is not straightforward, as we need to take special measures to insure that even subtle variations in the value of Hamiltonian show up where there might be an equilibrium. Therefore, we use a nonlinear mapping of values into colors; details are given in [4]. Furthermore, we would like to show adjacent orbits—particularly in the neighborhood of an equilibrium—with high contrast. Therefore, we repeat bands of color. Figure 9.2 shows the picture for the critical inclination; however, viewing the picture in color improves visibility.[8] More detail and other dynamical features are discussed in [1]. Figure 9.2 is a view looking at the north pole of the spherical phase space, so that the center represents circular orbits and the angle around the center is the argument of perigee. At the center is a stable equilibrium; around it are the four critical inclination equilibria apparent at second order, two $(g = 0, \pi)$ and two unstable $(g = \pi/2, 3\pi/2)$. As additional orders are computed or parameters such as H are changed, the picture can change radically.

One advantage of this method is that while it does involve a substantial amount of numerical computation in the form of a function evaluation, it is the same function being evaluated over and over again, with different arguments depending on the position in phase space. It is thus a ripe candidate for the "data-parallel" model of computation pioneered by the Connection Machine and most recently embodied in the Fortran-90/High-Performance Fortran language. The nearly linear speedup of such an "embarrassingly parallel" problem encourages easy exploration by playing with parameters to see how the dynamics changes, or to make a movie by having a single parameter stepped through a range of values and see where bifurcations occur and how equilibria move. By seeing the quick snapshot of phase space, we can respond by trying new and potentially interesting values of the parameters, or other equations,

---

[8]A color version of this figure is available from the SIAM WWW server at http://www.siam.org/books/astfalk/.

FIG. 9.2. *A view of the northern hemisphere of the reduced phase space for the main problem (J_2 only) in satellite theory. The central point represents a circular satellite orbit, with increasing distance from the center representing increasing eccentricity of the orbit and decreasing inclination. The angle around the central point is the argument of perigee.*

or we can go back to our other tools to investigate more thoroughly.

Parallel computation can also bring great benefit to the practical problem of following orbits of many satellites or processing of large numbers of satellite observations. These operations mean using the same algorithm on many pieces of data, which is precisely the data-parallel computational model. Much work has been done along these lines; for example, one may use parallelism to detect close approaches of satellites and thus assess potential collision hazards [5]. Therefore, many aspects of satellite orbit dynamics lend themselves to parallel computation.

9.4. Conclusion

The availability of powerful computers with novel architectures and software environments has revolutionized research about perturbed Keplerian systems. By rethinking problems to match the new capability, and by applying a variety of techniques, we may gain insight and improve our understanding of these systems. New software allows us to deal with the problem at an abstract level, to think about and see the mathematics in a way more like that to which humans are accustomed and less like that traditionally demanded by computers.

Already, the tools have combined to provide substantial progress in the main problem in the theory of the artificial satellite and other problems involving perturbed Keplerian systems. Even outside the realm of celestial mechanics these techniques have applicability. In atomic physics, the Stark–Zeeman problem, that of combined electric and magnetic fields, may be better understood to high orders. In many areas of the physical sciences, people have shied away from tackling these problems principally because of the complexity of the algebra. Perhaps this will now change.

Advancing the understanding of the dynamical systems and advancing the tools go hand in hand [2]. Given the obvious benefit of using the Connection Machine for the numerical computation necessary for the graphics, we naturally wonder if similar benefits could be derived in computer algebra. Already, steps have been taken to advance work in this area [3]. Meanwhile, improvements in the graphical techniques are contemplated to expand the amount of information in a picture, to increase accessibility of the graphics at remote sites, and to better locate bifurcations and equilibria.

9.5. Acknowledgments

This is a report on the joint work of Shannon Coffey, Etienne Deprit, and the author at the Naval Research Laboratory, and André Deprit and Bruce Miller at the National Institute of Standards and Technology.

References

[1] S. COFFEY, A. DEPRIT, E. DEPRIT, AND L. HEALY, *Painting the phase space portrait of an integrable dynamical system*, Science, 247(1990), pp. 769–892.

[2] S. COFFEY, A. DEPRIT, E. DEPRIT, L. HEALY, AND B. MILLER, *A toolbox for nonlinear dynamics*, in Computer-Aided Proofs in Analysis, K.R. Meyer and D.S. Schmidt, eds., The IMA Vol. Math. Appl., 28(1991), pp. 97–115.

[3] A. DEPRIT AND E. DEPRIT, *Processing Poisson series in parallel*, J. Symbolic Comput., 10(1990), pp. 179–201.

[4] L. HEALY AND E. DEPRIT, *Paint by number: Uncovering phase flows of an integrable dynamical system*, J. Comput. Phys., 5(1991), pp. 491–496.

[5] L. HEALY, *Close conjunction detection on a parallel computer*, J. Guidance Control Dynamics, 18(1995), pp. 824–829.

Chapter 10

Parallel Methods for Systems of Ordinary Differential Equations

Kevin Burrage

Editorial preface

Partial differential equations are often the central focus for high-performance computing. However, ordinary differential equations (ODEs) also occupy an important role in scientific calculations. This chapter surveys a number of approaches using parallelism to solve ODEs. The author also illustrates implementation methods for single-instruction multiple-data (SIMD) architectures as well as multiple-instruction multiple-data (MIMD) architectures. This chapter is an excellent starting point for those people who need to understand the issues involved in solving ODEs in parallel.

This article originally appeared in *SIAM News*, Vol. 26, No. 5, August 1993. It was updated during the summer/fall of 1995.

Considerable attention has been devoted recently to the development of efficient parallel algorithms for the numerical solution of initial value ordinary differential equations (ODEs) of the form

$$(10.1) \qquad y' = f(t, y), \quad y(t_0) = y_0, \quad f : \mathbb{R} \times \mathbb{R}^m \Rightarrow \mathbb{R}^m.$$

An example that gives an idea of the magnitude of some of the problems involved is the modeling of long-range transport of air pollutants in the atmosphere [23]. A relatively simple model generates a system of 267,264 ODEs; to study seasonal variations in the pollutants, the system must be solved over a long timescale. Clearly, such problems cannot be solved in reasonable time without some exploitation of concurrency.

In attempts to solve (10.1), three types of parallelism have been identified:

 (i) parallelism across the method,

 (ii) parallelism across the system (space),

(iii) parallelism across time.

It is highly likely that efficient parallel algorithms will take elements from all three of these categories, so that such algorithms will lie in a three-dimensional space (Figure 10.1).

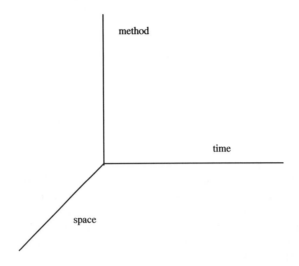

FIG. 10.1.   *The parallelism space.*

In this article two algorithms are introduced, and numerical results are given to indicate the efficiency of these approaches. One of the algorithms is based on parallelism across the method and is suitable for implementation on a single-instruction multiple-data (SIMD) architecture—the MasPar for the work described here. The other algorithm is based on parallelism across the system and is suitable for implementation on a multiple-instruction multiple-data (MIMD) architecture—in this case the Intel Paragon.

10.1. Parallelism Across the Method

One way to exploit parallelism across the method is to perform several function evaluations concurrently on different processors. This is possible with multistage methods, such as Runge–Kutta. In general, there is little advantage in the direct approach, although extrapolation techniques, with the work evenly balanced across the processors, are well suited for parallel implementation when the problem is large or function evaluations are costly [5]. However, indirect methods, such as prediction–correction techniques, can prove efficient in a parallel setting.

10.1.1. Prediction–Correction Techniques.   A popular technique for exploiting parallelism across the method [18] is based on the concept of a block method, in which a block of values is predicted concurrently by some explicit method from a set of previously computed values, which are then corrected a number of times by an implicit method with a fixed-point approach.

An example of this approach is the block method in which an explicit Euler predictor is used to update a set of k values concurrently at equidistant points $t_{n+1}, \ldots, t_{n+k}$; the values are then corrected twice by a trapezoidal corrector. This method can be computed in three steps:

$$(10.2) \quad \begin{aligned} y_{n+j}^{(0)} &= y_n + jh\, f(t_n, y_n), & j &= 1, \ldots, k, \\ y_{n+j}^{(1)} &= y_n + \tfrac{h}{2} f(t_n, y_n) + \tfrac{h}{2} f(t_{n+j}, y_{n+j}^{(0)}), & j &= 1, \ldots, k, \\ y_{n+j}^{(2)} &= y_n + \tfrac{h}{2} f(t_n, y_n) + \tfrac{h}{2} f(t_{n+j}, y_{n+j}^{(1)}), & j &= 1, \ldots, k. \end{aligned}$$

Although this method is very simple, it is illustrative of a much more general technique in which a block of k values, with components $y_1^{(n)}, \ldots, y_k^{(n)}$, are computed concurrently from step to step based on an Hermite predictor,

$$(10.3) \qquad\qquad Y^{(0)} = A_0 \otimes Y_n + hL_0 \otimes F(Y_n),$$

and an implicit corrector (with $Z_n = A_1 \otimes Y_n + hL_1 \otimes F(Y_n)$),

$$(10.4) \qquad Y^{(j)} = Z_n + hL_2 \otimes F(Y^{(j-1)}), \quad j = 1, \ldots, r,$$

where $F(Y_n)$ denotes the vector with components $f(y_1^{(n)}), \ldots, f(y_k^{(n)})$.

In general, however, such methods can have the disadvantages of poor stability and/or large error coefficients unless a large number of corrections are performed [4]. As a simple rule, each time a correction of the form (10.4) is performed, the order of the method increases by one until the order of the corrector is reached [2]. Further corrections do not increase the order of the method but do smooth out successively higher and higher truncation coefficients in the local error expansion. For large block sizes, however, the extrapolation error in the predictor can be very large, and it can take many corrections before acceptable accuracy is guaranteed. Nevertheless, the efficiency of this approach can be dramatically improved by the use of splitting techniques (which can be interpreted as a preconditioning) applied directly to the underlying corrector in (10.4).

This approach gives rise to a general iteration scheme of the form

$$(10.5) \quad M_{k,n} Y_{n+1}^{(k+1)} = (M_{k,n} - I) Y_{n+1}^{(k)} + Z_n + hL_2 \otimes F(Y_{n+1}^{(k)}), \quad k = 0, 1, \ldots.$$

In the case that

$$(10.6) \qquad\qquad M_{k,n} = I \quad \forall k, n,$$

then (10.5) gives the standard prediction–correction approach, which is just fixed-point iteration, while if

$$(10.7) \qquad\qquad M_{k,n} = I - hL_2 \otimes J_n \quad \forall k,$$

where J_n is the Jacobian of the problem evaluated at some point y_n, then (10.5) represents a modified Newton approach.

The $M_{k,n}$ can be chosen intermediate to the choices in (10.6) and (10.7) in an attempt to obtain both good convergence properties and cheap implementation in a parallel environment.

If, by definition,

$$(10.8) \qquad e_{n+1}^{(k+1)} = Y_{n+1}^{(k+1)} - Y_{n+1},$$

then a linearization of the problem gives

$$(10.9) \qquad e_{n+1}^{(k+1)} = R_{k,n}e_{n+1}^{(k)}, \quad R_{k,n} = I - M_{k,n}^{-1}(I - hL_2 \otimes J_n).$$

Another way of viewing this is to apply the underlying corrector in (10.4) to the linear problem

$$(10.10) \qquad y'(t) = J(t)y,$$

which gives

$$(10.11) \qquad PY^{(k)} = Z_n, \quad k = 1, \dots, r, \quad P = I - hL_2 \otimes J_n.$$

Thus, the choice of $M_{k,n}$ in (10.6) and (10.7) represents a preconditioning of the matrix P, which will make possible an acceleration of (10.5). If the eigenvalue structure of the underlying problem (10.1) is known (as is often the case, for example, with problems arising from the solution of parabolic partial differential equations by the method of lines), then polynomial preconditioning is a well-known procedure for accelerating the convergence. For example,

$$(10.12) \quad \begin{array}{l} \text{TYPE I}: \quad M_k^{-1} = \alpha_k I \Rightarrow R_{k,n} = I - \alpha_k P; \\ \text{TYPE II}: \quad M_k^{-1} = \alpha_k I - \beta_k P \Rightarrow R_{k,n} = I - \alpha_k P + \beta_k P^2. \end{array}$$

If the eigenvalues of J_n are real and lie in the interval $[-q, 0]$, $q > 0$, the rate of convergence over p iterations can be maximized by minimizing $\rho((\prod_{k=1}^p R_{p-k,n})^{1/p})$, where $\rho(H)$ denotes the spectral radius of H.

Analysis with Chebyshev polynomials [4] shows that the spectral radii of the amplification matrices, as functions of $z = hq$, are minimized with

$$(10.13) \quad \begin{array}{l} \alpha = \frac{2}{1+v}, \quad \rho(R) = 1 - \alpha, \\ \alpha = \frac{8(1+v)}{1+6v+v^2}, \quad \beta = \frac{\alpha}{1+v}, \quad \rho(R) = 1 - \alpha + \beta, \\ \alpha_k = \frac{4}{4v+s_k^2(1-v)^2}, \quad \beta_k = \frac{\alpha_k}{1+v}, \quad s_k^2 = \sin^2\left(\frac{(2k-1)\pi}{4p}\right), \quad k = 1, \dots, p. \end{array}$$

Here $v = \det(I + zL_2)$, which in the case of the trapezoidal rule is $1 + z/2$.

The advantage of this approach is that the implementation properties are similar to those of explicit methods, while the stability properties are similar to those of A-stable implicit methods. The computational savings arise from the fact that the corrections can be calculated from simple matrix–vector operations. No solutions of linear systems, which can be difficult to program efficiently in a parallel environment, are required. This approach is particularly appropriate for SIMD computers, such as the MasPar, and has been programmed on a 4K-processor MasPar MP-1 at the University of Queensland.

10.1.2. The MasPar. An array computer is a large collection of processing elements (PEs) arranged in a mesh topology or some close derivative. Typically, the PEs operate in a synchronized fashion, with all the PEs performing the same instruction in lockstep but on different data.

Since many modeling problems in, for example, fluid mechanics, stress analysis, and spatial modeling can be approximated by a spatially discretized mesh, there is a natural processor topology that allows automatic parallelization through the use of Fortran-90 constructs. These constructs are, in particular, the BLAS routines (see below).

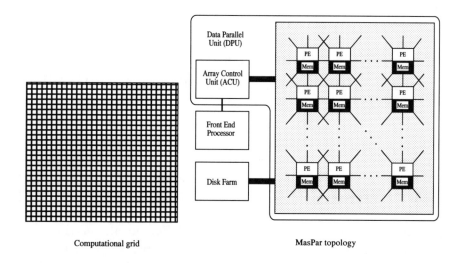

FIG. 10.2. *The computational space and network topology.*

The MasPar MP-1 at Queensland consists of a front-end UNIX workstation, which performs the serial part of the computation, and a back-end data parallel unit (DPU). The DPU consists of an array control unit (ACU) and an array of PEs; see Figure 10.2. The ACU handles all the program code for the scalar variables. The processing element consists of a 4-bit processor with 64 Kbytes of memory; sustainable performance is about 50 Kflops per PE.

Communication is via the XNET, which is a lockstep interprocessor communication protocol, through the eight nearest neighbors. Alternatively, a global router allows for arbitrary processor-to-processor communication via a three-stage switch router.

The preconditioning approach described in the previous section has been coded in MPFortran, which is based on the new Fortran-90 standard. Because the only operations required are BLAS-type operations, full advantage is taken of the parallelization by the Fortran compiler, which automatically partitions arrays and vectors on the DPU. Comments and numerical results for this implementation are given later in this article.

10.2. Parallelism Across the System

Perhaps the simplest way to exploit parallelism across the system is through the concept of Picard iteration and, more generally, iteration in the function space. Picard's method for obtaining global approximations to the solution of (10.1) is based on the solution of a sequence of functional iterations of the form

$$(10.14) \qquad y^{(k+1)\prime}(t) = f(t, y^{(k)}(t)), \quad y^{(k+1)}(t_0) = y^{(k)}(t_0).$$

In this case the problem can be split into m independent quadrature problems at each iteration level, an approach that appears to be appropriate for obtaining massive parallelism. Unfortunately, the convergence of the iterates $\{y^{(k)}(t)\}$ to $y(t)$ is very slow. For example, for the standard linear test problem

$$(10.15) \qquad\qquad y' = \lambda y, \quad t \in [0, T], \quad \lambda < 0,$$

it can be shown that the iterates $y^{(k)}(t)$ satisfy the following global error bound:

$$(10.16) \qquad\qquad \left| y(t) - y^{(k)}(t) \right| \le \frac{(|\lambda| t)^{k+1}}{(k+1)!}, \qquad t \in [0, T],$$

so that there is no convergence until $k \ge |\lambda| T$. Thus, one way to improve the convergence is by the technique of windowing, in which the region of integration is split into a series of windows, with the iterative process then taking place on each window.

The first extensive study of more general functional iteration schemes, which were given the name "waveform relaxation," was done by the electrical engineering group at the University of California, Berkeley in the early 1980s [16, 22]. With this technique, standard iteration schemes for the solution of linear systems can be applied directly to the differential system to create a sequence of differential systems, each of which can be solved by some discrete method, that converge to the solution of (10.1). Thus, Jacobi and Gauss–Seidel waveforms are described, respectively, as

$$(10.17) \quad y_i^{(k+1)\prime}(t) = f_i(y_1^{(k)}, \dots, y_{i-1}^{(k)}, y_i^{(k+1)}, y_{i+1}^{(k)}, \dots, y_m^{(k)}), \quad i = 1, \dots, m$$

and

$$(10.18) \quad y_i^{(k+1)\prime}(t) = f_i(y_1^{(k+1)}, \dots, y_i^{(k+1)}, y_{i+1}^{(k)}, \dots, y_m^{(k)}), \quad i = 1, \dots, m.$$

It is easy to prove results analogous to (10.16) for the nonlinear case, but again the implication is that convergence can be very slow. Nevertheless, for certain classes of problems, such as the integrated circuit design problems studied by the Berkeley group, waveform relaxation techniques can work very well indeed. This is because the physicality of the model suggests how the components can be grouped in tightly coupled subsystems, with the coupling occurring only over very short time intervals.

In general, however, difficulties arise both in choosing the components to be grouped and in reordering the equations that are crucial to the efficiency of waveform relaxation (see [11] and [13]). It was observed that convergence of the iterations can be slow in the case of strong coupling between subsystems. Gear and Juang [13] have examined the rate of increase of the order of accuracy of the iterates, and a speed of convergence can be defined in terms of the average number of additional terms in the power series expansion that are correct in each iteration. In the case of waveform Jacobi the accuracy increase is one, while for waveform Newton the accuracy increase doubles per iteration. For waveform Gauss–Seidel the increase is greater than one and depends on the ordering of the system components and the cycles in the directed dependency graph of the differential system (see also [4]).

Recently, multigrid acceleration techniques have been applied directly to linear problems arising from the solution of linear parabolic differential equations by the method of lines. It has been shown [17] that these multigrid techniques can dramatically accelerate the convergence behavior of such iteration schemes. This work has been extended in [21] to nonlinear problems.

A preconditioning approach is also used for accelerating the convergence of the waveform process. This process is completely analogous to those techniques used in the static case for linear systems of equations of the form $Qy = b$. Thus Burrage et al. [8, 9] have considered the acceleration of the waveform process by preconditioning on the left or right. In the case of right preconditioning this consists of transforming the original equation by an appropriate time-dependent transformation that will accelerate the waveform convergence of the transformed problem.

Multigrid acceleration or preconditioning will not be used for the numerical work presented here, but we will consider the use of overlapping the subsystems so that some components can appear in two neighboring blocks. This is a generalization of the work by O'Leary and White [19] for algebraic systems of equations and was adapted by Jeltsch and Pohl [15] to systems of ODEs. The effect of this is to reduce the coupling effects and allow faster convergence (see Burrage, Jackiewicz, and Renaut [8]).

10.2.1. Distributed Computing. The waveform approach is suitable for implementation in a distributed environment, because it allows a decoupling of the original problem into subproblems that can then be solved more or less independently of one another on different processors (depending, of course, on the nature of the coupling of the components in the original problem). This approach allows the programmer to take existing sequential codes that are known to be efficient and robust in a sequential environment and to apply them to the set of subproblems. One such code is VODE [1], which is based on the Adams and backward difference formula (BDF) methods and is suitable for both stiff and nonstiff problems. Through this type of decoupling we guarantee a portability and robustness of the parallel code through the same properties of the sequential solver that has been under continual development through its offspring EPISODE, LSODE, etc. for approximately the last twenty years.

VODE has a user interface that is almost identical to LSODE but improves on the efficiency of problems that require frequent and large changes in step size. In the solution of the nonlinear systems for the update point, VODE offers a choice of either functional iteration or modified Newton iteration in which the Jacobian is either user supplied or computed internally. VODE also caters to full or banded problems. Since most of the computational work in any differential equation solver involves the solution of the nonlinear equations defining the update, VODE uses a number of techniques to reduce this work. These techniques include repeatedly using the LU factors of the amplification matrix until convergence properties deteriorate as well as accelerating the convergence of the iterations within the Newton step by relaxing the amplification matrix based on estimates of the extreme eigenvalues of the problem. At the end of each step within VODE, the local error is estimated and a step size change is considered for the current step or subsequent steps depending on the magnitude of the error.

VODE uses direct linear algebra techniques for solving the linear systems associated with implicit methods. Recent advances in iterative techniques for linear systems based on preconditioned Krylov generalized minimum residual (GMRES) methods [20] have meant that these approaches can now be incorporated into ODE solvers, and this has been done within VODEPK. VODEPK has a very similar structure to VODE except that a preconditioned GMRES technique based on a scaled preconditioned incomplete version of GMRES (SPIGMR) is used for the linear solver.

In the tests presented in this paper, however, VODE is used as the core solver in the waveform relaxation code that will henceforth be known as PWVODE.

As a consequence of the use of a standard ODE integrator in PWVODE, the programmer now has to focus only on the communication protocols, which can be programmed in some generic message-passing environment, such as PVM (Parallel Virtual Machine) [14] or P4 (portable programs for parallel processors) [10]. These software environments are suited for Fortran-77 or C programs consisting of subtasks that offer highly granular parallelism. Both environments are based on the message-passing model, allowing message transmission, barrier synchronization, and broadcast. An essential difference between PVM and P4 is that PVM uses a *pvmd* daemon to control the status of the processes, whereas P4 does not, so that P4 communication on distributed-memory machines is done by native message-passing, whereas PVM utilizes the daemon.

A code has been implemented that uses VODE as the basic integrator and a standard message-passing environment such as the Message-Passing Interface (MPI) for the message-passing "glue." The code is based on a block Jacobi multisplitting technique. The first version of PWVODE uses a static load balancing in which the work is split more or less evenly among the processors at the beginning of the integration, and this loading then remains fixed. This version also used a fixed window length that could be chosen by the user. Within PWVODE there is a default setting that gives the number of output points in a given window. This is set to 40. An option in VODE is used to calculate an approximation to the solution at exactly these points. The waveform for that sweep is then calculated by a piecewise linear approximation.

Care must be taken in choosing the convergence criteria for the problem. In fact there are two criteria: the waveform tolerance and the tolerance used by VODE. The waveform tolerance (ϵ_{wr}) is used to determine whether successive waveforms are sufficiently close. The maximum absolute difference between successive waveforms over all components and all timepoints is computed. The iterates are deemed to have converged once this difference is less than ϵ_{wr}. The VODE tolerance ϵ_{vo} controls the local error and the choice of step-size.

Numerical results are presented in the next section. The advantage of these general message-passing environments is that a code can be debugged and tested on a network of workstations and then ported to a larger message-passing MIMD computer such as the Intel Paragon with no additional changes.

10.3. Numerical Results

To demonstrate some of the material presented earlier, two test problems, both involving two-dimensional partial differential equations, are presented. Although they can be considered as test problems rather than "real life" problems, they do allow a number of parallel aspects to be examined such as load balancing, communication protocols, static versus dynamic scheduling, etc.

The first problem is the linear diffusion equation defined on the unit square:

$$(10.19) \qquad \frac{\partial u}{\partial t} = \frac{\partial^2 u}{\partial x^2} + \frac{\partial^2 u}{\partial y^2},$$

with Dirichlet boundary conditions given by

$$u(t, x, 0) = u(t, x, 1) = u(t, 0, y) = u(t, 1, y) = 1,$$

which can be converted to a system of ODEs by the method of lines. If the second-order spatial derivatives are replaced by central finite differences on a uniform grid, with the grid discretization parameter given by $h = \frac{1}{N+1}$, the resulting linear system of differential equations of size N^2 will be of the form

$$(10.20) \qquad u' = (N+1)^2 Qu, \quad u(0) = 1.$$

Here Q is a block tridiagonal matrix of the form (I_N, T, I_N), where I_N is the identity matrix of order N and T is the tridiagonal matrix $(1, -4, 1)$ with -4 on the diagonal entries and 1 on the upper and lower subdiagonal entries. For this problem, q can be $8(N+1)^2$, so that the eigenvalues of Q lie in the interval $[-q, 0]$.

The second problem, a reaction–diffusion equation known as the diffusion Brusselator equation [21], takes the form

(10.21)
$$\frac{\partial u}{\partial t} = B + u^2 v - (A+1)u + \alpha\left(\frac{\partial^2 u}{\partial x^2} + \frac{\partial^2 u}{\partial y^2}\right),$$
$$\frac{\partial v}{\partial t} = Au - u^2 v + \alpha\left(\frac{\partial^2 v}{\partial x^2} + \frac{\partial^2 v}{\partial y^2}\right),$$

with initial conditions

$$u(0, x, y) = 2 + 0.25y, \quad v(0, x, y) = 1 + 0.8x, \quad A = 3.4, \quad B = 1, \quad \alpha = 0.002$$

and Neumann boundary conditions

$$\frac{\partial u}{\partial n} = 0, \quad \frac{\partial v}{\partial n} = 0.$$

Here u and v denote chemical concentrations of reaction products, A and B are constant concentrations of input reagents, and α is a constant based on a diffusion coefficient and a reactor length.

Again, central differencing leads to a system of coupled nonlinear equations of order $2N^2$ (with $\hat{\alpha} = \alpha(N+1)^2$) of the form

$$u'_{ij} = B + u^2_{ij}v_{ij} - (A+1)u_{ij} + \hat{\alpha}(u_{i+1,j} + u_{i-1,j} + u_{i,j+1} + u_{i,j-1} - 4u_{ij}),$$
$$v'_{ij} = Au_{ij} - u^2_{ij}v_{ij} + \hat{\alpha}(v_{i+1,j} + v_{i-1,j} + v_{i,j+1} + v_{i,j-1} - 4v_{ij}).$$

10.3.1. SIMD Implementation.

The linear problem and the one-dimensional form of the Brusselator have been solved on the 4K MasPar by a fixed step-size scheme based on the trapezoidal corrector. The computational results can be summarized as follows:

1. In the case of the linear problem of dimension N^2, some care must be taken in choosing N. If $N = 65$ and the number of available processors is $64 \times 64 = 4096$, for example, the computational time will be approximately twice as long as in the case of $N = 64$. The reason is that the MasPar automatically "layers" the computational grid into memory, so that two layers are required when $N = 65$. This layering is done automatically and does not require programmer intervention. On the other hand, the time required to solve a problem of any dimension N^2, with $N \leq 64$, should be approximately the same, although it can depend on the machine load.

2. Although the implementation described in §10.1 requires only BLAS operations of the form Qv, where Q is as in (10.20) and v is a vector defined on all the elements of the computational grid, it is important to structure the problem so that this is done efficiently. Accordingly, v is represented as an $N \times N$ matrix, and Qv is formed as a sequence of EOSHIFTs:

```
EOSHIFT(v,SHIFT=-1,BOUNDARY=f1,DIM=1) +
EOSHIFT(v,SHIFT=-1,BOUNDARY=f2,DIM=2) - 4.0*v +
EOSHIFT(v,SHIFT=+1,BOUNDARY=f3,DIM=2) +
EOSHIFT(v,SHIFT=+1,BOUNDARY=f4,DIM=1).
```

Here SHIFT represents a shift up or down the computational grid, DIM represents column or row shifts, and $f1$, $f2$, $f3$, $f4$ are the boundary conditions.

3. For the one-dimensional Brusselator, two coupled vectors, each of dimension N, are automatically layered as two row vectors onto the MasPar topology. For a type II implementation (see (10.12)) a Jacobian matrix has to be evaluated at each timestep. Because the Jacobian matrix has a simple block tridiagonal structure, with the identity matrix as the off-diagonal blocks, the vector product of the Jacobian and each of the two vectors representing the components of the problem is easily formed as a sequence of two EOSHIFTs columnwise for each vector.

4. Equation (10.20) has also been solved by a block method of size 2 based on a two-stage Radau corrector of order 3. In this case, two approximations (one a third of the way along the integration step and the other at the end of the integration step) are computed per processor. The computational time is, as expected, approximately twice that for the trapezoidal corrector.

10.3.2. MIMD Implementation.   The two-dimensional form of the Brusselator has been solved on a distributed cluster of SPARC-2 workstations and on the 96-node Intel Paragon, at ETH Zürich. Each node consists of two 50-MHz i860XP processors, with 32 Mbytes of memory and a 16-Kbyte cache. Only one processor per node is accessible for computation, and this processor has a peak rating of 75 Mflops. Before presenting the results we briefly discuss a number of issues relating to the effectiveness of any parallel waveform implementation.

10.3.3. Ordering.   Burrage and Pohl [6] observed that there are two natural ways of ordering the components of the two-dimensional Brusselator.

The first ordering is

$$u_{11}, \ldots, u_{1N}, u_{21}, \ldots, u_{2N}, \ldots, u_{NN}, v_{11}, \ldots, v_{1N}, v_{21}, \ldots, v_{2N}, \ldots, v_{NN}$$
(10.22)

and the second ordering for the components is

(10.23) $$u_{11}, v_{11}, u_{12}, v_{12}, \ldots, u_{NN}, v_{NN}.$$

If the first ordering is used, the Jacobian of the problem has the structure

(10.24)
$$\begin{array}{cc} T_1 & D_1 \\ D_2 & T_2 \end{array},$$

where each of the matrices are of dimension N^2, and D_1 and D_2 are diagonal. For the second ordering, the Jacobian is banded with a half-bandwidth of $2N$. If we ignore the effect of the ordering on the convergence rate of the waveforms and consider only the storage requirements of each ordering, it is apparent that the second choice is far superior. The bandwidth is a factor of $N/2$ smaller, which yields a substantial saving in storage and the linear algebra requirement. In the case of a parallel implementation, however, the situation is more complicated.

Using the ordering in (10.22), if the number of subsystems L satisfies

(10.25) $$\frac{2N^2}{3} \geq L \geq 2N,$$

then a block Jacobi approach leads to the solution of a tridiagonal problem on each processor. If $L < 2N$, then portions of the upper diagonal and lower diagonal blocks of T_1 and T_2 must be included in each subsystem. On the other hand, for the ordering in (10.23), if L satisfies

(10.26) $$\frac{2N^2}{5} \geq L \geq N,$$

then a block Jacobi approach leads to the solution of a banded problem with bandwidth 5 on each system. It is obvious that in a parallel environment the first ordering appears to be more appropriate if $L \geq 2N$, while the second ordering is preferable if $L < 2N$.

10.3.4. Communication. This analysis, however, is still incomplete because it does not address overlapping or communication issues. In the case of a dense problem, a waveform algorithm requires communication between all subsystems to update the waveform at the end of each iteration and to compute the input for the next step. Rather than perform multibroadcasts (with possible risk of deadlock) a master–slave model can be used. In this model, at the end of each waveform iteration the master collects and sends all the necessary information in order to proceed with the next iteration. This requires that each node send the computed waveform for its subsystem as well as error diagnostics to the host at the end of each iteration. This option is used as the default option if the problem is dense or if the user does not wish to exploit any structure within the problem. In this model it does not matter whether there is overlap or no overlap; nor does it particularly matter which subset of processors is allocated to a task or what the connection topology is.

If the first ordering is used it is not possible to write the problem in a block-tridiagonal form to allow for local communication in an efficient manner, whereas in the case of the second ordering this is possible, and, hence, local communication can be exploited. This means that when the waveform is to be communicated to the relevant processors for the next sweep only the processors to the left and right of a processor need communicate their results to that processor. Burrage [4] solves the two-dimensional Brusselator on 64 processors of an Intel Paragon using both a master–slave model and local communication for various overlaps. The problem dimension is $m = 5000$, and the interval of integration is $t = [0, 6]$. The second ordering of components is chosen. The executions times (in seconds) are given in Table 10.1.

TABLE 10.1

Overlap and communication times.

Overlap	0	1	2	3	4	5
Local	211.8	207.7	207.7	208.7	211.7	260.7
Master–slave	470.1	463.9	458.7	462.9	462.7	508.7

These results show that a master–slave implementation is at least twice as slow as a local communication implementation that exploits the tridiagonal structure of the problem. Although it is claimed that long messages can be sent efficiently by wormhole routing in a global manner, local communication is clearly important if the locality of data can be guaranteed.

Despite the previous comments on ordering, it has been found (see [4]) that the second ordering is much more efficient, in that the number of iterations needed to achieve satisfactory convergence is considerably less than for the first ordering. The reason for this is that there is a natural coupling between the u_{ij} and v_{ij}, which the second ordering exploits, while the first ordering breaks this coupling by putting corresponding elements u_{ij} and v_{ij} in separate blocks far removed from one another. However, the partitioning of the system for the second ordering must always be such that there is no splitting between the components u_{ij} and v_{ij}. Thus in this paper numerical results are based on the second ordering of the components.

10.3.5. Load Balancing. In spite of the apparent uniformity of the subsystems in (10.22), it has been observed that, through a study of the space–time diagram created by ParaGraph executed on the Paragon on a run of PWVODE, there are still considerable load-balancing difficulties (see Figure 10.3). Thus a dynamic load-balancing approach has been implemented by exploiting the use of overlap.

ParaGraph is an extremely useful software tool for analyzing the parallel performance of a code. The results given in Figure 10.3 show a concurrency profile for a 16-processor implementation of the code using static load balancing or dynamic load balancing, respectively, by controlling the overlap.

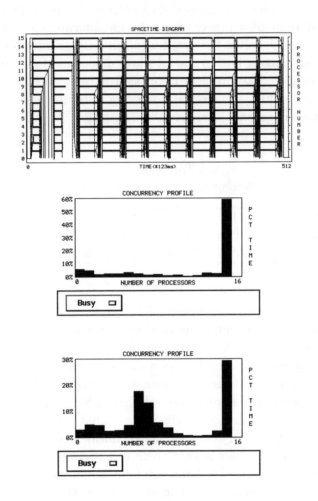

FIG. 10.3. *Space–time diagram created by ParaGraph.*

For the static load balancing it appears the full 16 processors are busy for approximately only 30% of the execution time and, for example, for 20% of the execution time only seven processors are fully busy. This suggests a poor load balancing. In the case of dynamic load balancing, however, the profile indicates that all 16 processors are busy for 60% of the time and there is an even distribution of busy time in general. This shows the success of dynamic load balancing by controlling the overlap. These results have been confirmed by analyzing a space–time diagram created by ParaGraph over the entire execution of the program. Here dynamic load balancing was enabled for the first two iteration sweeps and then turned off. The space–time diagram for the first two iterations shows that the processors are much busier than in later iterations.

Dynamic load balancing works on the assumption that initially the workload is distributed as above, with a small initial overlap θ_0 of, say, two or three. After a few waveform sweeps on a particular window, each processor i computes the average time T_i taken for one waveform sweep for its own subsystem. A native global routine is then used to compute the maximum ($T_{\max}$), minimum ($T_{\min}$), and average (T_{av}) of these times over all the processors. These numbers now reside on all the processors. Each processor now computes a load balancing factor

$$(10.27) \qquad\qquad B_i = C\frac{T_{av} - T_i}{T_{\max} - T_{\min}},$$

where C is a safety factor chosen to be 0.5 that mitigates against changing the dimensions of the subsystems by too much. This factor is now used to increase or decrease the dimension of the ith subsystem (see [4] for more details). Periodically these quantities are updated after a suitable number of sweeps.

10.3.6. Adapting the VODE Tolerances. The default settings for PWVODE implement VODE with the tolerance set at 10^{-8}. Furthermore, the initial guess is merely a matrix of constants. Therefore, it is highly unlikely that the first few iterations will be particularly accurate, and thus it seems that demanding the VODE tolerance to be 10^{-8} is a little unrealistic. A feature of PWVODE, then, is the ability to change the VODE tolerances so that initially they are relaxed and then they are progressively tightened. The user can thus specify a sequence of VODE tolerances and a corresponding sequence of iteration numbers that describes for how many iterations a given VODE tolerance is imposed. Such a modification generally results in a decrease in the total execution time by at least a factor of 2.

10.3.7. Adapting the Window Size. As we have seen, a feature of the waveform approach is that as t increases toward T, the waveforms are slower to converge; hence, smaller time windows are needed. If n windows are specified by the user, then each window is of length $(T - t_0)/n$. Although this strategy does allow some control over the length of each window, the control does not take into account the development of the waveforms. If it is expected that convergence deteriorates toward the end of the window of length $T - t_0$, then it may be ideal to use a nonuniform division of the windows (see Dyke [12]). Burrage and Dyke [7] have considered how to automate the choice of the number and length of time windows. Thus if $y^{(i)}(c,t)$ represents an approximation to the solution component c of y at time t during iteration i, and assuming that an approximation to the solution has already been calculated for iterations $i-2$ and $i-1$, then the behavior of the ratio

$$(10.28) \qquad \frac{||y^{(i)}(c,t) - y^{(i-1)}(c,t)||}{||y^{(i-1)}(c,t) - y^{(i-2)}(c,t)||}$$

may be monitored. If there are over k successive timepoints, $t, t+1, \ldots, t+k-1$, this ratio is greater than some convergence tolerance (tol) specified, in advance, by the user; then the window is rescaled from $[t_0, T]$ to $[t_0, t]$. If there are p timepoints in the original window, and the problem occurs at the kth timepoint, the new window created from $[t_0, t]$ will contain k timepoints, leaving the second window from $[t, T]$ with $p - k$ timepoints.

The window is rescaled by interrupting the integration in other subsystems. As soon as a troublesome timepoint t is encountered, integration on the involved processor ceases. The point t is sent to the master. Then the master sends emergency interrupt messages to all other subsystems. Once a subsystem has received the point t, there are two options available. First, if the timepoint t has not already been attained, then the integration proceeds until t is reached. However, it should be noted that in PWVODE a limit is placed on how many steps VODE can take in order to get to a specified point, and if this point is not reached in that number of steps, the window is shortened to the current point.

Second, if t has already been reached, then integration stops immediately. Meanwhile, the master is calculating the new number of timepoints required in each window and rescaling the window to the new endpoint. Integration then commences on each subsystem for the new window $[t_0, t]$, using the last iterate as the starting solution. In some applications it is possible that this resizing process will be necessary more than once. If this is the case, then suppose that the window has already been rescaled to the interval $[t_0, t_1]$ and that further rescaling to $[t_0, t_2]$ is necessary, where $t_2 < t_1$. Then on reaching t_2 the new window is $[t_2, t_1]$. This represents a global approach to windowing, in which the largest possible window size is chosen initially and is only reduced if appropriate.

10.3.8. Results. The results of adapting the window automatically are not given here (see Burrage and Dyke [7] for more details). The results are not conclusive. In some cases adapting can be more efficient than a fixed window implementation and sometimes it is not. It is expected that as the communication properties of the underlying architecture improve adapting would be the preferred option.

Table 10.2 shows the timings obtained for the (B) ordering compared with VODE running on a single processor on a problem of dimension 5000. In the parallel execution, L denotes the number of subsystems (with $L-1$ processors)

TABLE 10.2

Timings (secs), $m = 5000$.

VODE	L	Time
1348.66	3	13555.1
	7	6493.2
	15	4956.6
	31	991.3
	63	64.8

and the results are only given for the optimal window size and by adapting the tolerance.

It can be seen here that only when the number of processors exceeds 30 is there any improvement over VODE running on one processor. However, if 64 processors are used then the speedup over VODE is more than 20, which is very respectable. One of the reasons for this sudden improvement is due to the fact that only when $L \geq N$ (which in this case is equivalent to $L \geq 50$) does the (B) ordering give a small bandwidth (5) on each subsystem. Otherwise the bandwidth is $2N$. Caching can also have an effect for large systems with a small number of processors. These results are illustrated in Figure 10.4 in terms of speedup versus number of processors.

The final graph of results (Figure 10.5) gives a summary of speedups obtained for differing window sizes. The three curves show, respectively, the effects of both adapting the tolerance and using local communication (the solid line), or just using local communication (the dashed line), or just adapting the tolerance (the dotted–dashed line). As can be seen, the improvements are quite significant. For this problem, clearly a small window size is an efficient choice.

10.4. Conclusions

The results obtained for the MasPar suggest that large stiff problems can be solved without recourse to the solution of large systems of linear equations (possibly at each timestep) and that these techniques are ideally suited to massively parallel machines.

In a MIMD environment, the approach emphasized here suggests that, where possible, parallel algorithmic development should make use of existing sequential packages that have been fine-tuned over a number of years and that have proved to be robust and efficient. Such an approach will not only provide some robustness to the parallel algorithms, but will also allow the programmer to focus only on the interprocessor communications. The use of packages such as PVM or P4, moreover, will provide significant portability. In addition, it

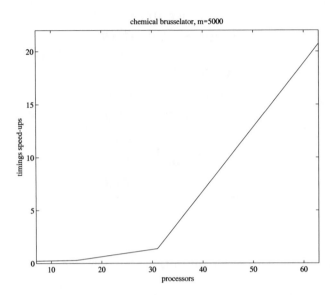

FIG. 10.4. *Speedup on the Paragon.*

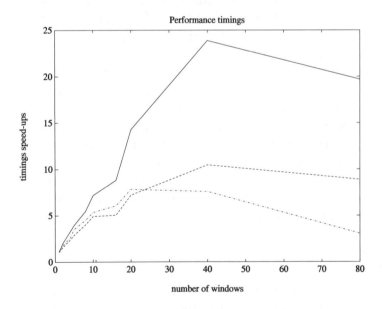

FIG. 10.5. *Effect of adapting tolerances and local communication.*

appears that waveform relaxation can be effective for solving large systems of equations in a parallel environment, especially if some form of preconditioning is used along with various adaptive strategies including adapting the tolerances and adapting the window size.

Lack of space precludes a more detailed discussion of the wide variety of parallel algorithms that have been developed for the solution of ODEs. It is clear that the various algorithms will not fare equally well on all architectures. For the immediate future, it seems likely that the variety of parallel codes, both problem-dependent and architecture-dependent, will have to be greater than in the sequential case. However, given the apparent trend among vendors toward massively parallel MIMD machines (possibly with shared memory, such as the SGI Power Challenge), it may well be that this situation is temporary and that we can expect uniformity and portability of codes across a large set of parallel machines, not only in the area of differential equations but in all areas of scientific computation.

10.5. Acknowledgments

The author thanks Pamela Burrage and Bert Pohl for their assistance in programming on the MasPar and the Intel Paragon, respectively. The author also wishes to thank the staff at ETH in Zürich for allowing access to the 96-node Paragon.

References

[1] P.N. BROWN, G.D. BYRNE, AND A.C. HINDMARSH, *VODE: A variable-coefficient ODE solver*, SIAM J. Sci. Statist. Comput., 20 (1989), pp. 1038–1051.

[2] K. BURRAGE, *The error behaviour of a general class of predictor-corrector methods*, Appl. Numer. Math., 8 (1991), pp. 201–216.

[3] ——, *Parallel methods for initial value problems*, Appl. Numer. Math., 11 (1991), pp. 5–25.

[4] ——, *Parallel and Sequential methods for Ordinary Differential Equations*, Oxford University Press, 1995.

[5] K. BURRAGE AND S. PLOWMAN, *The numerical solution of ODEIVPs in a transputer environment*, In Applications of Transputers2, D.J. Pritchard and C.J. Scott, eds., IOS Press, 1990, pp. 495–505.

[6] K. BURRAGE AND B. POHL, *Implementing an ODE code on distributed memory computers*, Comput. Math. Appl., 28 (1994), pp. 235–252.

[7] K. BURRAGE, C.T. DYKE, AND B. POHL, *On the performance of parallel waveform relaxations for differential systems*, Appl. Numer. Math., 20(1996), pp. 39–55.

[8] K. BURRAGE, Z. JACKIEWICZ, AND R.A. RENAUT, *The performance of preconditioned waveform relaxation techniques for pseudospectral methods*, Numer. Methods Partial Differential Equations, 12(1996), pp. 245–263.

[9] K. BURRAGE, Z. JACKIEWICZ, S.P. NORSETT, AND R.A. RENAUT, *Preconditioning waveform relaxation iterations for differential systems*, BIT, 36(1996), pp. 54–76.

[10] R. BUTLER AND E. LUSK, *User's Guide to the p4 Programming System*, Argonne National Laboratory Report ANL-92/17, 1992.

[11] C.H. CARLIN AND C. VACHOUX, *On partitioning for waveform relaxation time-domain analysis of VLSI circuits*, in Proc. of Int. Conf. on Circ. and Syst., Montreal, 1984.

[12] C.T. DYKE, *The Solution of Differential Equations in a Parallel Environment*, Ph.D. thesis, Mathematics Dept., Univ. of Queensland, 1995.

[13] C.W. GEAR AND F.L. JUANG, *The speed of waveform methods for ODEs*, In Applied and Industrial Mathematics, R. Spigler, ed., Kluwer, Netherlands, 1991, pp. 37–48.

[14] A. GEIST, A. BEGUELIN, J. DONGARRA, W. JIANG, R. MANCHEK, AND V. SUNDERAN, *PVM 3.0 User's Guide and Reference Manual*, Report ORNL/TM–12187, Mathematical Sciences Section, Oak Ridge National Laboratory, Oak Ridge, TN, 1987.

[15] R. JELTSCH AND B. POHL, *Waveform Relaxation with Overlapping Systems*, Research Report 91-02, Seminar für Angewandte Mathematik, ETH Zürich, 1991.

[16] E. LELARASMEE, *The Waveform Relaxation Method for the Time Domain Analysis of Large Scale nonlinear Dynamical Systems*, PhD thesis, University of California, Berkeley, CA, 1982.

[17] C. LUBICH AND A. OSTERMANN, *Multi-grid dynamic iteration for parabolic equations*, BIT, 27 (1987), pp. 216–234.

[18] W.L. MIRANKER AND W. LINIGER, *Parallel methods for the numerical integration of ordinary differential equations*, Math. Comp., 21 (1967), pp. 303–320.

[19] D.P. O'LEARY AND R.E. WHITE, *Multi-splittings of matrices and parallel solution of linear systems*, SIAM J. Alg. Disc. Meth., 6 (1985), pp. 630–640.

[20] Y. SAAD AND M. SCHULTZ, *GMRES: A generalized minimal residual algorithm for solving nonsymmetric linear systems*, SIAM J Sci. Statist. Comput., 7 (1986), pp. 856–869.

[21] S. VANDEWALLE AND R. PIESSENS, *Numerical experiments with nonlinear multigrid waveform relaxation on a parallel processor*, Appl. Numer. Math., 8 (1991), pp. 149–161.

[22] J. WHITE, A. SANGIOVANNI-VINCENTELLI, F. ODEH, AND A. RUEHLI, *Waveform relaxation: Theory and practice*, Trans. Soc. for Computer Simulation, 2 (1985), pp. 95–133.

[23] Z. ZLATEV, *Treatment of some mathematical models describing long-range transport of air pollutants on vector processors*, Parallel Comput., 6 (1988), pp. 87–98.

Chapter 11

Parallelizing Computational Geometry: First Steps

Isabel Beichl

Francis Sullivan

Editorial preface

Triangulation ought to be quite familiar to finite-element practitioners. In addition to this well-known application, triangulation can also be used in visualization. In each case the triangulation is viewed as an adjunct task to the computation itself; however, it is also true that the triangulation itself can be a challenging computation. This chapter details the data structures and algorithms used to completely parallelize the triangulation task. The development is done in two dimensions, but the extension to three dimensions and the performance on three-dimensional problems is discussed.

This article originally appeared in *SIAM News*, Vol. 24, No. 6, November 1991. It was updated during the summer/fall of 1995.

Great progress has been made in devising parallel algorithms for classes of problems in which the core computation is a convergent iterative method. In the case of linear systems, results have been extremely encouraging. Good results have also been obtained for some nonnumeric problems, such as sorting and graph traversal. Methods for those problems that combine numeric and combinatoric features, however, are less well developed. An especially important and interesting subarea in this class is computational geometry and, in particular, geometric methods that will work on real machines and real problems. The questions themselves are often deceptively simple to describe. However, the numeric problems are different (some would say much harder) because there is no guarantee of convergence as there is, for example, in Newton's method. Instead, we are given a finite number of coordinates and asked to determine things like interior and exterior, where there is no real notion of an approximate answer.

We will outline a new algorithm for triangulation that exploits the parallelism of a SIMD machine. We will describe a two-dimensional version because it is simpler, but at the end we'll see that things really aren't too different in three dimensions.

11.1. What is Triangulation?

Triangulation has been used in finite-element calculations for many years. It has been used to determine the structure of crystals, given only the coordinates of the points (which can number in the thousands). Other researchers are interested in triangulation for free Lagrangian calculations, geometric modeling, global climate models, and particle simulations. Recently, investigators at NIST in collaboration with a group at Clark University have been using triangulation in molecular dynamics calculations to study dense, two-component liquids and glasses. The object is to relate the geometric properties of glasses to their dynamic behavior by examining defect structures and "holes." The typical simulation involves 500 particles, and one would like to generate a picture of the geometric structure after every few time integration steps. For this to be a practical possibility, the triangulation should require only a few floating-point operations per particle. Our new algorithm will make this possible in two and three dimensions.

Suppose we start with a set of input vertices given by their coordinates. We want to fill out the convex set that these points span using triangles. Geometrically speaking, the triangles must fit together meeting only at edges, and we are not allowed to add any extra vertices. A *triangulation* of these input vertices is this set of triangles. Of course the triangulation is *really* the set of triples defining the triangles.

How could we find the triangles? A naive approach would be to try all triples of points, but this would give us at best an $\mathcal{O}(n^3)$ algorithm even in two dimensions—not a very satisfactory solution. In fact, there are well-known methods for finding triangulations in three dimensions in $\mathcal{O}(n^2)$ operations. Our algorithm is more efficient than this.

In addition, there is still the question of determining that a triangulation is complete and correct. A good way to ensure completeness is to have the triangles enumerated in *shelling* order. This means that we start somewhere, find triangle number one, and then add triangles successively so that at each stage, the set of already enumerated triangles is always simply connected. Intuitively, step k of a shelling is a choice of triangle k in the triangulation so that no holes or bridges are formed with the already chosen set of triangles 1 through $k - 1$. Figure 11.1 is an example of a triangulation; the numbers indicate a shelling order. Note that if we had placed triangle 44 immediately after triangle 22 it would have made a bridge separating the area containing triangles 23–43. This is illegal. Specifically, a new triangle may intersect previous ones only in something homeomorphic to a ball, which in the two-dimensional case means that it can intersect in one edge or two edges and nothing else.

Shelling has applications in physical problems involving a moving front,

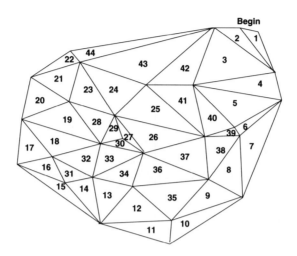

FIG. 11.1. *An example of a triangulation; the numbers indicate the shelling order.*

such as simulation of self-avoiding random surfaces. It also reduces the amount
of calculation needed in a sequential version of the algorithm, because we don't
need to consider points that are behind the moving front when building new
simplices. Since shelling checks topological properties, it helps to ensure that
numerical results are logically consistent. We use it as a check of correctness
at every stage of a triangulation.

11.2. Empty Spheres: The Central Idea

Not all triangulations are the same. We would like to find the Delaunay
triangulation. This means that the circle determined by the three vertices
making up a triangle contains none of the other vertices [3].

Here is the procedure we use to make a single triangle. Suppose that we
already know $\langle a,b \rangle$ is an edge of a triangle, and we are looking for the third
point, c. We know that if we already had c, the center of the circle determined
by $\langle a,b,c \rangle$ would be somewhere on the perpendicular bisector of $\langle a,b \rangle$ (see
Figure 11.2). So we search along the perpendicular bisector for the center of a
circle that goes through a, b, and one other point, c, and that does not contain
any other vertices in its interior. We start by picking *any* other point d and
calculating ξ, the center of the circle that passes through a, b, d. Then, using
a nearest neighbors algorithm (for the moment pretend that we just try every
point, but there are faster ways), we find the nearest neighbor of ξ. If the
nearest neighbor isn't a, b, or d, then that circle is not a Delaunay circle and
$\langle a,b,d \rangle$ is not a legal triangle. Suppose the nearest neighbor is point c. We
then just repeat the center calculation with d replaced by c.

Here is the whole procedure:

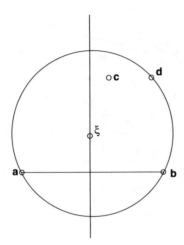

FIG. 11.2. *Procedure for making a single triangle.*

0. Choose any d different from a or b.

1. Find ξ, the center of the circle determined by $\langle a,b,d \rangle$.

2. Find the input point c closest to ξ (using a nearest neighbors algorithm).
   If $c \in \{a, b, d\}$, we're done. If not, repeat step 1 with d replaced by c.

We must also keep track of an orientation of the points $\langle a,b \rangle$ and consider only points c, d on the positive side of $\langle a,b \rangle$. This technicality is necessary because there are usually two triangles that attach to an edge, and we need to find both of them. In addition, we must determine when edge $\langle a,b \rangle$ is on the boundary. This happens when there are no points on its positive side. How we do this in practice is slightly more complicated than we want to be in this article.

Two important facts make this an efficient method. First, there is a modified nearest neighbors algorithm that works in $\mathcal{O}(\log(n))$ time, assuming there are n input points. That is what step 2 in the above procedure costs us. Second, the step 1-step 2 cycle can be refined using a binary search so that it will converge in $\mathcal{O}(\log(n))$ steps. When we combine these two ideas, we get an $\mathcal{O}(T * \log^2(n))$ algorithm for triangulation and shelling, where T is the number of triangles created. The algorithm is more stable than classical methods because distances rather than angles are compared.

11.3. Putting Triangles Together

We still haven't explained how to put the triangles together so that we don't get repetitions and we are sure to stop correctly. To do this, we use a concise data structure to represent the triangulation: *t-list*'s. We number the triangles according to the order in which we make them. We call these shelling numbers. The *t-list*'s are an array of lists, one list for each point a; *t-list[a]* is an ordered list of the shelling numbers of triangles that contain a. If at any point *t-list[a]* is empty, it means that point a has not been seen before. We can also tell the number of triangles that contain an edge $\langle a,b \rangle$ by finding #(*t-list[a]* $\cap$ *t-list[b]*), that is, the number of elements in this intersection. Because the average number of triangles that contain a point is six, these intersections don't usually become unwieldy. In pathological cases we use the fact that since the lists are ordered, we can do the intersections using yet another binary search.

Here's how things fit together: we start with any point and its nearest neighbor as an initial edge $\langle a,b \rangle$ and find the potential triangles $\langle a,b,c \rangle$ and $\langle a,b,c' \rangle$ that may be made on either side of $\langle a,b \rangle$. We create a stack of potential triangles to be processed, initially containing $\langle a,b,c \rangle$ and $\langle b,a,c' \rangle$. (We get both if $\langle a,b \rangle$ is not a boundary edge. We will not give the details of boundary conditions in this article.) We remove the top element, $\langle a,b,c \rangle$, from the stack. Intersecting *t-lists* will tell us if edge $\langle a,b \rangle$ is already in two triangles. In that case we do nothing and continue to the next stack element. Otherwise, we check for the legality of $\langle a,b,c \rangle$ by checking *t-lists*. If $\langle a,b,c \rangle$ passes the legality tests, we make the triangle and then, for each of the edges $\langle a,c \rangle$ and $\langle c,b \rangle$, we use the empty spheres method to tell us if either edge is on the boundary and, if not, what point it should connect to. We then put these new potential triangles on the stack. We continue in this way until the stack is empty.

Note that because of the shelling order, the checking tells us when a vertex is completely surrounded by triangles. In two dimensions, for example, this occurs when #(*t-list(c)*) $\neq$ 0, but no bridge is present. In that case, either point a or b is surrounded. Such points are "dead" because they need not be considered in later determinations of potential triangles. In the sequential case, recognition of dead points decreases running time significantly.

We can streamline the shelling process greatly by associating with each triangle a floating-point number $\mu = p^2 + q^2 - r^2$, where (p,q) is the center and r is the radius of the circle determined by the three points of the triangle. The connection between shelling and μ was discovered by Bruggesser and Mani [1], and we use it in our program. Shelling by μ ordering starts at the origin in the middle of the points and orders triangles by distance from the origin and the size of the circles. We get μ "free" as a side effect of the way in which we determine triangles. Our procedure for finding a new triangle always ends by determining the center and radius of the circle. Figure 11.1 is a strict ordering; Figure 11.3 is not. Bruggesser and Mani proved that if we follow strict μ ordering, we will be guaranteed that there will be no bridges and no wasted steps. But the theorem applies under ideal (i.e., infinite-precision arithmetic) conditions, which in practice occasionally do not happen. Thus, the legality checking is still needed. The stack ordered by μ is now a heap, and at each stage we remove the potential triangle with the smallest μ.

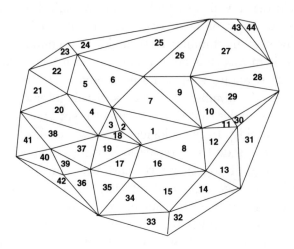

FIG. 11.3. *A triangulation in which the shelling order is a strict μ ordering.*

11.4. Parallelizing

We can identify the major components of the algorithm:

(1) Getting a next point for defining a new triangle, which requires determination of the nearest neighbors of the successive centers.

(2) Saving triangles (*t-list*) and potential triangles (the stack).

(3) Using the *t-lists* to avoid repetitions and bridges, i.e., to check the legality of the potential new triangle $\langle a,b,c \rangle$.

We parallelize each of these operations. The Connection Machine is arranged with a front end that has its own memory and a large number of parallel processors, which we call CM processors, each that has its own local memory. The CM processors all perform the same instruction at the same time, but it is possible to turn individual processors off and on. When we do this we call it changing the context. There will be three ways that we look at the CM processors, i.e., three basic contexts: (1) Each CM processor represents a point. The coordinates of the point as well as its index are stored there. (2) The processor is a potential triangle. In this context each processor holds the indices of the vertices of the potential triangle and its associated μ. (3) Each CM processor represents a triangle as a *t-list*, with slots in memory representing each vertex. The vertices that actually make up the triangle represented by this processor have 1's in the appropriate vertex slot; the others have 0's. Thus, a vertex is a bit plane across the processors, while the processor itself represents the triangle (see Figure 11.4).

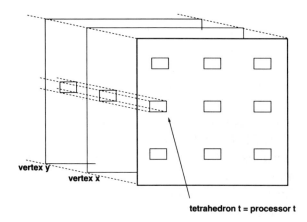

vertex y
vertex x

tetrahedron t = processor t

FIG. 11.4. *Context 3 for the CM processors: each processor represents a triangle as a t-list.*

We select the next potential triangle to test by switching to potential triangle context 2 and carrying out a global minimum over all processors to find the one with the smallest μ. The indices of these three points are brought back to the front end. We then switch to *t-list* context (context 3) to check for legality. We can check if a point has been seen before by doing a global **or** over all bit planes. We can find how many times an edge $\langle a,b \rangle$ has been used before in the triangulation by doing an **and** of the bit planes for a and b and then a global **or** over all those that are true followed by a **count** of the number of "trues."

To find new points for the next potential triangles, we switch to the point context 1. We broadcast the new edge $\langle a,b \rangle$ to all the processors. Since each processor represents a point d, we can check if d is on the positive side of $\langle a,b \rangle$. For those that are, we compute the center of the circle determined by $a, b,$ and d. The center at the minimum distance from the center of the previous triangle containing $\langle a,b \rangle$ is the desired one. (The very first triangle must be computed on the front end as described above for the sequential version.)

Note that in all dimensions the parallel complexity is $\mathcal{O}(M(n)T(n))$, where $M(n)$ is the complexity for determining the minimum or maximum of n items and $T(n)$ is the number of simplices made. The $M(n)$ term appears because of the determination of the minimum μ value and the minimum distance between centers.

11.5. How Are Things Different in Three Dimensions?

We have written a three-dimensional CM-2 program. It is very similar to the two-dimensional version described here. The average number of tetrahedra that contain a point is 24 instead of six. In order to check legality, we need to find the number of times a face has occurred. This requires three intersections in the sequential case and two local **and**'s and one global **or** in the parallel case. The empty spheres method does indeed find spheres rather than circles, so the linear algebra is slightly more complicated, but this doesn't change the complexity. Instead of potential triangles, we have potential tetrahedra. Last, the overall complexity is the same, but $T(n)$ will be the number of tetrahedra made rather than the number of triangles.

The sequential version with 1000 three-dimensional input vertices takes 220 seconds on an IBM RS/6000 workstation. The parallel version with 1000 points takes 106 seconds on the CM-2. If we don't do any checking for shelling, but keep enough logic to prevent infinite loops, it takes half as long. Because standard algorithms are $\mathcal{O}(n^2)$ and the parallel algorithm is $\mathcal{O}(M(n)T(n))$, the CM program eventually beats any standard program. However, one can really see the price paid for doing the legality checking, and the communication it requires, on the relatively slow processors of the CM-2. A back-of-the-envelope estimate says that the floating-point arithmetic accounts for only 1% of the time.

11.6. Alternative Parallel Methods

Because we construct the triangles in shelling order, there will always be a $T(n)$ term in the complexity. However, if we give up shelling order, we could triangulate in parallel by another method that first assigns a vertex to each processor and then gives all possible triangles for the given vertex [4]. In this article, in order to get the fastest time, we keep the list of available edges from which tetrahedra may grow. This list is distributed over the processors using load-balancing techniques, and separate triangulation processes are run at each node. A binning technique is used so that not every vertex is examined to find one tetrahedron, and this gives on the average an $\mathcal{O}(\text{polylog}(n))$ algorithm. The worst case, however, is $\mathcal{O}(n)$. This algorithm has been implemented on the CM-2 and CM-5, and running times can be found in the paper.

In effect, what we have is a concordance of the triangulation: for each point p we have a complete record of all occurrences. We can then compute μ and use the μ values as a key variable for a giant merge–sort step that eliminates the duplicates and assembles the triangles in shelling order. Doing this without additional logic checking is fast, but believing the result would require deeper faith in floating-point arithmetic. For example, if some triangle is found by points p_1 and p_2, we would have to trust the arithmetic to tell us that μ from the p_1 search is the same as μ from the p_2 search.

Another interesting algorithm and implementation, for two dimensions, is described in [2]. This algorithm divides areas to be triangulated among processors.

11.7. Acknowledgments

We thank the University of Maryland (UMIACS) for the use of their Connection Machines for this project.

References

[1] H. BRUGGESSER AND P. MANI, *Shellable decompositions of cells and spheres*, Math. Scand., 29(1971), pp. 197–205.

[2] G. MILLER, D. TALMOR, S. TENG, AND N. WALKINGTON, *A Delaunay based numerical method for* 3D, in Proc. 27th Annual ACM Symposium on Theory of Computing, ACM Press, 1995, pp. 683–692.

[3] F.P. PREPARATA AND M.I. SHAMOS, *Computational Geometry: An Introduction*, Springer-Verlag, New York, 1985.

[4] A. TENG, F. SULLIVAN, I. BEICHL, AND E. PUPPO, *A data parallel algorithm for the three dimensional Delaunay triangulation and its implementation*, in Proc. Supercomputing '93, IEEE Computer Society Press, 1993, pp. 112–121.

Chapter 12

Parallel Inverse Iteration for Eigenvalue and Singular Value Decompositions

Richard J. Hanson
Glenn R. Luecke

Editorial preface

The determination of eigenvalues and their eigenvectors and the singular value decomposition (SVD) is a rather common linear algebra operation. A novel approach to both tasks on a single-instruction multiple-data (SIMD) machine is considered here. Also illustrated is how the choice of language can strongly affect the performance of the algorithm's implementation, and how the use of additional memory can help gain enhanced performance.

This article originally appeared in *SIAM News*, Vol. 27, No. 5, May/June 1994. It was updated during the summer/fall of 1995.

In this article we describe new algorithms we developed for carrying out two commonly used mathematical techniques in scientific computing on the MasPar MP-1 and MP-2 single-instruction multiple-data (SIMD) parallel computers. In particular, the symmetric, real matrix eigenvalue–eigenvector and singular value decompositions (SVDs) are discussed.

Parallel implementations of the serial Fortran-77 versions of the symmetric matrix eigenvalue–eigenvector and SVDs—the routines _SYEV and _GESVD from LAPACK [2]—had performed poorly on these computers.

Parallel program execution on the MasPar computers is achieved by using Fortran-90 array syntax and some of the new Fortran-90 intrinsics. (For a description of Fortran-90, see [8] and especially [1].) Parallel execution can also be achieved by writing in a lower-level language called MPL, a C language extension available on MasPar computers. Although better performance can often be achieved with MPL [7] than with Fortran-90, it is much easier to write in Fortran-90 than in MPL.

12.1. Background

The eigenvalue problem for a real, symmetric matrix A requires the computation of an orthogonal matrix Q such that $AQ = QD$, where D is an $n \times n$ diagonal matrix. This decomposition is obtained by first reducing A to tridiagonal form, $AH = HT$, where H is orthogonal and T is tridiagonal. This is followed by the decomposition $TG = GD$, where G is orthogonal and D is diagonal. Thus, $Q = HG$. An algorithm that uses the product of $N - 2$ Householder transformations to reduce a symmetric matrix to tridiagonal form, yielding the matrix T, can be found in [5]. This same algorithm also produces the H matrix.

The SVD of a rectangular, real matrix $A \in \mathbb{R}^{m \times n}$ is $AV = US$, where $U \in \mathbb{R}^{m \times m}$ and $V \in \mathbb{R}^{n \times n}$ are orthogonal matrices and $S \in \mathbb{R}^{m \times n}$ is a rectangular, diagonal matrix. This decomposition is obtained by first reducing A to upper bidiagonal form, $AX = YB$, where X and Y are orthogonal and $B \in \mathbb{R}^{m \times n}$ is upper bidiagonal. This step is followed by the decomposition $BW = ZS$, where W and Z are orthogonal and S is diagonal. Thus, $V = XW$ and $U = YZ$. The reduction to bidiagonal form and the computation of X, Y, and the bidiagonal matrix B can be found in [4].

12.2. Eigenvectors of Tridiagonal Matrices

In this work we have focused on computing the eigenvalues and eigenvectors of the real tridiagonal matrix

$$(12.1) \qquad T = \begin{bmatrix} a_1 & b_1 & 0 & & \cdots \\ b_1 & a_2 & \ddots & & 0 \\ 0 & \ddots & & a_{n-1} & b_{n-1} \\ \vdots & & 0 & b_{n-1} & a_n \end{bmatrix}$$

and the SVD of the real upper bidiagonal matrix

$$(12.2) \qquad B = \begin{bmatrix} d_1 & f_1 & 0 & \cdots & & 0 \\ 0 & d_2 & f_2 & \ddots & & \vdots \\ & & \ddots & \ddots & \cdots & 0 \\ \vdots & & & 0 & d_{n-1} & f_{n-1} \\ 0 & \cdots & & & 0 & d_n \end{bmatrix}.$$

Our methods for computing the SVD of B use the observation [5] that the singular values of B are opposite-signed eigenvalue pairs of the $2n \times 2n$ symmetric tridiagonal matrix

$$(12.3) \qquad \hat{T} = \begin{bmatrix} 0 & d_1 & 0 & & \cdots & 0 \\ d_1 & 0 & f_1 & & & \vdots \\ 0 & f_1 & 0 & \ddots & & \\ & & \ddots & & f_{n-1} & 0 \\ \vdots & & & f_{n-1} & 0 & d_n \\ 0 & \cdots & & 0 & d_n & 0 \end{bmatrix}.$$

The equivalence relations between the left and right singular vectors for B, in terms of eigenvalues and eigenvectors for $\hat{T}$, are

$$(12.4) \qquad \begin{aligned} \hat{T}q &= \sigma q, \\ \left. \begin{aligned} w_i &= q_{2i-1} \\ z_i &= q_{2i} \end{aligned} \right\} \, & i = 1, \ldots, n, \\ Bw &= \sigma z, \\ B^T z &= \sigma w. \end{aligned}$$

Since the time required to compute the eigenvalues is a small part of the time required to compute both the eigenvalues and the eigenvectors, the increased efficiency resulting from the use of a "very fast" method for eigenvalue calculation will have little effect on overall performance. Thus, we use the Pal, Walker, and Kahan (PWK) method [9, §8.15] because of its accuracy, simplicity, and reasonable efficiency. Other efficient methods are available (see [11] and its bibliography). The eigenvalues of T, $\lambda_1, \ldots, \lambda_n$, are sorted so that $|\lambda_1| \geq \cdots \geq |\lambda_n|$.

The eigenvectors of T are computed with two applications of inverse iteration [9], $(T - \lambda_i I)g_i = p_i$, $i = 1, \ldots, n$, where a uniform random number generator is used to initialize each p_i. Indeed the set of p_i must not be pathologically ill conditioned for clusters of λ_i. In the extreme case in which the clusters of values are multiple, then choosing the same initial value of p_i may result in a deficient set of eigenvectors.

Since the matrix elements for the n problems represent the lattice points of a three-dimensional $n \times n \times n$ box, each $n \times n$ rack, or slice, of the box is the matrix $T - \lambda_i I$, $i = 1, \ldots, n$. Let $\square M$ denote this box, and $\mathbf{p}$ the corresponding right-hand-side vectors associated with inverse iteration. For the SVD problem the n systems $(\hat{T} - \sigma_i I)q_i = p_i$ are solved; the problem size is $2n \times 2n \times n$. The operations we perform are designated $G = \square M^{-1} \mathbf{p}$ for the eigensystem case and $Q = \square \hat{M}^{-1} \mathbf{p}$ for the SVD case. The eigenvectors and singular vectors are then orthogonalized by the method presented by Björk and Paige [3, §7]. Their approach is based on Householder triangularization, in which G is augmented with a zero matrix.

For efficient parallel implementation of inverse iteration, it is critical that the operations performed to solve a system in each rack of $\square M$ be identical. By contrast, the LAPACK code SSTEQR computes the eigenvalues and accumulates the vectors during the decomposition of the tridiagonal matrix. Similar remarks hold for the decomposition of the bidiagonal matrix in the SVD with **SBDSQR**. Considerable care is taken to apply the products of plane rotations in consecutive planes, using routine **SLASR**, and to enhance the efficiency of vector accumulation. This inherent dependence on immediate past results prevents parallelization and will result in poor parallel performance.

12.3. The Parallel Eigenvector/Singular Vector Computation

Our approach can be superficially summarized as follows:

> Start with a fast method, i.e., cyclic reduction. If this fails, use a more reliable and slower method, i.e., Gaussian elimination.

This approach is effective if the fast method is not likely to fail and if a failure can be cheaply discovered and fixed. Typically, parallel cyclic reduction plus checking for a failure was roughly twice as fast as parallel Gaussian elimination on the test problems used for this study. Moreover, very few failures of cyclic reduction were encountered when we used matrices whose elements had been randomly generated. Thus, at least for the eigenvalue and singular value test cases we have run, this approach was efficient.

In developing our cyclic reduction code, we began with the description given in [10, Kershaw]. The parallel cyclic reduction code added a second rank or subscript to the single system code. It is well known that for systems of equations that are not positive definite, cyclic reduction is not necessarily numerically stable. As indicated earlier, this was not a problem for the test cases we ran. However, it is necessary to evaluate the quality of the results produced by cyclic reduction, as shown later in this article.

An example of a tridiagonal matrix for which cyclic reduction fails to give a complete set of eigenvectors is given by W_{2n+1}^+ [12], even for modest values ($n \geq 10$). This matrix has subdiagonals, all with a value of 1. The diagonals decrease from the value n to 0 and then increase to n. Some pairs of eigenvalues agree to working precision for $n = 10$, even though the values of the off-diagonals imply that eigenvalues are distinct.

Our advice to users is to try parallel cyclic reduction first, and to keep using it unless it has clearly failed to compute a solution that is accurate in the direction of the eigenvectors. As an alternative, the solver code will compute the LDU factorization, using parallel Gaussian elimination with partial pivoting. This algorithm delivers numerically stable results for the eigenvectors.

In developing our parallel eigenvector and parallel singular value implementations, which are summarized below, we wrote the parallel Gaussian elimination and parallel cyclic reduction routines in MPL in an effort to achieve very high performance.

To compute the eigenvectors of T perform the following steps:

1. Compute the eigenvalues $\lambda_1, \ldots, \lambda_n$ of T by the PWK method. This portion is executed serially.

2. Solve in parallel the multiple system $G = \Box M^{-1} \mathbf{p}$, using $\mathbf{p} \leftarrow G$ on the second iteration. Parallel cyclic reduction is used to solve each of the above systems.

3. Normalize G so that all columns have unit Euclidean length. This is executed in parallel.

4. If the residuals satisfy $\|TG - GD\|/(n \times epsilon(T)) \leq \phi \|D\|$ (we use $\phi = 4$), in which case the answers computed by using cyclic reduction are acceptable, proceed to step 5.

   If this condition is not satisfied, then repeat step 2 using parallel Gaussian elimination in the solve step, normalize G, and then proceed to step 5. The function $epsilon(T)$ is the Fortran-90 intrinsic that gives the value of working precision of the data for T.

5. Orthogonalize the matrix G.

6. Compute the matrix multiplication $Q = HG$.

To compute the singular vectors of B perform the following steps:

1. Compute the singular values $\sigma_1, \ldots, \sigma_n$ of B, using PWK. This portion is executed serially.

2. Solve in parallel the multiple systems $Q = \Box \hat{M}^{-1} \mathbf{p}$, using $\mathbf{p} \leftarrow Q$ on the second iteration. Parallel cyclic reduction is used to solve the systems.

3. Normalize Q so that all columns have unit Euclidean length. This is executed in parallel.

4. If the residuals satisfy $\|\hat{T}Q - QS\|/(n \times epsilon(\hat{T})) \leq \phi \|S\|$ (we use $\phi = 4$), in which case the answers computed by cyclic reduction are acceptable, proceed to step 5.

   If this condition is not satisfied, repeat step 2 using parallel Gaussian elimination, normalize G, and then proceed to step 5.

5. Extract the $n \times n$ matrices W and Z from Q using (12.4). Orthogonalize both W and Z. This is executed in parallel.

6. Compute the matrix products $V = XW$ and $U = YZ$. This is executed in parallel.

In each case, the residuals are computed in parallel in step 4. A classic "storage-required-versus-time" tradeoff is evident in both cases at step 2. The storage required for computing the systems equals twice the number of nonzero grid points in the box. The total storage used by our solver is $8n^2$ for the tridiagonal problem and $16n^2$ for the SVD. Traditional approaches based on the accumulation of plane rotations typically require n^2 additional storage for eigenvalues and $2n^2$ for the SVD. The storage requirement for our cyclic reduction algorithm can be attributed to its use of twice the dimension of the problem size for each diagonal, codiagonal, and right-hand-side vector to avoid storage-accessing conflicts. This additional storage requirement can restrict the size of the problem computed, and we consider this the major defect of the algorithm.

On the up side, cyclic reduction is inherently "fast": the computational complexity is $\log_2(n)$ array arithmetic operations of size no larger than $(n/2) \times n$. In fact, both of the algorithms sketched here are rich in $\mathcal{O}(n^3)$ and $\mathcal{O}(n^2)$ array arithmetic operations.

12.4. Performance Results

TABLE 12.1

CPU times, in seconds, for the eigenexpansion on the 16K processor MasPar MP-1. Speedup is the ratio of the Fortran-90 algorithm to the MPL algorithm.

n	MPL new alg	F90 new alg	Speedup
128	1.88	2.89	1.5
512	11.23	19.52	1.7
896	32.03	67.06	2.1
1280	92.39	158.31	1.7

Tables 12.1–12.5 support our claims that these methods for computing eigenvectors and singular vectors have merit in terms of computational efficiency. The front-end workstations for the MasPar MP-1 and MP-2 are DEC-Stations 5000/240 running ULTRIX V4.3A (rev .146). The Fortran compiler used on the DEC-Stations was the DEC Fortran X3.2-430. All runs were made on a 16K MasPar MP-1 and on the newer 4K MasPar MP-2 with version 2.2.52 of the MasPar Fortran compiler and version MP3.2.1 of the MPL compiler. The MP-1 has a peak theoretical speed of 1137 Mflops (16K processors, 0.0694 Mflops/processor) for 32-bit floating-point arithmetic and half that speed for 64-bit. The MP-2 has a peak theoretical speed of 1528 Mflops (4K processors, 0.373 Mflops/processor) for 32-bit floating-point arithmetic and half that speed for 64-bit. All timings reported in this paper use 32-bit. The speed of the communication network is the same on the MP-2 and MP-1 machines. All MasPar computers use a two-dimensional toroidal grid of processors and are SIMD machines.

All times reported are wallclock timings measured in seconds. All matrices were initialized with the pseudorandom number generator **rand_gen** [6]. In Tables 12.1–12.4 "F90 new alg" refers to the algorithm described earlier, written in Fortran-90, with the exception that cyclic reduction was written

TABLE 12.2

CPU times, in seconds, for the eigenexpansion on the 4K processor MasPar MP-2. Speedup is the ratio of the Fortran-90 algorithm to the MPL algorithm.

n	MPL new alg	F90 new alg	Speedup
128	2.06	2.93	1.4
512	10.74	24.72	2.3
896	31.38	97.26	3.1
1280	66.11	246.75	3.7

TABLE 12.3

CPU times, in seconds, for the SVD on the 16K processor MasPar MP-1. The speedup is the ratio of the Fortran-90 to the MPL implementation.

n =size	MPL new alg	F90 new alg	Speedup
128	1.95	5.38	2.8
384	11.01	34.79	3.2
640	32.43	101.43	3.1
896	62.15	218.28	3.5
1152	111.13	517.68	4.7

in MPL. "MPL new alg" refers to the algorithm described earlier, written in Fortran-90 with computationally intensive portions written in MPL to achieve high performance. The Gaussian elimination algorithm was also written in MPL, although, to our knowledge, the branch that executes Gaussian elimination never occurred in our test problems.

Since all LAPACK routines are written in Fortran-77, these routines will execute only on the front-end workstation and are not able to take advantage of the parallelism of the MasPar machines. The MasPar will provide parallel execution only for those portions of a program written in Fortran-90 array syntax. It is our experience that simply translating the Fortran-77 LAPACK code for eigensystems and for the SVD to Fortran-90 array syntax produces parallel code that actually runs *slower* than the original Fortran-77 code! This observation underscores the need to develop new algorithms to achieve high performance for these operations on MasPar computers.

As expected, the larger the problem size, the greater the advantage we could achieve by using parallel processing; for a problem size of 128, parallelism really offers no advantage, and it is just as effective to use LAPACK routines on the front-end workstation. For an absolute comparison, we include sample times for eigenexpansion and SVD, restricting computations to the front-end

TABLE 12.4

CPU times, in seconds, for the SVD on the 4K processor MasPar MP-2. The speedup is the ratio of the Fortran-90 to the MPL implementation.

n	MPL new alg	F90 new alg	Speedup
128	1.77	5.30	3.0
384	9.83	40.37	4.1
640	27.67	132.22	4.8
896	59.31	305.92	5.2
1152	107.92	589.00	5.5

TABLE 12.5

CPU times, in seconds, for the eigenexpansion and SVD on the DEC-Stations 5000/240 with F77 LAPACK.

n	Eigenexpansion	n	SVD
128	2.98	128	3.39
512	111.38	384	142.26
896	557.14	640	655.07
1280	1650.69	896	1802.74
1664	3488.96	1152	3946.78

workstation (Table 12.5). This shows the relative merit of using the MasPar and a new algorithm for these calculations.

What are the benefits of writing computationally intensive portions of these routines in MPL? For the eigenvalue routine, use of MPL increased performance by a factor of 1.5–1.7 on the 16K MP-1 and 1.4–3.7 on the 4K MP-2, as the problem size increased from 128 to 1280. Because of memory limitations, 1280 is about the size of the largest eigenvalue problem that could be run on either the MP-1 or the MP-2. For SVD, use of MPL increased performance by a factor of 2.8–4.7 on the 16K MP-1 and 3.0–5.5 on the 4K MP-2 as the problem size increased from 128 to 1152. Because of memory limitations, 1152 is about the size of the largest SVD problem that could be run on either the MP-1 or the MP-2.

12.5. Conclusions

Parallel execution on MasPar computers can be achieved only by using Fortran-90 array syntax or MPL, an extension of C. As an example of the greater difficulty of writing programs in MPL than in Fortran-90, a 60-line Fortran-90 code required approximately 800 lines of MPL! As mentioned earlier, simple modifications for converting the Fortran-77 LAPACK code for eigensystems and SVD to Fortran-90 array syntax produce code that actually runs slower than the original Fortran-77 code. The new parallel algorithms for eigenvalue and SVDs presented here achieved high performance on MasPar computers. The good performance achieved with Fortran-90 array syntax improved even further when the computationally intensive portions of the code were written in MPL.

12.6. Acknowledgments

We thank Youngtae Kim for writing the MPL programs used in this project and are grateful for the help provided by Suraj Kothari. We also thank the Scalable Computing Laboratory at Ames Laboratory for allowing us to use their MasPar computers for this project.

References

[1] J.C. ADAMS, W.S. BRAINERD, J.T. MARTIN, B.T. SMITH, AND J.L. WAGENER, Fortran-90 Handbook, Complete ANSI/ISO Reference, McGraw-Hill, New York, 1992.

[2] E. ANDERSEN, ET AL., LAPACK Users' Guide, Society for Industrial and Applied Mathematics, Philadelphia, PA, 1992.

[3] Å. BJÖRCK AND C.C. PAIGE, Loss and recapture of orthogonalization in the modified Gram–Schmidt algorithm, SIAM J. Matrix Anal. Appl., 13(1992), pp. 176–190.

[4] J. DONGARRA, S.J. HAMMARLING, AND D. SORENSEN, Block Reduction of Matrices to Condensed Forms for Eigenvalue Computations, LAPACK Working Note # 2, Argonne National Laboratory, MCS TM No. 99, September 1987.

[5] G.H. GOLUB AND C. VAN LOAN, Matrix Computations, John Hopkins University Press, Baltimore, MD, 1989.

[6] IMSL MP Library, Version 1.1, Visual Numerics, Inc., part number 3026, Visual Numerics, Inc., Houston, TX, 1993.

[7] MasPar Parallel Application Language (MPL) Reference Manual, MasPar Computer Corporation, Sunnyvale, CA, 1990.

[8] M. METCALF AND J. REID, Fortran-90 Explained, Oxford Science Publications, Oxford, UK, 1990.

[9] B.N. PARLETT, The Symmetric Eigenvalue Problem, Prentice-Hall, Englewood Cliffs, NJ, 1980.

[10] G. RODRIGUE, Parallel Computation, Academic Press, New York, 1982.

[11] P.N. SWARZTRAUBER, A parallel algorithm for computing the eigenvalues of a symmetric tri-diagonal matrix, Math. Comp., 60(1993), p. 651.

[12] J.H. WILKINSON, The Algebraic Eigenvalue Problem, Clarendon Press, Oxford, UK, 1965.

Chapter 13

Parallel Branch-and-Bound Methods for Mixed Integer Programming

Jonathan Eckstein

Editorial preface

Mixed integer programming (MIP) problems are quite often very highly leveraged problems in industry. This means that their solution offers, or leverages, large savings or manufacturing efficiencies. Therefore, the ability to solve larger MIP problems, and to solve existing MIP problems faster, is of more than academic interest. This chapter details how this can be achieved by mapping the branch-and-bound MIP solution algorithm to parallel machines.

This article originally appeared in *SIAM News*, Vol. 27, No. 1, January 1994. It was updated during the summer/fall of 1995.

Mixed integer programming (MIP) is an important class of mathematical problems that arise in industrial and operational planning. MIP, like linear programming, involves maximizing or minimizing a linear function subject to linear equality and inequality constraints, but with the restriction that some subset of the variables can take only whole-number values. Its applications run the gamut from production planning and freight routing to the design of fiber optic networks and the development of schedules for sports leagues.

MIP is a very general problem class that essentially subsumes all of combinatorial optimization, and the problems are therefore $\mathcal{NP}$-hard. In practice, MIP problems with identical numbers of total variables, integer variables, and constraints can vary from trivial to intractable.

Many special cases of MIP have been extensively studied in their own right, and some have been solved by efficient algorithms. Among algorithms for *general* MIP, however, the traditional *implicit enumeration*, or *branch-and-bound*, class is by far the most common in practice. These methods effectively search the entire space of possible solutions and select a solution that is provably the best; however, they try to use information gleaned from related linear programs to limit the amount of detail in which they examine portions of the search space. With luck, the vast majority of the space proves "uninteresting" and need not be looked at very closely.

Often, however, one is not so lucky, and this is where parallelism enters the picture. The branch-and-bound process subdivides the search space into successively smaller pieces. Each piece must be examined in some aggregate manner (*bounding*) and, possibly, further subdivided (*branching*). The bounding and branching tasks for any given region of the search space are largely independent of those for other regions and can thus be performed simultaneously. As the space is divided into ever-finer pieces, the potential for parallelism grows.

Parallel branch-and-bound algorithms have existed since the 1970s— see [4, 5, 8] for some survey material—but most earlier work tested specialized versions used for very specific problem classes, such as traveling salesman or vertex cover. There are some partial exceptions to this observation [1, 3, 4], but results are available for only a handful of processors, and the parallel architectures used were not very advanced.

This article describes the implementation, called "CMMIP," of a general MIP branch-and-bound algorithm on a large-scale parallel computing system, the Thinking Machines Corporation CM-5. The CM-5 implementation shows that, on a reasonably up-to-date parallel system and without extensive specialization or tuning, efficient use of at least a hundred or so processors is now a reality for a variety of hard, "real-world" MIP problems. The implementation, which uses a single "hub" processor to control the search, is described in considerably more detail in [5]. Central control has its limitations, and [5] begins to address the possibilities for more distributed control strategies, drawing on the prior literature.

As usual, parallelism is not a panacea that makes it possible to use carelessly selected or inappropriate algorithms to solve arbitrary problems. What the work described here does show is that, in a fairly broad range of practical contexts, if a sequential branch-and-bound algorithm is within a few orders of magnitude of running within some required timeframe, it is likely that parallel processing can "close the gap."

13.1. General Branch-and-Bound Algorithms for MIP

Formally, we consider optimization problems of the form

$$
\begin{align}
(13.1) \quad & \text{minimize} \quad & c^{\mathsf{T}} x \\
(13.2) \quad & \text{subject to} \quad & A_{(1)} x = b_{(1)}, \\
(13.3) \quad & & A_{(2)} x \le b_{(2)}, \\
(13.4) \quad & & l \le x \le u, \\
(13.5) \quad & & x_j \in \mathcal{Z} \quad \forall j \in D \, .
\end{align}
$$

$\mathcal{Z}$ denotes the set of integers, $A_{(1)}$ and $A_{(2)}$ are $m_1 \times n$ and $m_2 \times n$ matrices, respectively, and c, $b_{(1)}$, and $b_{(2)}$ are conformally sized vectors. The vectors l and u are sets of lower and upper bounds on the n decision variables x; elements of l may be $-\infty$, and elements of u may be $+\infty$. $D \subseteq \{1, \ldots, n\}$ is a nonempty set of indices identifying the *discrete* variables, which may take only integer values.

The standard branch-and-bound method for MIP, which is described in most introductory operations research textbooks, starts by considering the linear programming (LP) relaxation (13.1)–(13.4) of the original problem, (13.1)–(13.5).

This "root" problem, which is denoted by R, has only continuous variables and can be solved efficiently by such techniques as the classical simplex method. The optimal objective value of R can be denoted by $z(R) = c^\top x(R)$, where $x(R)$ is some optimal solution. This quantity provides a lower bound on the solution z^* of the original problem (13.1)–(13.5). If $x(R)$ happens to satisfy (13.5), then $x(R)$ solves the entire problem.

Usually, of course, this is not the case, and one selects some $j(R) \in D$ such that $x_{j(R)}(R) \notin \mathcal{Z}$. R is then separated into two *subproblems*: one, the "up child," $C^+(R)$, has the additional constraint that $x_{j(R)} \geq \lceil x_{j(R)}(R) \rceil$, and the other, the "down child," $C^-(R)$, has $x_{j(R)} \leq \lfloor x_{j(R)}(R) \rfloor$. These restrictions can be enforced by modifying the l and u vectors. Since $x_{j(R)}$ is supposed to be integer, any solution to (13.1)–(13.5) must be feasible for at least one of these children. Furthermore, the value $z(C^+(R))$ of $C^+(R)$ is a lower bound on all solutions to the original problem with $x_{j(R)} \geq \lceil x_{j(R)}(R) \rceil$, and the optimal value $z(C^-(R))$ of $C^-(R)$ is a lower bound on all others. If the solution to either child meets (13.5), then its value is also an *upper* bound on z^*.

The algorithm continues by treating subproblems in much the same way as the root. At any time t, there is a *pool* of active subproblems $\mathcal{P}(t)$. Each member of the pool resembles (13.1)–(13.4), except that some of the bounds l_j and u_j are more restrictive. There is also an *incumbent* value $\bar{z}(t)$, which is the best objective value of a feasible solution to (13.2)–(13.5) seen up to time t. If no such solution has been encountered, $\bar{z}(t) = +\infty$. Thus, $\bar{z}(t)$ is a decreasing upper bound on z^*.

The algorithm removes some problem Q from the pool and calculates an optimal solution $x(Q)$ with objective value $z(Q) = c^\top x(Q)$. If $z(Q) \geq \bar{z}(t)$, then Q can be "fathomed" (discarded), because any feasible solution to Q, including those meeting (13.5), must have an objective value less desirable than $\bar{z}(t)$. Therefore, the portion of the search space corresponding to Q need not be considered in any further detail.

In the alternative case, $z(Q) < \bar{z}(t)$. If $x(Q)$ meets (13.5), there is no need to examine Q's portion of the search space any further, and $z(Q)$ becomes the new incumbent, $\bar{z}(t+1) = z(Q)$. For any subproblem S, let $P(S)$ denote its parent. All outstanding subproblems S for which $z(P(S)) \geq z(Q)$ can be immediately fathomed by deleting them from the pool. If, on the other hand, $x(Q)$ does not meet (13.5), the corresponding piece of the search space must be examined more closely. Some $j(Q) \in D$ is chosen such that $x_{j(Q)}(Q) \notin \mathcal{Z}$, and Q is separated into two children, $C^+(Q)$ and $C^-(Q)$, just as with the root problem. The children are placed in the pool, and the algorithm then repeats. When the pool becomes empty, the method terminates with $\bar{z}(t)$ proved equal to z^*.

Because the method's hierarchy of subproblems can be thought of as a tree emanating from the root problem, fathoming is also called *pruning*. At any time t, define $\underline{z}(t) = \min\{z(P(Q)) \mid Q \in \mathcal{P}(t)\}$. With the convention that $z(P(R)) = -\infty$, $\underline{z}(t)$ is an increasing lower bound on z^*. The goal of the algorithm is to "squeeze" $\bar{z}(t)$ and $\underline{z}(t)$ together until they meet.

Commercial MIP codes tend to contain a number of refinements on the basic algorithm. One such refinement concerns the choice of branching indices, $j(Q)$. In theory, an arbitrary member of D with $x_{j(Q)}(Q) \notin \mathcal{Z}$ will suffice, but a more careful choice is required in practice. In this project, the branching index was chosen to maximize a "score" computed from an optional set of user-assigned "priorities," the values of the $x_j(Q), j \in D$, and the relevant *pseudocosts*. Each $j \in D$ has an "up" and a "down" pseudocost; see, for example, [9]. The up pseudocost of x_j, which estimates the average rate at which the optimal objective value rises as x_j is forced upward, is computed by comparing all problems with branching index j to their up children. The down pseudocost is similar but estimates the objective change as the variable is forced lower. Both are continually updated as the algorithm proceeds. For further particulars of the implementation, see [5] and references therein. Generally speaking, variables with large pseudocosts tend to be the most desirable branching variables, as they are the most likely to cause an increase in $\underline{z}(t)$.

The selection of problems from the pool is another important detail. Two popular general search strategies are depth-first, in which the task pool is treated as a stack, and breadth-first, in which the pool is a queue. In branch-and-bound algorithms, a common alternative to breadth-first is "best-first," in which the subproblems Q with the loosest bounds $z(P(Q))$ are processed first, on the grounds that they are the least likely to be pruned later. Such strategies usually involve organizing the pool as a heap.

A typical approach in commercial serial codes is to pursue a depth-first search until an incumbent (or a value of "reasonably good" quality) is encountered, and then to switch to best-first. Finding any valid incumbent might require probing quite deeply into the search tree, something depth-first search can accomplish without massive consumption of time or memory. Later in the search, $\bar{z}(t)$ may be at or near z^*, and the primary goal should be to increase $\underline{z}(t)$; in this case best-first search is most appropriate. In the CMMIP runs presented here, the switch from depth-first to best-first occurs when $\bar{z}(t)$ and $\underline{z}(t)$ come within some fixed percentage, τ (typically 50%), of one another.

Alternative methods for generating incumbents are another important detail. Rather than waiting for some $x(Q)$ to meet (13.5), additional, heuristic methods can be used to generate integer-feasible solutions. Solutions to (13.2)–(13.5) obtained in this way are just as valid for pruning the tree and updating $\bar{z}(t)$ as any other and can result in very aggressive fathoming if found early enough. The possibilities range from simple rounding off of the $x_j(Q)$, $j \in D$, to a variety of more sophisticated or specialized heuristics. CMMIP's current heuristic is general and somewhat ad hoc, using linear programming with large cost perturbations in an attempt to modify $x(Q)$'s to meet (13.5) without violating (13.2)–(13.3). For further details, see [5].

Some useful enhancements, such as special techniques for branching on *groups* of variables, and polyhedral methods for strengthening linear programming lower bounds, have not yet been incorporated into CMMIP. However, they are not ultimately incompatible with a parallel implementation of the type described here.

13.2. Implementation on the CM-5

The CM-5 is a MIMD multiprocessor that has $p = 2^k$ processing nodes (PNs), each with a single SPARC-2 microprocessor and four optional vector floating-point units. The basic approach of the implementation, as in most prior parallel branch-and-bound work, is to evaluate multiple subproblems simultaneously, each on its own PN. Due to the short vector lengths typically encountered in commercial mixed integer models, there was no attempt to take advantage of the vector units. For the work described in this article, then, the CM-5 can be viewed as a tightly coupled set of SPARC-2 microprocessors with distributed memory.

Each node runs its own copy of the CPLEX implementation of the simplex method (from CPLEX Optimization, Inc.). CPLEX was used strictly as a linear program solver; there was no attempt to use the serial mixed integer capabilities *within* CPLEX, except for one problem-input routine.

The processors of the CM-5 are interconnected by two user-accessible networks, the *data network* and the *control network*. The control network is dedicated to global computation and synchronization operations. The data network is a point-to-point message system whose architecture makes it possible for many arbitrarily chosen pairs of nodes to exchange data packets without "blocking" one another. Thus, the CM-5 can often be programmed as if it were a densely connected set of processors in which the available raw bandwidth between any pair of processors is about five megabytes per second in each direction.

CMMIP is implemented in the C language, using the CMMD 3.0 message-passing programming environment. CMMD incorporates a facility called *active messaging* [11]. An active message is a single data network packet, the first four bytes of which contain the address of a "handler" function f. Such messages behave like asynchronous remote procedure calls: the arrival of one of these messages immediately causes suspension of the current program and execution of f, with the remainder of the packet supplied to f as arguments. Once the handler has run, control returns to the suspended program. This mechanism permits communication-related code to operate in the "background," while "foreground" code (such as CPLEX) may be completely oblivious to being used in a parallel environment.

13.2.1. Central Control. A key issue in parallel search is whether tasks should be assigned to processors by a single "master" processor or shared data structure, or in some more decentralized way. As the number of processors grows, a centralized approach risks bottleneck or contention problems involving common resources. With more distributed approaches, on the other hand, scheduling decisions must be made on the basis of limited information; there is a risk that the search process might evolve very differently from serial versions of the underlying algorithm, perhaps resulting in a larger search tree.

Initially, a decision was made to implement a centralized scheme and then assess the seriousness of the resulting bottlenecks. Therefore, one processor was set aside to be the "hub" that controls the search, with the others designated as "workers." The hub maintains the pool of active subproblems and assigns subproblems to workers in an asynchronous manner, attempting to give each worker a new task as soon as possible after it becomes idle. On receiving a subproblem Q, a worker uses CPLEX to calculate the lower bound $z(Q)$, which it then compares with the incumbent value. If the bound is worse, the worker immediately reports itself idle. Otherwise, if $x(Q)$ meets (13.5), the worker asynchronously broadcasts $z(Q)$ as the new incumbent value (see below), saves $x(Q)$ in a local buffer, and then reports itself idle. If $x(Q)$ violates (13.5), the worker creates a new pair of child subproblems, reports the pair to the hub, optionally executes the incumbent heuristic, and then reports itself idle.

The workers use a feature of CPLEX that allows subproblem solutions to be aborted if their outcomes are provably above some "cut-off" value. Just before entering CPLEX, this cutoff is set to the current value of the incumbent.

As with the basic serial algorithm, the search is divided into two phases. In the initial, depth-first phase, the hub treats the pool as a stack. Due to parallelism, however, the search tree can evolve in a way very different from that of a pure serial depth-first search. When the relative gap between $\bar{z}(t)$ and $\underline{z}(t)$ drops below τ, the hub switches to a best-first strategy, storing the pool as a heap.

13.2.2. The Nature of the Active Pool. In the operations research community, it is common to view the branch-and-bound process as acting on a pool of unsolved problems, each inheriting a lower bound from its parent. In the computer science community, it is more common to work on a pool of *solved* subproblems. In this alternate version of the method, the basic work of evaluating a subproblem is *first* to separate it into two children, and then to calculate a lower bound for each child. Children that violate (13.5) and survive comparison with the incumbent are returned to the pool.

The tradeoffs between the two approaches are fairly involved: the solved-pool approach may require less memory, because some of the leaf nodes of the fully developed search tree need not pass through the pool, but these savings may be offset by the need to attach extra data to each subproblem entry. The solved-pool approach makes branching decisions as late as possible, presumably benefiting from more accurate pseudocost information.

In a parallel setting, however, the unsolved-pool approach has an advantage that would be of little concern in a serial context—the task granularity is finer, because evaluating a member of the pool involves the calculation of only one LP lower bound. The consequence of this difference is greater potential parallelism, at the cost of additional communication. Because the CM-5 scales to potentially large configurations, interprocessor communication is fairly rapid, and the LP work required to calculate a lower bound is generally substantial, a decision was made to stick with the unsolved-pool approach. To save memory and communication, however, the two children of a subproblem are reported to the hub as a single data structure, with a "tag" indicating the branching variable. A single message reports the existence of both problems, and the pool consists of subproblem *pairs*. A pair is deleted from the pool only after it has been dispatched twice, once for each child.

13.2.3. Quasi-Distributed Pool Storage. When a worker reports a subproblem pair to the hub, it sends only a small "token" consisting of the parent's calculated lower bound, the worker's processor number, and a memory address a. All the remaining information associated with the pair, including the parent's l and u bound vectors, a concise description of the parent's optimal simplex basis, and branching variable information, are stored in the *worker's* memory, at address a. This scheme dramatically reduces communication and memory requirements at the hub, and the tokens fit into low-overhead active messages. The hub stores a "scale model" of the active pool, containing just enough information to control the search, while most of the related data is distributed essentially randomly among the workers.

Active messages figure prominently in the way CMMIP transfers subproblem information to idle workers. Suppose that the hub learns that worker B is idle. According to the best-first rule, some subproblem Q should be evaluated next. Chances are, however, that this problem is stored on some other worker, say A, which is busy using CPLEX to bound some other subproblem. To deliver all of Q's data to B, the hub sends an active message to A. This message interrupts A and instructs it to begin an asynchronous, background transfer of Q's data, including all bound and basis information, to B. Worker A then returns to its foreground task. When the transfer is complete, B begins work on the newly arrived problem, while A is again interrupted, and decrements problem Q's count of unevaluated children. If this count reaches zero, A deletes Q from its memory.

Figure 13.1 outlines this quasi–distributed-memory scheme, which allows

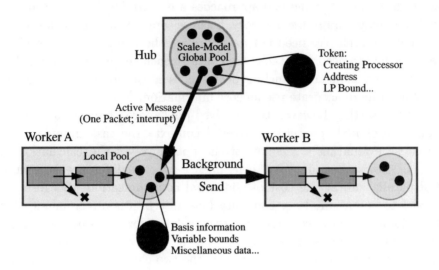

FIG. 13.1.   *Hub-initiated transfer of subproblem data from a busy worker (A) to an idle worker (B). The hub stores a "scale model" of the active pool (a small "token" for each pair of subproblems), containing just enough information to control the search. Most communications do not pass through the hub; rather, the data are transmitted between random pairs of worker processors. Use of this scheme reduces communication and memory requirements dramatically.*

virtually all communication involving the hub (including some bookkeeping functions glossed over here) to be low-overhead, single-packet, active messages. More crucially, the bulk of the code's communications do not pass through the hub at all. Instead, the vast majority of data is transmitted in relatively large batches between random pairs of worker processors. Many such transfers may take place at any one time. Random point-to-point communication of this type is well suited to the CM-5 data network architecture.

Rayward-Smith, Rush, and McKeown [10] have implemented a similar combination of central control and distributed storage for more specialized branch-and-bound algorithms on "transputer" multiprocessors. Their approach employs multiple processes per computing node, communicating via intraprocessor "messages." While overhead is probably higher than for active messages, the general principles of the implementations are similar.

13.2.4. Distribution of Incumbent Values.   Active messages also figure in the distribution of incumbent information. Essentially, the code implements an emulation of a shared, global, monotonically decreasing memory register.

Each processor has a memory location called `incumbent`. Whenever a worker constructs a candidate incumbent, it compares its objective value with `incumbent`. If the candidate is better, the candidate solution is copied to a local buffer, its value replaces `incumbent`, and the worker initiates an asynchronous broadcast operation. The worker starts this broadcast by sending an active message to the k processors whose addresses differ from its own in exactly one bit.

Consider now a receiver of one of these messages, whose address differs from the sender at bit number $l \leq k$. The receiver interrupts and compares the contained incumbent value with its own instance of `incumbent`. If `incumbent` is better, the receiver does nothing. If the message's value is better, the handler updates `incumbent` and forwards the message to the $l - 1$ processors whose addresses differ in lower order bits than l. These messages are then processed and/or forwarded by those processors, and so on. Thus, all processors learn the new incumbent value within $O(k) = O(\log p)$ time, or distribution of the value is stopped through collision with a better, simultaneously spreading value. The initial value of `incumbent` is $+\infty$.

Because this scheme is interrupt driven, the foreground code merely sees a memory location that occasionally changes to reflect improving values for the global incumbent. When the algorithm terminates, the worker originating the final value of `incumbent` has an optimal global solution in its local buffer. Interrupt-driven incumbent distribution is described in [4], although on a smaller scale and in a more cumbersome form.

13.2.5. Storage of the Pseudocosts. Like the incumbent value, the pseudocost tables are global information. It appears that no prior parallel branch-and-bound implementation has attempted to maintain any global data of this sort. Because production of new pseudocost information is much more frequent than creation of new incumbents and the amount of data involved is much greater, global memory emulation seems less practical.

As with the search control strategy, a decision was made to start with a centralized approach, assess the severity of the resulting bottlenecks, and then move to a more distributed method if necessary. Another processor, the *pseudocost server*, was set aside to store the pseudocost tables.

Whenever a worker computes the LP bound of a subproblem, it sends an active message packet to the pseudocost server. This packet identifies the problem's branching variable, states whether it is an "up" or a "down" child, and gives the lower bound difference between it and its parent. When the message arrives, the server immediately updates the master pseudocost tables.

When a worker needs to make a branching decision, it sends a list of the integrality-violating variables to the server. While the worker is waiting for a reply, it attempts to occupy itself with other tasks. The server collects the corresponding pseudocosts and returns them to the sender.

13.2.6. Starting and Stopping.   Starting the method is quite straightforward: the hub creates a description of the root LP relaxation (13.1)–(13.4) and sends it to a worker. This is the only message involving the hub that exceeds one packet.

When the hub detects termination, it uses the control network to simultaneously inform the pseudocost server and all the workers. All processors then enter a largely synchronous clean-up phase in which they jointly calculate statistics about the run, and one processor outputs the optimal solution to a file. To select this processor, the code uses the global reduction capabilities of the control network to compute the global minimum of `incumbent`. Then, from the group of workers whose local buffers correspond to the minimum, they elect the one with the lowest network address. This step requires one additional control network global reduction.

13.3. Sample Computational Results

The CM-5 implementation was tested on problems culled from "real-world" industrial applications, which are all available in the MIPLIB public collection [2]. Full details appear in [5]. It should be kept in mind that for any particular problem, there may be specialized algorithms that perform much better than the *general* branch-and-bound method described here. In other cases, a general method enhanced by cutting plane techniques might be most appropriate. However, to the degree that they incorporate branch-and-bound-style search procedures, such alternative algorithms might benefit from parallel implementations resembling CMMIP.

Some care was exercised in choosing test problems. Some MIP problems have relatively small search trees, consisting of tens or hundreds of nodes. In such cases, little parallelism is inherent in the tree, and parallel branch-and-bound is unlikely to perform well. To apply parallel computing to such cases would require the exploitation of some kind of concurrency *within* each LP bound calculation. Parallel LP methods do exist, but they depend on the structure and sparsity of the constraint matrices $A_{(1)}$ and $A_{(2)}$.

On the other hand, some MIPLIB problems are specifically designed to require specialized solution methods and produce gargantuan search trees when solved with standard branch-and-bound. In such cases, ample parallelism is available in the search tree, but an astronomical number of processors would be needed for CMMIP, as currently constituted, to take sufficient advantage of it.

The trees in the problems tested in [5] have on the order of 1000 to 100,000 search nodes, although it turns out that even larger trees would probably prove tractable. Another criterion for problem selection was that the LP relaxations be solvable within a few minutes on a single processor.

Figure 13.2 summarizes the results for four of the most difficult problems in the study, on configurations ranging from three to 128 processors. Since message latency in the CM-5 data network is not entirely deterministic, search

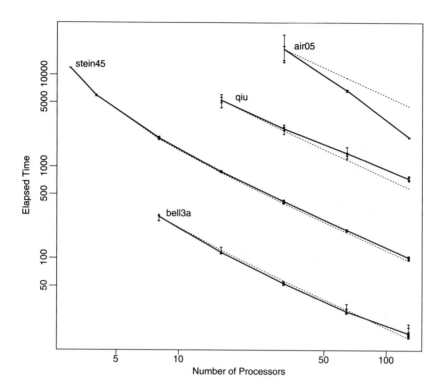

FIG. 13.2. *Results for four difficult MIP problems. Both axes are logarithmic;
the x axis plots the total number of processors, and the y axis the total runtime in
seconds. The solid lines trace the average runtime for each problem as the numbers
of processors vary, and the dashed lines show how average time would scale if it were
perfectly linear in the number of workers.*

tree sizes and runtimes can vary between runs of the same problem on the
same number of processors.

Problem `air05` is the difficult "kernel" portion of a hard airline crew
scheduling problem. The search tree stays fairly stable at about 5000
subproblems, each of which takes about two minutes to solve using CPLEX's
dual steepest-edge simplex algorithm. Runtime scales superlinearly—as the
number of workers grows, it appears that CPLEX is able to abort a larger
proportion of problems, as it can prove they are above the incumbent cutoff.
This effect appears in turn to be a result of the earlier discovery of better
incumbents obtained with the initial parallel depth-first search than with a
purely serial depth-first search. `Air05` has "set partitioning" structure and (to
the author's knowledge) has so far been solved only by specialized techniques.[9]

---

[9]In subsequent research, it was found that some adjustment of algorithm parameters, principally

`Qiu` is a fiber optic network design problem whose solution initially required about 90 hours on a desktop workstation. In parallel, it is possible to solve it in under 20 minutes. Runtime scales very slightly worse than linearly, partly because the root problem is quite time consuming. There are about 20,000 to 30,000 search nodes.

`Stein45` has more than 100,000 search nodes, and the optimal solution tends to be discovered very early in the search. Runtimes scale almost perfectly linearly with the number of workers. `Bell3a`, another fiber optic network design problem, has a search tree of more than 40,000 nodes. Again, average performance is nearly linear. Interestingly, configurations of fewer than eight processors could not solve the problem at all, because they never discovered an initial incumbent. Apparently, pure depth-first search is not a good strategy for this problem, and the more randomized search obtained by using multiple workers is much more effective.

Tests of 15 additional problems, most considerably easier than those shown in Figure 13.2, are contained in [5]. As a very rough rule, it was found that for problems with T nodes in their serial search trees, at least $T/100$ processors could be used very efficiently. Harder problems showed good efficiencies through at least 64 processors, while runtimes for easier problems could generally be reduced to roughly the order of seconds, after which gains from parallelism tended to become minimal.

For some problems, central control did pose some limitations. For example, `stein45` and `bell3a` both have relatively easy LP relaxations, solvable in a tenth to a few hundredths of a second. In the 128-processor configuration, the hub and pseudocost server occasionally fall behind in servicing the workers, explaining some small deviations from linear performance. For problems with extremely easy LP relaxations, such limitations can become severe even with as few as 32 processors.

Forthcoming papers will describe versions of CMMIP that have partially and fully distributed storage and control options capable of eliminating such bottlenecks.

While there are certainly a number of open questions, the results here and in [5] suggest that parallel branch-and-bound solution of real MIP problems is now a reality. For repeated use in a specific application, of course, it would probably be advisable to specialize the method somewhat. Such specializations could include some of the cutting–plane-based techniques for improving bounds that have become popular during the last decade or so. (It is also possible that these techniques will help in general settings.) There is no reason such techniques could not be combined with parallel search, providing even greater strides toward the solution of difficult MIP problems.

---

disabling the incumbent heuristic, reduced the difficulty of `air05` to somewhere in between the next two problems, `qiu` and `stein45`. Runtime then scaled approximately linearly with the number of processors. Specialized cutting plane techniques should also significantly reduce runtimes for `air05` (see, e.g., [6]).

13.4. Acknowledgments

The author would like to thank the following people for their help and advice on this project: Doug Wiedemann, Xiru Zhang, Adam "Moose" Greenberg, Alan Mainwaring, and Bradley Kuszmaul—all currently or formerly affiliated with Thinking Machines Corporation; Todd Lowe, Mary Finelon, Bob Bixby, and Irv Lustig—all affiliated with CPLEX Optimization, Inc.; and Jerry Shapiro (MIT), Michael Saunders (Stanford), Yuping Chiu (USWest), and Andy Boyd (Texas A&M).

References

[1] T.S. ABDEL-RAHMAN, *Parallel Processing of Best-First Branch and Bound Algorithms on Distributed Memory Multiprocessors*, PhD thesis, Computer Science and Engineering, University of Michigan, Ann Arbor, MI 1989.

[2] R.E. BIXBY, E.A. BOYD, AND R.R. INDOVINA, *MIPLIB: A test set of real-world mixed-integer programming problems*, SIAM News, 25(1992), p. 16.

[3] R.L. BOEHNING, R.M. BUTLER, AND B.E. GILLETT, *A parallel integer linear programming algorithm*, European J. Oper. Res., 34(1988), pp. 393–398.

[4] T.L. CANNON AND K.L. HOFFMAN, *Large-scale 0–1 programming on distributed workstations*, Ann. Oper. Res., 22(1990), pp. 181–217.

[5] J. ECKSTEIN, *Parallel branch-and-bound algorithms for general mixed integer programming on the CM-5*, SIAM J. Optim., 4(1994), pp. 794–814.

[6] K.L. HOFFMAN AND M. PADBERG, *Improving LP-representations of zero-one linear programs for branch-and-cut*, ORSA J. Comput., 3(1991), pp. 121–134.

[7] R.M. KARP AND Y. ZHANG, *A randomized parallel branch-and-bound procedure*, in Proc. ACM Symp. Theory Comput., 20(1988), pp. 290–300.

[8] G.A.P. KINDERVATER AND J.K. LENSTRA, *Parallel computing in combinatorial optimization*, Ann. Oper. Res., 14(1988), pp. 245–289.

[9] A. LAND AND S. POWELL, *Computer codes for problems of integer programming*, in Discrete Optimization II, Annals of Discrete Mathematics 5, P. L. Hammer, E.L. Johnson, and B.H. Korte, eds., North-Holland, Amsterdam, 1979.

[10] V.J. RAYWARD-SMITH, S.A. RUSH, AND G.P. MCKEOWN, *Efficiency considerations in the implementation of parallel branch-and-bound*, Ann. Oper. Res., 43(1993), pp. 123–145.

[11] T. VON EIKEN, D.E. CULLER, S.C. GOLDSTEIN, AND K.E. SCHAUSER, *Active messages: A mechanism for integrated communication and computation*, in Proc. 19th International Symposium on Computer Architecture, Gold Coast, Australia, 1992.

Chapter 14

Optimal Scheduling Results for Parallel Computing

Håkan Lennerstad

Lars Lundberg

Editorial preface

Load balancing is one of many possible causes of poor performance on
parallel machines. If good load balancing of the decomposed algorithm or
data is not achieved, much of the potential gain of the parallel algorithm is
lost to idle processors. Each of the two extremes for load balancing—static
allocation and dynamic allocation—has advantages and disadvantages.
This chapter illustrates the relationship between static and dynamic
allocation of tasks.

This article originally appeared in *SIAM News*, Vol. 27, No. 7, Au-
gust/September 1994. It was updated during the summer/fall of 1995.

It is a common situation to have several tasks to be executed, some of
which may be dependent on others, and several available executors of tasks.
Clearly, some kind of scheduling is needed: Which executor is to do which task,
and when? A frequent goal is to minimize the global execution time, i.e., the
time from the start to the termination of the last task. One central scheduling
question is whether to allow transfers of tasks from one executor to another.
The two extremes are to allow any transfer at any time, usually called dynamic
allocation of tasks, and to allow no transfers at all, which is static allocation.

A situation of this type occurs when a parallel program is executed on a
parallel computer. In this case the processors are most often identical, and
the tasks themselves are usually not affected by the way they are scheduled.
The formulas we present in this article compare the performance of optimal
static and optimal dynamic allocations for scheduling problems of this type.
The functions give the ratio of the execution times in the worst case: it
is a maximum of the ratio over all parallel programs with any interprocess
dependency structure, with the only exception of deadlock. Immediate
applications include the design of parallel computers and the evaluation of

static allocation algorithms. Our results arose in the context of parallel computing, and it is in that terminology that they are presented here.

14.1. Parallel Program Scheduling

Consider a parallel computer with k identical processors and a parallel program P with n processes. In the usual and worst case, the program P has many dependencies among its processes—at several points in time some processes cannot execute unless certain other processes have completed certain calculations. If $n > k$, then some processor must execute at least two processes. If $n \leq k$, there is no scheduling problem; in this case each process can occupy a processor of its own.

An essential performance issue is how to distribute the processes over the processors in a way that minimizes the total execution time. A parallel program P can obviously be allocated to the processors in many different ways, with varying total execution times. There is a basic set consisting of all possible allocations. For some of these allocations, no processes are transferred; this is the subset of all static allocations. Within this subset, an allocation for which the total execution time is minimal represents an optimal static allocation. The execution time for the program P with the optimal static allocation is denoted by $T_s(P)$. The execution time for the program P with an optimal dynamic allocation is analogously denoted by $T_d(P)$.

Static allocation is a more restricted scheduling scheme than dynamic allocation. Clearly, the execution time for a parallel program P with optimal static allocation is never shorter than that for an optimal dynamic allocation. However, the transfer of processes allowed in dynamic allocation is sometimes complicated and can be time consuming. This suggests that the adverb "clearly" may be too definitive. How much worse can it actually be to apply static allocation than to apply dynamic allocation is an issue of real concern to parallel computer designers and the subject of this article.

14.2. NP-Hard Scheduling Problems

The problems of finding optimal static and optimal dynamic allocations are both known to be NP-hard. Nevertheless, we can compare the performance of the two allocation schemes; the ratio is given by the values obtained from the formulas presented in this article. The execution times for calculating these values increase slowly with n but faster with k. Experiments show that the computation of the values can be parallelized, allowing values for large n and k to be computed. It is not known whether the execution time increases exponentially with k.

We make two standard assumptions. First, we neglect the execution time of dependency signals between processes, which usually is negligible. Second, we neglect the cost of transferring processes, which often is not negligible. With these two assumptions, we have a situation where the adverb "clearly" of our earlier statement is fully valid, and where, given an executable parallel program P, the execution times with optimal static and dynamic allocations, $T_s(P)$ and $T_d(P)$, respectively, are well defined.

In this notation we present an explicit formula for the function

$$(14.1) \qquad g(n,k) = \max_P \frac{T_s(P)}{T_d(P)},$$

where the maximum is taken over all parallel programs P with n processes, executed on a multiprocessor with k processors.

14.3. The Process-Dependent Function

If $n \geq k$, we define $g(n,k) = 1$. Otherwise, if $w = n/k$ is an integer,

$$(14.2) \qquad g(n,k) = \frac{1}{\binom{n}{k}} \sum_{l=1}^{\min(w,k)} l\pi(k,w,k,l),$$

where $\pi(k,w,q,l)$ is defined below. If $w = n/k$ is not an integer, we let $w = \lfloor n/k \rfloor$ and denote the remainder of n divided by k by n_k; i.e., $n_k = n - k\lfloor n/k \rfloor$. Then

$$(14.3) \qquad g(n,k) = \frac{1}{\binom{n}{k}} \sum_{l_1=\max(0,\lceil \frac{k-(k-n_k)w}{n_k}\rceil)}^{\min(w+1,k)} \sum_{l_2=\max(0,\lceil \frac{k-l_1 n_k}{k-n_k}\rceil)}^{\min(w,k-l_1)} \max(l_1,l_2)$$

$$\times \sum_{i=\max(l_1,k-l_2(k-n_k))}^{\min(l_1 n_k,k-l_2)} \pi(n_k,w+1,i,l_1)\pi(k-n_k,w,k-i,l_2).$$

The function $\pi(k,w,q,l)$ is 0 if $\min(q,w) < l$ or if $q > kl$; otherwise, if $k = 1$ then $\pi(k,w,q,l) = \binom{n}{q}$. In all other cases it is given by

$$(14.4) \quad \pi(k,w,q,l) = \binom{w}{l} \sum_I \binom{w}{i_1} \cdots \binom{w}{i_{k-1}} \frac{k!}{\prod_{j=1}^{b(\{l\}+I)} a(\{l\}+I,j)!}.$$

Here the sum is taken over all sequences of nonnegative integers $I = \{i_1,\ldots,i_{k-1}\}$ that are decreasing ($i_j \geq i_{j+1}$ for all $j = 1,\ldots,k-2$), bounded by l ($i_1 \leq l$), and have the sum $q - l$ ($\sum_{j=1}^{k-1} i_j = q - l$). By definition, $b(I)$ is the number of distinct integers in I, and $a(I,j)$ is the number of occurrences of the jth distinct integer in I. The notation $\{l\} + I = \{l, i_1, \ldots, i_{k-1}\}$ is used.

The maximum of $g(n,k)$ for n and k up to 50 is 2.543. It follows that for any parallel program consisting of at most 50 processes, the execution time with optimal static allocation is never more than a factor of 2.543 longer than that with optimal dynamic allocation. The ratio itself is optimal; there are programs having this value. Computer scientists studying parallelism have found this ratio surprisingly low.

The cost of transferring processes is probably strongly application dependent. In any case, taking this into account favors static allocation even more. On the other hand, it is easier in practice to find close-to-optimal dynamic allocations than close-to-optimal static allocations. For the static case, the formula gives quantitative information on how close to optimal the allocation is.

14.4. The Mathematical Formulation

We briefly and intuitively sketch the mathematical reformulation; for a full presentation see [2, 6]. Formulas (14.2)–(14.4) can be deduced by reformulating the problem about parallel programs into a problem about 0,1 matrices. A 0,1 matrix represents a program P with optimal dynamic allocation. The matrix consists of n column vectors representing the processes; the rows represent time intervals of equal lengths. In this matrix, 1 denotes an active process and 0 an idle process for a certain time interval.

For a given program, the time is discretized in time intervals of equal length in such a way that each dependency signal and process activity change occurs at an end-point of a time interval. Each row has k 1's, one for each processor, since the worst-case programs keep all processors constantly busy. Each static allocation is represented by a partition of the set of column vectors. The execution time with a specific static allocation is computed by adding the column vectors in each partition group, resulting in k column vectors representing the load on the k processors for each time interval. Next the componentwise maximum is taken, since at each time interval all processors must wait for the processor with maximal load. This results in a single remaining column vector, representing the multiprocessor load for each interval. The sum of its entries is the execution time with this static allocation. In this formulation, because the execution time with optimal dynamic allocation is the number of rows, the ratio can be described as the arithmetic mean of the entries in the multiprocessor load vector. It is proved [2, 6] that a matrix whose rows are exactly all $\binom{n}{k}$ permutations of the k 1's in n positions represent extremal programs, those that maximize the ratio of static to dynamic allocation. In addition, there is always an optimal static allocation from which the sizes of the sets in the partition differ as little as possible. When this situation is achieved, a formula for $g(n, k)$ can be deduced.

14.5. Optimal Topography

The function $g(n, k)$, when plotted as a function of two variables (see Figure 14.1), has a surprising and interesting topography. The main parts of the plot can be described as plateaus and the transitions between them.

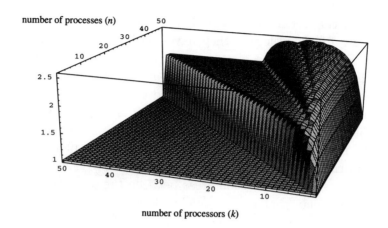

number of processes (n)

number of processors (k)

FIG. 14.1. *The optimal worst-case function $g(n,k)$, defined as the worst-case ratio of static versus dynamic allocations: $g(n,k) = \max_P \frac{T_s(P)}{T_d(P)}$. Here $T_s(P)$ and $T_d(P)$ denote the execution times for a parallel program P with optimal static and optimal dynamic allocations, respectively, executed on a multiprocessor with k identical processors. The maximum is taken over all parallel programs P consisting of n processes.*

The plateau $g = 1$ is immediately visible; this is the trivial case in which there are more processors than processes. In this case no process ever needs to be transferred, there is no difference between static and dynamic allocations, and the ratio is thus identically 1.

The plateau $g = 2$ is also visible within $n, k \leq 50$. When moving from the first plateau, from, say, the point $k = n = 15$ to $k = 15, n = 16$, we can observe a sharp jump. In the 0,1 matrix formulation, we have here a worst-case matrix (parallel program) whose rows are the 16 permutations of 15 1's and a single 0. In this time interval the execution time with optimal dynamic allocation is 16. In the partition (static allocation), two columns (processes) will belong to the same partition set (execute on the same processor). When adding the column vectors, we will get 14 2's and 2 1's (in most time intervals this processor will have two busy processes), giving $g(16,15) = (14 \times 2 + 2)/16$, which is almost 2.

The plateau $g = 3$ is not visible in the domain $n, k \leq 50$. It is established that for each positive integer w, $g(n,k)$ has a plateau $g = w$; i.e., $|g(n,k) - w|$ is arbitrarily small for all points (n,k) in an unbounded domain sufficiently far away between the "straight lines" $n = (w-1)k$ and $n = wk$. The main unknown geometrical feature of the function $g(n,k)$ is perhaps the distance from the origin to each plateau.

14.6. Optimal Ridge Mountains

Valleys and ridges can be observed in the plot of Figure 14.1. Before the plateaus are fully developed, they have the form of "mountain ridges." As shown in Figure 14.2, the plateau $g = 2$ actually has a ridge, although very flat, in the middle. Observed from the n-axis, and perpendicular to it, this is the highest ridge until $n = 10$. The second ridge starts at $n = 10$, taking over as the highest from $n = 11$ to $n = 23$, where it is surpassed by the third ridge. At $n = 36$ the first sign of the fourth ridge appears. The last two ridges will eventually become the plateaus $g = 3$ and $g = 4$. The plots suggest that each subsequent ridge rises to the plateau level much more slowly than the previous one. Accordingly, the transition areas between the plateaus become larger.

The statements that can be made when the plot is viewed from the k-axis are even more definite. The function $g(n, k)$ is increasing as a function of n; when the number of processes increases, it is possible to find parallel programs with larger ratios of optimal static to optimal dynamic allocations. This is an obvious truth for computer scientists and can be proved within this mathematical framework. It turns out that it is also possible to calculate an explicit formula for the limit $\lim_{n \to \infty} g(n, k) = G(k)$. In the plot of $g(n, k)$, the function $G(k)$ represents the horizon when the landscape of $g(n, k)$ is viewed from the k-axis, in the direction of the n-axis. In parallel computing terms, the function $G(k)$ is the worst-case ratio of optimal static to dynamic allocations for a parallel computer with k processors, where a worst-case program is found in the domain of all parallel programs of any number of processes.

14.7. The Process-Independent Function

With the same notation used above, the formula for the horizon function $G(k)$ [3, 7] is

$$(14.5) \quad G(k) = \frac{(k!)^2}{k^k} \sum_{l=1}^{l=k} \frac{1}{(l-1)!} \sum_{I} \left(\prod_{j=1}^{k-1} i_j! \prod_{j=1}^{b(\{l\}+I)} a(\{l\} + I, j)! \right)^{-1}.$$

The sum over I is taken over the same sequences as in (14.2) and (14.3), with $q = k$. Here we have a factor, $(k!)^2 k^{-k}$, that tends rapidly to infinity, multiplied by a sum over certain decreasing sequences. The number of terms in this sum tends to approach infinity rapidly. On the other hand, the terms of the sum are inverse values of products of factorials, i.e., numbers that are certainly very small and tend to zero very rapidly as k becomes large. These numerical aspects balance each other to result in moderate behavior for $G(k)$. $G(k)$ stays slightly larger than $\log_e k$ up to $k = 38$, but $G(39) = 3.65393 < 3.66356 = \log_e 39$.

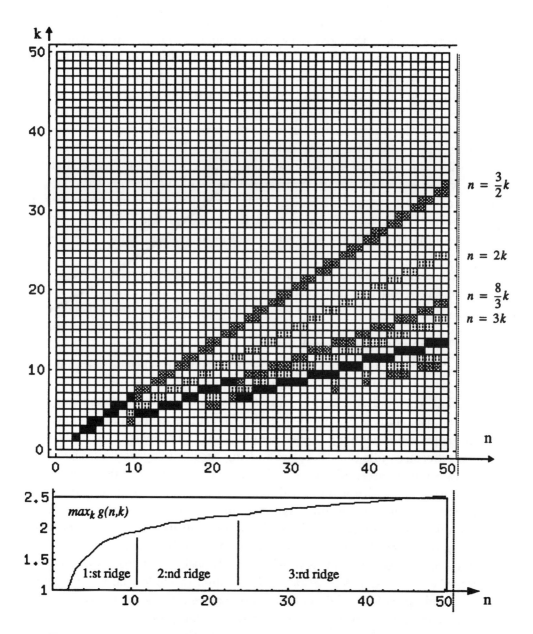

FIG. 14.2. *Ridges and valleys of* $g(n, k)$. *Details of the ridge/valley structure of the surface* $g(n, k)$ *are revealed by studying the function* $g(n, k)$ *as a function of* k, *for each constant* n. *The locations of global maxima* ■, *local maxima* ■ ▦, *and local minima* ⊞ *are plotted. Notes:* (1) $g(n, k)$ *is increasing as a function of* n. (2) *The straight line equations shown in the margin are simply constructed from the graph.* (3) *The function* $\max_k g(n, k)$ *gives the worst-case ratio for all parallel programs with* n *processes and for all multiprocessors with identical processors.*

14.8. Further Applications and Generalizations

By formulating a central parallel computing problem mathematically, we were able to derive formulas that give answers to the parallel computation scheduling problem. Since mathematics is in principle application independent, it is natural to look for related applications. From a general point of view, the formulas are optimal scheduling bounds, which represent a problem domain that appears in several other contexts. A practical example is the construction of a house; of the many tasks that need to be performed, some can be done only after others are fully or partially completed. The processes here are the work tasks; the processors are the workers. Static allocation means that the worker who starts a task always completes it, while in dynamic allocation a task can be taken over by another worker. The hypothesis of identical processors is usually not valid here—humans tend to be specialists. This, like further parallel computer problems, suggests one line of generalization: the case of nonidentical processors.

In telecommunications, scheduling problems are frequent. The performance of the algorithm that schedules incoming messages and other events on the available nodes is essential for telecommunication system efficiency.

14.9. Previous and Future Results

The only previous general results of a similar type are those of Graham [1]. In [3] it is established that, as expected, $\lim_{k\to\infty} G(k) = \infty$. In [1] it is proved that when only parallel programs with no interprocess dependency are considered, this limit is 2. Another result of [1] concerns a dynamic allocation scheme called self-scheduling. Self-scheduling has the property that any time a processor becomes idle and there are waiting executable processes, one of the processes is immediately allocated to this processor. In [1] it is also proved that the execution time with a self-scheduling algorithm is never more than twice the execution time with optimal dynamic allocation. This result is useful in, for example, the evaluation of static allocation algorithms, since the problem of finding self-scheduling allocations is certainly not NP-hard.

With similar techniques it is possible to establish results comparing cluster allocation with dynamic allocation [3, 7], optimal ratios for parallel programs with a certain parallelism [8], and the gain achieved by increasing the number of memory units in simultaneous memory-allocation problems [9].

The results of [10] come closer to the application. In this report one test execution of the program is done with some static allocation schedule, and this execution time is measured. By taking advantage of the extra information we obtain a tighter bound, optimal for the new situation.

The report [4] presents our first results on single processors. Optimal bounds comparing cache memory associativity are derived. Also in this article we finally arrive at a basic problem that is solved by similar arguments as in the other reports. However, the sequence of transformations leading to this basic problem here constitute the major part of the report. The report [5] is a survey article over all our results to this date.

References

[1] R.L. GRAHAM, *Bounds on multiprocessing anomalies*, SIAM J. Appl. Math., 17(1969), pp. 416–429.

[2] H. LENNERSTAD AND L. LUNDBERG, *An optimal execution time estimate for static versus dynamic allocation in multiprocessor systems*, SIAM J. Comput., 24(1995), pp. 751–764.

[3] ——, *Optimal Performance Functions Comparing Process Allocation Strategies in Multiprocessor Systems*, Research Report 3/93, Högskolan i Karlskrona/Ronneby, Sweden, 1993.

[4] ——, *Optimal Worst Case Formulas Comparing Cache Memory Associativity*, Research Report 5/95, Högskolan i Karlskrona/Ronneby, Sweden, 1995.

[5] ——, *Combinatorics for scheduling optimization*, in Proc. Conference of Combinatorics and Computer Science, Brest, France, Lecture Notes in Computer Science, Springer-Verlag, Berlin, New York, 1995.

[6] L. LUNDBERG AND H. LENNERSTAD, *An Optimal Bound on the Gain of Using Dynamic versus Static Allocation in Multiprocessor Computers*, Technical Report, Högskolan i Karlskrona/Ronneby, Sweden, 1992.

[7] ——, *An Optimal Performance Bound on the Gain of Using Dynamic versus Static Allocation in Multiprocessors with Clusters*, Technical Report, Högskolan i Karlskrona/Ronneby, Sweden, 1993.

[8] ——, *An optimal upper bound on the minimal completion time in distributed supercomputing*, in Proc. 8th ACM Conference on Supercomputing, Manchester, England, July 1994.

[9] ——, *Bounding the Maximal Gain of Changing the Number of Memory Modules in Multiprocess Computing*, Technical Report, Högskolan i Karlskrona/Ronneby, Sweden, 1994.

[10] ——, *An optimal lower bound on the maximal speedup in multiprocessors with clusters*, in Proc. First International Conference on Algorithms and Architectures for Parallel Processing, Australia, April 1995.

Chapter 15

Airline Crew Scheduling: Supercomputers and Algorithms

Jeremey Schneider
Theresa Hull Wise

Editorial preface

Integer programming (IP) problems are difficult to solve. Particularly difficult is the airline crew scheduling problem, a 0–1 set partitioning variant that requires special solution methods. The typically large size of the set-partitioning problem for the crew scheduling problem makes it even more challenging.

This chapter formulates the crew scheduling problem, summarizes some of the past solution methods, and details various current approaches. The application of a vector computer to this problem is presented.

This article originally appeared in *SIAM News*, Vol. 23, No. 6, November 1990. It was updated during the summer/fall of 1995.

"The crew scheduling problem has been 'solved' many times using many and varied techniques. It has never, however, been completely solved to the satisfaction of all airlines, and certainly not to the degree of rigor that the term 'solution' would imply to a mathematician." [2]

Stated more than 20 years ago, this passage is an equally valid assessment of the airline crew scheduling problem today. While advances in computer technology and operations research (OR) algorithms have made it possible to handle larger, more complex problems more quickly and more precisely, the size of the crew scheduling problem and the demand for precision have evolved even more rapidly. Mergers and acquisitions have brought larger and more complex problems to the crew scheduler. To help Northwest Airlines solve this problem, in 1988, we acquired a Convex supercomputer specifically and exclusively for this application. In the ensuing years, we have continued to research other forms of advanced computing as well. We discuss here our usage and on-going investigation into advanced computers and OR algorithms for the crew scheduling problem.

The crew scheduling problem is an economically significant topic for the airline industry. Crew costs are, in fact, one of the two highest components of direct operating cost, exceeded only by fuel costs. Therefore, the OR practitioner, equipped with today's hardware technology, has the potential to make a deep financial impact by attacking this problem. Even a 1% improvement could be worth millions of dollars per year. Whereas manual solutions once gave acceptable (if not optimal) results, the problem has so grown in difficulty and importance that it now warrants the very latest in OR and computing technology. While some organizations have had success developing this in-house, others are interfacing with the academic community for mathematical and technological expertise. Northwest Airlines' AirLine Pairing Planning System (ALPPS) development project has taken the latter approach.

15.1. Overview

The goal of the airline crew scheduler is to assign personnel optimally within a strictly specified set of constraints. At the very least, each flight leg (a nonstop flight between a pair of cities) in a given flight schedule must be covered once and only once with a full complement of pilots and flight attendants. Each leg then becomes part of a roundtrip, called a "pairing," for some set of crew members.

Each pairing originates at a crew-base station, traces a path through a network defined by the flight schedule, and returns to its original crew-base station. The airline's crew staffing plan dictates the maximum and minimum number of crew members who may start from each base. The duration of the trip varies with equipment type and other factors from as little as two days to as many as 14. The number of flight legs in the trip can also vary from trip to trip and among equipment types.

The description above transforms the task of legally assigning airline crews into a search for round-trip pairings that partition the legs of a given flight schedule. Now, by defining a cost for each pattern, the optimal assignment of personnel becomes equivalent to the minimum cost partition of the flight legs into pairings. In this way, the airline crew scheduling problem acquires the structure of a set partitioning problem (SPP), making it amenable to known OR techniques. As an SPP, the problem takes the form

$$\text{(SPP)}\quad \text{minimize}\quad \sum c_j x_k$$

$$\text{subject to}\quad \sum a_{ij} x_j = 1\quad \text{for } i = 1,\dots,m,$$

$$x_k \in \{0,1\}\quad \text{for } k = 1,\dots,n.$$

Each row of $\mathbf{A}$ (indexed by $i = 1,\dots,m$) represents a flight leg to be flown. Each column of $\mathbf{A}$ (indexed by $j = 1,\dots,n$) represents a round-trip pairing that a crew may legally fly.

As the matrix $\mathbf{A}$ is constructed, column-by-column,

$$a_{ij} = \begin{cases} 1 & \text{if flight leg } i \text{ is on pairing } j, \\ 0 & \text{otherwise.} \end{cases}$$

For example, the first pairing may include flight legs 1, 3, and 62, causing the first column to have a "1" in rows 1, 3, and 62 with "0" in all other rows. Then we have something like

$$\mathbf{A} = \begin{array}{c} \\ leg\#1 \\ leg\#2 \\ leg\#3 \\ leg\#4 \\ \\ leg\#61 \\ leg\#62 \\ leg\#63 \\ \\ leg\#m \end{array} \begin{pmatrix} pairing\#1 & pairing\#2 & \cdots & pairing\#n \\ 1 & a_{1,2} & \cdots & a_{1,n} \\ 0 & a_{2,2} & \cdots & a_{2,n} \\ 1 & a_{3,2} & \cdots & a_{3,n} \\ 0 & a_{4,2} & \cdots & a_{4,n} \\ \vdots & \vdots & \ddots & \vdots \\ 0 & a_{61,2} & \cdots & a_{61,n} \\ 1 & a_{62,2} & \cdots & a_{62,n} \\ 0 & a_{63,2} & \cdots & a_{63,n} \\ \vdots & \vdots & \ddots & \vdots \\ 0 & a_{m,2} & \cdots & a_{mn} \end{pmatrix}.$$

Now, supposing that this first pairing is legal and its cost to the airline is 210, then the corresponding (first) entry in the cost vector would be $c_1 = 210$. This means that if the first pairing is selected to be flown, its cost of 210 must be added to the objective function. Note also that once pairing 1 has been selected no other chosen pairing may cover legs 1, 3, or 62 as mandated by the equation $\mathbf{Ax} = \mathbf{1}$ in SPP. Here $\mathbf{1}$ is a vector of 1's (i.e., the row sums are identically 1).

The goal of the optimization phase is to select a "covering" or "partitioning" set of pairing columns at minimal cost; i.e., we must solve SPP for the assignment vector $\mathbf{x}$. The vector $\mathbf{x} = (x_1, x_2, \ldots, x_n)$ has one entry for each pairing column in matrix $\mathbf{A}$.

An element that greatly complicates this formulation of the crew scheduling problem is the complexity that arises from contractual rules and pay guarantees. To be considered, a pairing must conform to both Federal Aviation Regulations and union contract requirements. These include maximum flying and minimum rest times per duty period. Other company policies, generally described as "soft" constraints, are not steadfast rules but, rather, preferences honored by the scheduler if the cost of doing so is not too high. Such policy parameters could include minimum crew connect times. The airline pays no actual monetary penalty for violating policies of this type.

The airline also maintains a set of rules that pushes the dollar cost of a pairing higher than its actual operating cost. This difference (the "penalty") is driven by a collection of, again, well-defined parameters. Most are categorized as "credit time" paid to the crews. Because of minimum guarantees per duty period, for example, crew members who fly less than x hours are nonetheless paid for x hours. Likewise, there are average and minimum percentages of time away from home that must be paid.

"Deadheading," which occurs when pilots fly as passengers to position themselves to fly the next leg of their crew pairing, is another expense incurred as difficult pairings are completed. Although sometimes necessary, deadheading is an expensive option as a result of both the loss of revenue seats occupied by the crew and the pay provided to the crew to travel but not fly.

The length of a pairing can also create penalty costs in the form of per diem expenses and hotel room fees. While these are major expenses, they tend not to be significant in driving decisions. Other "soft" penalties arise from situations such as substandard crew connect times, pairings beginning at one coterminal station and ending at another, and trips that contain too much flying.

15.2. Methods for Solving SPP

A common *simplifying* assumption is that the crew scheduling problem has two more or less distinct phases: the first is the explicit generation of the pairings, also called column generation; the second is optimization to select a subset of those pairings meeting all requirements at a low, if not the lowest, cost to the airline. However, as will be discussed below, the generation of *all* legal pairings (or, equivalently, all columns for matrix **A**) is not practical from an operational standpoint. As a result, the solution methods addressed below are solutions to subproblems of the original crew scheduling set partitioning problem. Owing to the number of possible approaches, and the mathematical flavor of the problem, we elaborate on several of the existing methods we have used or considered and several of the emerging methods in which we hold great promise.

15.2.1. Implicit Enumeration Techniques. The success of implicit enumeration, a widely used approach for solving crew optimization problems, apparently results from the special structure of the problem. Its performance depends on an appropriate selection of a *branching* scheme, which considers variables for assignment, and a *bounding* scheme, which assesses the quality of each assignment.

The required branching scheme may involve fixing particular variables (e.g., $x_j = 1$) to satisfy certain constraints, where the decision is based on whether or not a particular pattern should be forced to cover a given leg. The verdict is then fixed on that branch of the solution process and on all subproblems extending from it. An analogue of this approach (developed in the 1970s and used in an early version of Northwest Airlines' ALPPS optimizer [10]) is to group the pattern columns into classes and then to select the class to cover a given row. This early heuristic was considered to be well suited for the larger crew optimization problems of the time.

A good bounding scheme enhances the performance of the implicit enumeration technique as it reduces the size of the enumeration tree. Early bounding strategies, such as that of Marsten [14], computed the bounds via linear programming (LP) relaxation at each step. This later proved quite costly for the larger problems. The next set of strategies, such as that of Marsten and Shepardson [15], sought good approximations of the LP bounds without actually computing them. This was used in the early versions of the ALPPS optimizer that remained in use through the late 1980s. Bounds were approximated with Lagrangian relaxation and subgradient optimization. This approach is also described by Etcheberry [8].

While it, too, is sufficient for smaller problems, this approach has several inherent drawbacks. The algorithm requires an initial bound as well as several parameter settings, all of which may vary from problem to problem and affect overall performance. Also, because convergence is asymptotic and not monotonic, an oscillating lower bound may be observed. Finally, convergence is demonstrably slow for larger problems. This originally prompted Northwest Airlines' search for new optimization algorithms for our crew scheduling system.

More recently, Chan and Yano [6] introduced the multiplier adjustment method to find lower bounds. This algorithm is a hybrid involving both Lagrangian relaxation and LP. By quickly identifying a good set of Lagrangian multipliers, they are able to initiate their linear program at a more advanced state. Their computational results show this method to be superior to subgradient optimization. In addition, convergence is monotonic and neither user-specified parameters nor an initial upper bound are required as input. Further studies also found this method to be superior to early use of LP bounds.

Another bounding approach considered for the Northwest Airlines' crew optimization problem was that of Fisher and Kedia [9] who used the "greedy heuristic" followed by an improvement heuristic, followed (only if necessary) by subgradient optimization.

Table 15.1 displays the relative results of a 1990 benchmark at Northwest Airlines, using four different solution techniques to solve the same series of 100 small subproblems. While exact problem sizes are not available, the number of rows was approximately 30 and the number of columns approximately 700. Selected for the study were an implicit enumeration approach using Lagrangian relaxation and subgradient optimization (as was found in our crew optimization software at that time), Kedia's SPMINC heuristic [13], PAR4, an implementation of the interior-point method by Lustig, Marsten, and Shanno [16], and the simplex-based CPLEX-mip[10] developed by Bixby [3]. All arrived at the same solution values but at greatly differing rates. Note that while the problems were too large for the Lagrangian relaxation method to solve efficiently, they were perhaps too small to give an advantage to an

---

[10]CPLEX is a trademark of CPLEX Optimization Inc.

interior-point method. Note that the simplex method CPLEX was superior in terms of computation time.

TABLE 15.1

Running time on DC9-daily batch job (100 subproblems).

	Lagrangian	SPMINC	PAR4	CPLEX
Total time (sec.)	341.57	94.30	68.54	10.10

15.2.2. Elastic Set Partitioning Techniques. This model for the crew scheduling subproblem is described by Graves et al. [11]. Rather than the usual set partitioning model, this group considers the *elastic* set partitioning problem

$$
(15.1) \qquad
\begin{aligned}
\text{minimize} \quad & c^t x + p^t s^- + p^t s^+ \\
\text{subject to} \quad & Ax + I s^- - I s^+ \; = 1, \\
& x \qquad\qquad\quad \text{binary,} \\
& s^-, s^+ \qquad\quad\ \geq 0,
\end{aligned}
$$

which allows each constraint i to be violated at a cost of penalty p_i. They indicate that performance is again constrained by the size of the problems to be solved.

15.2.3. Probabilistic Relaxation Techniques. Wedelin [19] describes a probabilistic relaxation algorithm that can be used as a general purpose approximation for combinatorial optimization. This fundamentally different approach has been applied to airline crew scheduling problems by the Volvo Transportation Group in Sweden. In combinatorial optimization problems, such as this one, probabilistic relaxation seeks a good integral approximation to the optimal solution.

The procedure is iterative. Each variable, x_i, is initially assigned between 0 and 1. Later iterations, in sequence or parallel, adjust the variable subject to the constraints until they converge to "yes" or "no" in the final assignment network.

In terms of computational results, our benchmarks in 1990 indicated that while probabilistic relaxation did produce high quality approximations to very large problems, the computational effort outweighed the benefit to Northwest Airlines. However, on crew optimization problems with up to 2,000,000 columns, Desrosiers et al. [7] later found that this method produced high quality approximations in less time than was required by standard LP packages.

15.2.4. LP Relaxation. An intuitive approach to the set partitioning formulation is to "relax" the IP problem to an LP problem. This is achieved by replacing the integrality condition $x_i = \{0, 1\}$ with the nonnegativity condition $x_i \geq 0$ for all i. While the relaxed problem is much easier to solve, we now face the possibility of fractional solutions—an unacceptable alternative in the crew scheduling application.

Interestingly, some computational experience has indicated that for crew optimization subproblems, the LP relaxations often yield natural integral optimal solutions. While some believe this behavior to be characteristic of the airlines' set partitioning problems in a certain size range, our recent research contradicts this claim [20]. As such, one cannot rely on an LP relaxation alone.

Nonetheless, LP relaxation has been a primary tool in the development of solution methods for IP problems. And, as such, knowing that we will be making use of large LP relaxations we digress here to discuss their behavior on supercomputers. The combination of new algorithms for solving LPs, improvements to classical methods, and the availability of modern computer architectures has worked to permit solutions to problems of sizes previously thought impractical. Today's vector processors, which are capable of efficient indirect addressing, help the math programmer take full advantage of the sparse matrices encountered in LP and in particular in the crew scheduling model.

The simplex method has great success with such large systems. Vector processing enhances the simplex method allowing larger problems to be solved with reasonable computer resources.

Research into interior-point methods also improved the prospect of solving very large LP problems. The major computational step in this method is amenable to vector processing and loop-level parallelization. While using technology too new to be as solid as the simplex codes, benchmarks on large crew scheduling problems have proven very successful using interior-point methods. Problems with a few hundred rows and several million columns can now be handled by modern LP codes such as CPLEX, OB1, and OSL. For further details, refer to Bixby et al. in [4].

15.2.5. Integerization Techniques. As mentioned above, one difficulty in crew scheduling is arriving at an optimal integral solution. While the LP relaxations may be a first step, they do not always deliver the desired integrality. In some cases, the problem structure results in "highly integral" answers that are then resolved with simple heuristics. In other cases, however, stronger integerization procedures must be employed.

As an example of an integerization procedure, consider the branch-and-cut method used by Hoffman and Padberg [12]. In this case, branch-and-bound is combined with the generation of cutting planes based on the polyhedral structure of the integral polytope. The process has, in fact, been specialized to solve general 0–1 problems and performs particularly well on integer linear programs with a large number of branching nodes. Hoffman and Padberg [12] have shown very promising results in their application of this method to airline crew optimization problems. Trotter [18] has also demonstrated success on the Hoffman and Padberg data set using a set of related heuristics.

The sequential implementation of the branch-and-cut approach has been adapted to allow parallel processing by Cannon and Hoffman [5]. In that work they describe an implementation on a collection of commercial workstations connected by a local area network.

Cannon and Hoffman note that the speedup of their approach over others is especially prevalent on large-scale 0–1 IP problems, particularly those with a large number of branching nodes. This describes the crew scheduling set partitioning problem. Studies have been performed on both distributed- and shared-memory environments.

15.3. The Challenge of Column Generation

As stated above, a common simplifying assumption made during crew optimization is that the problem may be partitioned into the two more or less distinct phases of column generation and optimization. Our recent research indicates, however, that such an assumption is not generally valid [20]. We outline first the magnitude of the problem of explicit column generation, followed by several approaches designed to avoid this difficulty.

The following is an example that emphasizes the complexity of the column generation problem in crew scheduling. For most airlines, the scheduling task is viewed from a 30-day perspective, and every takeoff and landing in that month-long period must be covered by some round-trip crew pattern. Fortunately, each type of airplane (747, DC9, DC10, for example) has its own uniquely qualified pool of crew members. Thus, the crew scheduling problem may be decomposed into a set of subproblems, one for each fleet type. While this greatly reduces the complexity, the problems associated with some fleets remain too large to handle efficiently. Further note that these subproblems are *not* necessarily fully independent, because in some situations the pilots deadhead on a fleet type other than their own. Crews deadhead, that is travel as passengers, to position themselves to fly the next leg of their pairing.

Computational experience has indicated that the number of columns that could be generated for a crew scheduling problem with r flight legs and rows is of the order

$$c = \mathcal{O}(C_0 e^r),$$

where c is the number of columns (pairings) that may be generated. To give an approximation of a "typical" problem size, consider that Northwest Airlines' DC9 fleet contains over 150 aircraft flying over 800 flight legs per day. Extending this to the perspective of a 30-day flight schedule, we now have a problem with 24,000 legs that could then induce on the order of $c = e^{24000}$ columns, a number far too large for explicit column generation.

In earlier versions of our crew scheduling software, we applied empirical rules to avoid considering such a problem. We did this by working first on so-called daily problems, knitting the solutions into weekly solutions, and then knitting the larger solutions into the monthly schedule. Even within the daily problem, which was done first, we did not attempt to solve the 800-row problem. Rather, a series of subproblems covering 50–80 flight legs were iteratively solved. The problems at all levels were cast as SPP. Even with this empirically reduced approach, it was necessary for us to make use of advanced architectures to approach the problem more efficiently.

15.3.1. Column Generation in Parallel. The route structure of Northwest Airlines is such that we have a small number of crew bases for each equipment type. If we view the production of the pairings for a given base as completely independent of all other bases, then it is a relatively straightforward task to allow the code to operate on each base simultaneously with regard to pairing generation. This parallel approach to the task of column generation is in some sense a brute-force approach, but the speedup achieved, weighed against the complexity of the pairing-generation code, warrants it use. The use of bases as parallel tasks is not only quite natural but also nicely amenable to the machine that we were using when it was implemented, a Convex C220, a shared-memory MIMD machine. The resulting "grain size" can be quite large.

The two major issues of concern with this approach were (1) synchronization of the updating of the data structure and (2) the handling of redundant pairings. Both of these were easily handled. Redundant pairings would only have been a concern if a large number of them were produced relative to the number of unique pairings. This did not happen. Before the pairings were passed to the SPP algorithm they were sorted and redundancies eliminated. The synchronization issue was handled by updating the global data structure after all the individual bases were completed.

At that time, the computational work involved in pairing generation alone was roughly comparable with that of solving the SPP. Thus we were taking approximately 50% of the runtime of our application and dividing it among a small number of tasks (i.e., bases).

15.3.2. The Pattern Sifting Approach. Even in parallel, the column generation technique described above for the original ALPPS is able to consider only small subproblems of the true crew scheduling problem. While improvement may be observed from iteration to iteration, this approach generally does not guarantee optimality. *Pattern sifting* is a more global approach to crew optimization, as a much larger set of crew patterns may be generated and evaluated using a *dual pricing* scheme. This approach is based on the SPRINT methodology introduced in the early 1970s [17].

Before sifting, many millions of crew patterns must be generated and stored. Next, a set partitioning problem such as those described above must be built. Dual pricing, based on the solution to set partitioning subproblems, is then used to "sift" through the millions of pregenerated crew patterns to determine which should be next considered for optimization. In this way, the many million-column SPP may be solved to optimality.

Promising results of the application of sifting to crew optimization problems is described by Anbil, Tanga, and Johnson [1], as well as by Bixby et al. [4]. Anbil, Tanga, and Johnson use the IBM Optimization Subroutine Library on the 3090E Vector Facility, while Bixby et al. use CPLEX and OB1, as well as a hybrid combining the simplex and interior-point approaches on a CRAY Y-MP supercomputer.

15.3.3. Delayed Column Generation. Even with the millions of columns that may be considered with a column sifting approach, we are still considering only a subproblem (a subset of columns) of the larger crew scheduling problems. We are not ensured that the *best* or even a *good* subset of columns is contained in this original set.

Unknown at the outset of pairing generation is the meaning of "good" in the context of the given schedule. A high cost pairing that completes an otherwise inexpensive crew schedule may be worth its price. *Reduced costs*, which can be generated from the LP solution to SPP, provide insight into the fit of newly generated patterns into the existing crew network. This is part of the rationale behind "delayed column generation," which has recently been incorporated into the ALPPS software at Northwest Airlines [20].

After solving SPP over an initial subset of legal crew pairings, we use reduced cost information to help us generate more legal crew pairings that, if added to the initial subset, could improve the overall cost. The SPP is then resolved, and the process is repeated until no more "improving" crew pairings can be generated. In this case the column generation phase is a variation of a shortest path problem, and it, too, is amenable to parallelization at several levels.

15.4. Conclusions

While still not solved to the complete satisfaction of either airlines or mathematicians, the crew scheduling problem has benefited from advanced computing and OR technology.

We have applied a shared-memory MIMD computer in which each CPU is a vector processor to large-scale optimization problems. Each phase of the application, column generation, and optimization has utilized a particular aspect of the architecture to advantage.

The scalar-parallel pairing-generation phase maps well to computing platforms such as the shared-memory MIMD Convex by treating crew bases as the parallel chores. This area is also one that is quite different among the different airlines. Owing to the rules and regulations of the airline and the setup of the schedule development organization, it is not likely that a single-column generation algorithm can be applied across all airlines. We suspect that at a high enough level, in whatever algorithm is used, the column generation phase will continue to utilize parallelism.

The solution of the large LP relaxations has benefited from the vector architecture of the Convex. This combination of supercomputing hardware and the newer algorithms for solving the SPP has permanently changed our view of the crew scheduling problem. The expected savings from the combination of hardware and software will certainly help Northwest Airlines.

Our near-term approach to the problem will continue to focus on the advances in the SPP algorithms as well as improved problem formulation and column generation techniques. All aspects of this problem are interesting blends of supercomputing, mathematical programming, and combinatorics.

A final point on our improved crew scheduling is that the crews themselves are happier with the newer trips. The crews want to fly when on duty and by producing more efficient, i.e., less costly, pairings they achieve that goal. Therefore, not only has supercomputing saved the corporation money, it has resulted in greater satisfaction on the part of the crews.

References

[1] R. ANBIL, R. TANGA, AND E.L. JOHNSON, *A global approach to crew-pairing optimization*, IBM Systems Journal, 31(1992), pp. 71–78.

[2] J.P. ARABEYRE ET AL., *The airline crew scheduling problem: A survey*, Transportation Sci., 3(1969), pp. 140–163.

[3] R.E. BIXBY, *Implementing the Simplex Method: The Initial Basis*, Technical Report TR90-32, Department of Mathematical Sciences, Rice University, Houston, TX, 1990.

[4] R.E. BIXBY, J.W. GREGORY, I.J. LUSTIG, R.E. MARSTEN, AND D.F. SHANNO, *Very Large-Scale Linear Programming: A Case Study in Combining Interior Point and Simplex Methods*, Technical Report J-91-07, School of Industrial and Systems Engineering, Georgia Institute of Technology, Atlanta, GA, 1991.

[5] T.L. CANNON AND K.L. HOFFMAN, *Large-scale 0-1 linear programming on distributed workstations*, Ann. Oper. Res., 22(1990), pp. 181–217.

[6] T.J. CHAN AND C.A. YANO, *A Multiplier Adjustment Method for the Set Partitioning Problem*, Technical Report, Department of Industrial and Operations Engineering, University of Michigan, Ann Arbor, MI, 1987.

[7] J. DESROSIERS, Y. DUMAS, M.M. SOLOMON, AND F. SOUMIS, *Time Constrained Routing and Scheduling*, Technical Report G-92-42, GERAD, Montreal, Canada, 1993.

[8] J. ETCHEBERRY, *The set covering problem: A new implicit enumeration algorithm*, Oper. Res., 25(1977), pp. 760–772.

[9] M.L. FISHER AND P. KEDIA, *Optimal solutions of set covering/partitioning problems using dual heuristics*, Management Science, 36(1990), pp. 674–688.

[10] R. GERBRACHT, *A new algorithm for very large crew pairing problems*, in Proc. 18th AGIFORS Symposium, Vancouver, BC, Canada, September 1978.

[11] G.W. GRAVES, R.D. MCBRIDE, I. GERSHKOFF, D. ANDERSON, AND D. MAHIDHARA, *Flight crew scheduling*, Management Science, 39(1993), pp. 736–745.

[12] K.L. HOFFMAN AND M. PADBERG, *Solving airline crew-scheduling problems by branch-and-cut*, Management Science, 39(1993), pp. 657–682.

[13] P.K. KEDIA, *Solver for Crew Scheduling Problems SPMINC*, Presentation to Northwest Airlines, 1990.

[14] R.E. MARSTEN, *An algorithm for large set partitioning problems*, Management Science, 20(1974), pp. 774–787.

[15] R.E. MARSTEN AND F. SHEPARDSON, *Exact solutions of crew scheduling problems using the set partitioning model: Recent successful applications*, Network, 11(1981), pp. 165–177.

[16] I.J. LUSTIG, R.E. MARSTEN, AND D.F. SHANNO, *On Implementing Mehrotra's Predictor-Corrector Interior Point Method for Linear Programming*, Technical Report SOR 90-3, Department of Civil Engineering and Operations Research, Princeton University, Princeton, NJ, 1990.

[17] SPERRY UNIVAC, *Series 1100 Functional Mathematical Programming System (FMPS) Programmer Reference*, UP-8198, 1974.

[18] L.E. TROTTER, *Private Communications*, September 1995.

[19] D. WEDELIN, *Probabilistic Networks and Optimization*, Report 49, Programming Methodology Group, University of Goteborg and Chalmers University of Technology, Goteborg, Sweden, May 1989.

[20] T.H. WISE, *Column Generation and Polyhedral Combinatorics for Airline Crew Scheduling*, PhD Dissertation, Cornell University, Ithaca, NY, January 1995.

Chapter 16

Parallel Molecular Dynamics on a Torus Network

Klaas Esselink
Peter A.J. Hilbers

Editorial preface

The parallelization of molecular dynamics enables scientists to study macroscopic properties from large collections of microscopic particles. Accomplishing this requires a very large number of molecules, but because the treatment is of discrete particles, there is inherent parallelism.

Particle interactions are shown to be problematic, and the authors develop a parallel algorithm for molecular dynamics that runs on a network of Transputers.

This article originally appeared in *SIAM News*, Vol. 26, No. 3, May 1993. It was updated during the summer/fall of 1995.

Computer simulations of physical systems are playing an important role in statistical mechanics. Macroscopic phenomena of physical systems are studied on the scale of microscopic particles, such as atoms, molecules, electrons, and nuclei. The capacity of available computers, however, places restrictions on the total number of particles that can be studied and on the total simulation time. In the past, considerable effort was devoted to optimizing algorithms for molecular dynamics simulations for both sequential and vector architectures. Today, because the computations involved are suited to parallelization, more and more articles are describing parallel implementations. The details of a parallel algorithm depend in general on the topology of the processor network. Therefore, it is important to be able to estimate a priori which mappings of an algorithm onto a processor network will yield optimal performance with respect to both the distribution of work over the processors and communication costs.

In this article, we explain some aspects of molecular dynamics simulations and show why the use of "geometric parallelism" leads to very efficient implementations. This approach yields optimal mappings with respect to both load balancing and communication costs. We also describe some aspects of the implementation of simulations involving multiparticle potentials. Timings of simulations performed on toroidal networks of Transputers demonstrate that molecular dynamics simulations can greatly benefit from parallel computing both in time and in cost. In effect, parallel machines make it possible to study very large systems and new problems that previously could not be dealt with in reasonable computing times.

16.1. Molecular Dynamics

Molecular dynamics is a simulation technique that can be used to study the dynamic behavior of many-particle systems. By integrating Newton's equation of motion at each timestep, trajectories of mutually interacting particles are calculated. Given a system of particles we solve

$$(16.1) \qquad\qquad m_i \frac{d^2 r_i}{dt^2} = f_i,$$

where m_i and r_i are the mass and the position of particle i, respectively, and where $f_i = -\nabla_i \mathcal{V}$ is the force on particle i due to the total potential $\mathcal{V}$. Usually, this potential is divided into terms depending on the coordinates of pairs, triplets, and quadruplets of particles. In this section we consider some potentials that are widely used in molecular dynamics codes. For readers who are interested in more details, we recommend [1].

The first and most important potential is the Lennard–Jones 12–6 potential for the interaction between pairs of nonbonded particles. It has the following form:

$$(16.2) \qquad LJ_{ij}(r_{ij}) = 4\epsilon \left[\left(\frac{\sigma}{r_{ij}} \right)^{12} - \left(\frac{\sigma}{r_{ij}} \right)^6 \right],$$

where r_{ij} is the length of $\vec{r}_{ij}$, the vector from the position of particle i to particle j, ϵ is an energy parameter, and σ is a length parameter. Because the Lennard–Jones potential rapidly decays to zero, it is usually truncated at a fixed cut-off radius R_c and then shifted to make the potential continuous. This means that a particle has nonbonding interactions only with the particles contained within the sphere of radius R_c of which it is the center.

Other forces within molecules also play a role. We take these intramolecular potentials from [10]. A chemical bond between two particles can be modeled by a harmonic spring potential:

$$(16.3) \qquad\qquad BO(r_{ij}) = \frac{1}{2} c_0 (r_{ij} - l_{BO})^2,$$

where l_{BO} is the equilibrium bond length and c_0 is the spring constant.

Bending forces in a chain of bonded particles are modeled by the three-particle potential, which maintains the valence angle between a successive pair of bonds close to the tetrahedral value θ:

$$(16.4) \qquad BE(\theta_{ijk}) = c_1(\cos\theta_{ijk} - \cos\theta)^2.$$

θ_{ijk} is the angle between $\vec{r}_{ji}$ and $\vec{r}_{jk}$ in the chain (i, j, k), and c_1 is an energy parameter.

A four-particle torsion potential can be associated with an angle τ_{ijkl} between the planes ijk and jkl:

$$(16.5) \qquad TO(\tau_{ijkl}) = c_2 \sum_{n=0}^{5} a_n \cos^n \tau_{ijkl},$$

where a_n, with $0 \leq n \leq 5$, and c_2 are energy parameters; see also [10]. The torsion potential models the molecular property that four particles preferably all lie in the same plane ($\tau_{ijkl} = \pm 0°$, the so-called trans conformation), or that $\tau_{ijkl} = \pm 60°$ (gauche conformation).

Lennard–Jones (truncated), bonding, bending, and torsion are computationally short-range potentials, since the distance between any two particles involved is at most a certain R_c. If constant density and a homogeneous distribution of particles are assumed, the amount of work done for each particle does not depend on the total number of particles in the universe. Implementations for which the time complexity scales linearly with the number of particles are therefore feasible.

For instance, for a typical system of 1000 decane molecules modeled by a linear chain of 10 particles, the number of interactions to be calculated each timestep and the execution time per potential evaluation (on a T800 Transputer) are shown in Table 16.1. The bulk of the computation consists of the evaluation of Lennard–Jones potentials. Torsion, although an expensive potential, has a contribution that involves four particles. The cost *per particle* is therefore roughly equal for all potentials.

Because of the large number of Lennard–Jones evaluations, we concentrate first on the efficient implementation of the nonbonding interactions in parallel.

16.2. Parallel Molecular Dynamics

Two phases can be distinguished in molecular dynamics simulations. In the first phase, the forces on each particle are determined, and in the second the displacements of the particles are determined from all the forces taken together, possibly along with some macroscopic properties of the system. The latter phase is trivial to parallelize. Parallelization of the former is usually more difficult, as the processors need to cooperate (exchange information) in order to compute the potentials.

TABLE 16.1

The number of evaluations to be performed at each timestep, and the execution times per potential evaluation on a single T800 *Transputer, for a system of* 1000 *chains of length* 10, *with a typical density of* $2.3\sigma^{-3}$, *and a cut-off radius of* 2.5σ (σ *is a length parameter*).

Potential	Evaluations	Time (μs)
Lennard–Jones	752292	110
Bonding	9000	125
Bending	8000	306
Torsion	7000	507

There are three main techniques for exploiting parallelism: processor farm, particle parallelism, and geometric parallelism. We briefly discuss each one, with reference to its appropriateness for the evaluation of the Lennard–Jones interactions and its scalability properties.

16.2.1. Processor Farm. In the processor farm approach to parallel processing, a group of independent work processors receive tasks from a single control processor. Because of the bottleneck at the control processor, the scalability is poor. Therefore, we do not consider this technique useful.

16.2.2. Particle Parallelism. The second technique for exploiting parallelism assigns particles to processors [6, 7]. Each processor continually calculates the forces and the new positions for the particles it "owns." The initial distribution of particles remains unchanged during the simulation and can be chosen such that the workload is evenly distributed. For bonding, bending, and torsion potentials, this technique is particularly convenient if particles from a bonding, bending, or torsion tuple are assigned to the same processor. However, in some cases (those involving large chains and small processor networks, for example), it is not possible to guarantee that all the particles involved will belong to the same processor.

Implementation of the Lennard–Jones potential poses a problem. It is quite possible for two particles, initially far apart, to approach each other closely during the course of the simulation. Despite the short-range nature of the potential, it is therefore necessary for each processor to communicate with all the others to determine whether any two particles have come close enough to each other (i.e., within R_c).

It is difficult to analyze the communication behavior of this technique a priori. In general, it will be difficult to achieve good scale-up properties if the processor network size increases. This disadvantage is quite serious, as the bulk of the computation consists of the evaluation of Lennard–Jones (or comparable) potentials.

16.2.3. Geometric Parallelism.   When geometric parallelism is applied, space, rather than particles, is assigned to processors [8, 9].   During a computation, a processor calculates the trajectories of all particles it finds in its space.   Because of the movement of the particles, some particles may enter or leave a processor's space during the computation.   For this reason, processors continually need to redistribute the particles to ensure that each one has the right subset.   The short-range nature of the Lennard–Jones potential can be used to determine which part of the universe should be assigned to a processor—because the interaction does not extend over distances larger than R_c, it is not necessary to exchange information over longer distances.   As in the "linked-list" method [1], we assume that the simulation box is divided into a number of cells, such that particles interact only with particles in the same cell or in neighboring cells.   Hence, we can associate with each cell a search space of cells in which particles to be investigated for interaction reside.   In [4] we considered several choices for cell shapes, such as octahedron, rhombic dodecahedron, and cube, to determine which shape results in the smallest search space.   The cube turns out to be the best, because it has both a modest volume and a modest number of neighbors (see Figure 16.1).

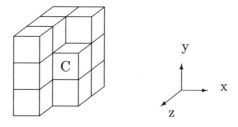

FIG. 16.1.   *Cell "C," with 13 neighbors in three dimensions. By using Newton's third law ($f_{ij} = -f_{ji}$) it suffices to search only half the space, and 13 neighbors therefore suffice.*

Having divided the simulation box into a regular cubic lattice of cells, and assuming a homogeneous distribution of particles, we can achieve good load balancing by assigning the same number of cells to each processor. To minimize communication costs, which, in general, depend on both the number of cells for which results must be communicated between processors and the distance between the communicating processors, cells must be assigned judiciously to processors. Suppose we have a square torus of processors. The most "natural" mapping (and also the most widely used) is the orthogonal projection of the universe onto the torus of processors. This mapping has the advantage that the wrap-around in the z-direction can be achieved without the need for extra communication. We can show, however, that this z-mapping is not always the best.

Suppose that the processor torus is of size Q^2. We divide this torus into subsquares of size q^2 (q divides Q) and define a Hamiltonian path in each subsquare, such that the endpoints are at a distance of at most 2. If the size of the universe is n^3 cells (Q divides nq; notice that there may be more than one cell per processor in either dimension), we map columns of size $n(nq/Q)^2$ to each subsquare. In this way each processor gets a subcluster of $(nq/Q) \times (nq/Q) \times n/q^2$ cells. For the special case of $q = 1$, we obtain the z-mapping; for larger q an implementation that also clusters in the z-direction is obtained. In this way it is possible to derive a formula for the communication cost as a function of q. In [4] we show that column mapping is very efficient for processor networks of up to size 32×32, but that for larger networks it is better to map more spherical subclusters of cells.

When the size of the cubic cell is R_c, each processor needs particle information from only four other processors. Figure 16.2 shows a situation in which the cells are not cubic; the edge lengths are $R_c/2$ in the x-dimension and $R_c/3$ in the y-dimension. In that case, processor P needs to communicate with 17 other processors. In the most general case, communication cost increases with the distance between the two communicating processors. Here, however, we can avoid high cost by using the following scheme. First processor P receives particle information from its first neighbor to the south (simultaneously sending its own particles to its first neighbor to the north, north being the positive y-direction). P should then get information from its second neighbor to the south, but this information has just been sent to its first neighbor to the south. Thus, although the distance between P and its second neighbor to the south is 2, the necessary information can be obtained at a communication cost of 1. This principle can be extended, with the result that all data are effectively obtained at a communication cost of 1. Consequently, the time needed for communicating particle information is roughly independent of the size of the processor network.

16.2.4. Multiparticle Potentials with Geometric Parallelism.

So far we have considered only the Lennard–Jones potential in combination with geometric parallelism. For a multiparticle potential, such as the torsion potential, the implementation seems to be rather complex. Four particles are involved, and they may reside on different processors. Moreover, during the course of the simulation the particles may change processors. Thus, we cannot predict which processor should evaluate the potential if no extra communication is allowed (see Figure 16.2). We can, however, prove the following theorem.

DEFINITION 16.2.1. *Consider a particle i, and let its Lennard–Jones list be the collection of particles j for which a Lennard–Jones potential evaluation has to be performed (the distance between i and j is thus at most R_c). Moreover,*

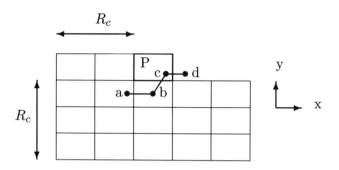

FIG. 16.2. *Processor P and its environment of* 17 *other processors. Which processor evaluates torsion potential (a,b,c,d)? It cannot be P, as it lacks information about particle d.*

for any pair of particles, say p and q, p is not in the Lennard–Jones list of q or q is not in the Lennard–Jones list of p.

THEOREM 16.2.1. *Let the quadruplet* (i, j, k, l) *of particles determine a torsion. Assume that the distance between any two particles from* $\{i, j, k, l\}$ *is at most* R_c *and that the search space of a cell is as shown in Figure* 16.1. *Then there exists exactly one particle* $p \in \{i, j, k, l\}$ *such that the other three are in its Lennard–Jones list.*

In [3] a constructive and more general proof is given; here we give only a sketch of a proof. Let (i, j, k, l) determine a torsion such that the distance between any two of the four particles is at most R_c. Then we can determine a cubic block of eight cells, each of size R_c, such that all four particles are in this block. Assume that one of the four particles is in the cell marked with a dot (Figure 16.3A; compare the dotted cell with cell C in Figure 16.1). Then the other three particles are in the search space of this cell and, hence, in the Lennard–Jones list of the particle residing in this cell. This particle will therefore have information for the other three particles (in fact, it is the only one that does so; see the definition of the Lennard–Jones list), and it should evaluate the torsion potential. To complete the proof, we next assume that none of the four particles is in the marked cell of Figure 16.3A and, hence, that all four particles lie in the remaining seven cells (Figure 16.3B). Assume that one of the four particles is in the cell marked with a dot (in Figure 16.3B). Then the other three particles are in the search space of this cell and, hence, in the Lennard–Jones list of the particles residing in this cell. Next we assume that none of the four particles is in the marked cell of Figure 16.3B and, hence, that all four particles lie in the remaining six cells, and so forth. By continually choosing an appropriate cell and a particle in that cell, proving that the other three particles are in the search space of that cell, and removing that cell, all possibilities can be checked. This proves the theorem.

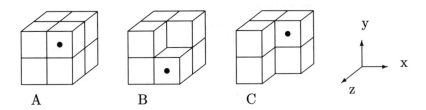

FIG. 16.3.   *Three blocks of cells.*

Thus, by using the information of the two-particle Lennard–Jones potential, we can determine which particle, and hence which processor, should evaluate the torsion potential. It is possible to prove a similar theorem for other
multiparticle potentials. For the evaluation of multiparticle potentials, no extra communication is needed, beyond that for the Lennard–Jones potential,
when geometric parallelism is used.

16.3.   Timing Results

Our simulations are run on two toroidal networks of T800 Transputers, one
with 36 and the other with 400 processors. The implementation (written in
Pascal) does not assume a relation between the size of a column of the universe
and the cut-off radius R_c. Furthermore, a processor can have any number of
columns in the x- and y-directions. The Lennard–Jones lists are used to keep
track of nearby particles for the Lennard–Jones potential. If the search space
of a cell is enlarged slightly, these lists can be used for several (typically 10)
iterations before updating.

Without using bonding, bending, or torsion potentials, we found the
following formula for the time τ per timestep [4]:

$$(16.6) \qquad\qquad \tau \approx 0.064 + 0.011\rho\frac{N}{P},$$

where ρ denotes the density, N the number of particles, and P the number of
processors. There is some performance degradation when N/P becomes very
small.

For an average system, the number of bonding, bending, and torsion
potentials is several orders of magnitude smaller than the number of Lennard–
Jones evaluations (see Table 16.1). We found that the Lennard–Jones potential
is responsible for 92–94% of the time needed to calculate all forces. This
includes the search for all three- and four-particle tuples constituting the
potentials.

16.4. Applications

The implementation described here has been used to study the behavior of rather large systems. The first application consisted of surfactants in water and oil solutions [11]. The water and oil were modeled by using particles interacting by a Lennard–Jones potential. The water–water and oil–oil interactions were truncated at 2.5σ, while the water–oil interaction was truncated at 1.12σ, so as to make this interaction completely repulsive. (The potential minimum of the Lennard Jones function can be found by solving "$\mathbf{d}LJ(r)/\mathbf{d}r = 0$." The result is $r = 2^{1/6}\sigma$, which is $\approx 1.12\sigma$. After the shift only a repulsive part remains.) Surfactants were introduced by using harmonic spring potentials to chain together water-like and oil-like particles. In this way, a surfactant could have a hydrophilic head and a hydrophobic tail. No specific three- or four-particle potential was used. A simulation was started with 39,304 particles, a density of $0.7\sigma^{-3}$, equal amounts of bulk water and oil, and surfactants randomly placed in the simulation universe at concentrations of 1.5 and 3%. After 250,000 timesteps ($\Delta t = 0.005$; we use reduced units [1]), the system had many interesting features.

The surfactants were aggregated into spherical micelles, a monolayer was formed, and there was a depletion layer void of surfactants close to the interface. The micelles were evident in the water phase; the oil phase was much less ordered. One of the novel aspects of these simulations is that the interface and the micelles are described with one model, whereas previous theoretical work was concerned mainly with either isolated micelles or a monolayer. In earlier simulations of much smaller systems (up to 1000 particles), the formation of micelles was not observed. A possible explanation is that those systems were too small for this phenomenon to be observed; finite size effects prevent cluster formation because of periodic boundary conditions. When a parallel machine is used, much larger systems can be studied in a cost-effective way, and simulations like the one reported here become possible. In this case, one iteration of the 39,304 particle system takes 0.82 seconds on the 400 processor network, while a vectorized version on a much more expensive single processor Cray X-MP takes 2.2 seconds.

The second application involves the simulation of universes consisting of chain molecules, like decane [5]. In this work, we simulated a universe of 1000 chains of length 10. Within a chain, bonding, bending, and torsion forces were used, as in [10], and a Lennard–Jones potential, cut off at 2.5σ, was applied between nonbonded atoms and atoms separated by more than three bonds. The density was approximately $2.3\sigma^{-3}$.

Starting from a high temperature, the system was cooled down to 210°K, using the loose coupling techniques of [2]. The temperature coupling $\Delta t/\tau_T$ was set to 0.0001, and the pressure was kept at 1 atmosphere by using a pressure coupling constant $\Delta t/\tau_P$ of 0.00001. The timestep was set to 10 femtoseconds. The system started to show crystallization behavior as soon as 210°K had been reached. The ordering process took approximately one million iterations (10 nanoseconds of simulated time), resulting mostly in two large layers of decane, wrapped around the periodic boundaries. Using a new algorithm, we are able to identify all nuclei in the system and thus to study the nucleation process in great detail. Furthermore, the size of the simulated systems is kept large to minimize size effects. One timestep in this simulation takes 0.85 seconds on the 400 processor network, while a comparable run on a Silicon Graphics Iris 4DW35 takes 15 seconds.

16.5. Conclusion

Parallel computers can be successfully applied to solve problems that are too large to be tackled by conventional workstations. Some phenomena depend critically on the size of the simulated system. We have presented our approach to an efficient parallel implementation that is being used by physicists and chemists at our laboratory.

References

[1] M. ALLEN AND D. TILDESLEY, *Computer Simulation of Liquids*, Oxford Science Publications, 1987.

[2] H. BERENDSEN, J. POSTMA, W. VAN GUNSTEREN, A. DiNOLA, AND J. HAAK, *Molecular dynamics with coupling to an external bath*, J. Chem. Phys., 81(1984), pp. 3684–3690.

[3] K. ESSELINK AND P. HILBERS, *Efficient parallel implementation of molecular dynamics on a toroidal network, Part* II: *Multi-particle potentials*, J. Comput. Phys., 106(1993), pp. 108–114.

[4] K. ESSELINK, B. SMIT, AND P. HILBERS, *Efficient parallel implementation of molecular dynamics on a toroidal network, Part* I: *Parallelizing strategy*, J. Comput. Phys., 106(1993), pp. 101–107.

[5] K. ESSELINK, P.A.J. HILBERS, AND B.W.H. VAN BEEST, *Molecular dynamics study of nucleation and melting of n-alkanes*, J. Chem. Phys., 101(1994), pp. 9033–9041.

[6] D. FINCHAM, *Parallel computers and molecular simulation*, Molecular Sim., 1(1987), pp. 1–45.

[7] J. LI, D. BRASS, D. WARD, AND B. ROBSON, *A study of parallel molecular dynamics algorithms for N-body simulations on a transputer system*, Parallel Comput., 14(1990), pp. 211–222.

[8] H.G. PETERSEN AND J.W. PERRAM, *Molecular dynamics on transputer arrays. I. Algorithm design, programming issues, timing experiments and scaling projections*, Molecular Phys., 67(1989), pp. 849–860.

[9] M. PINCHES, D. TILDESLEY, AND W. SMITH, *Large scale molecular dynamics on parallel computers using the link-cell algorithm*, Molecular Sim., 6(1991), pp. 51–87.

[10] D. RIGBY AND R.-J. ROE, *Molecular dynamics simulation of polymer liquid and glass. I. Glass transition*, J. Chem. Phys., 87(1987), pp. 7285–7292.

[11] B. SMIT, P. HILBERS, K. ESSELINK, L. RUPERT, N. VAN OS, AND A. SCHLIJPER, *Computer simulations of a water/oil interface in the presence of micelles*, Nature, 348(1990), pp. 624–625.

Chapter 17

A Review of Numerous Parallel Multigrid Methods

Craig C. Douglas

Editorial preface

The study of multigrid methods is a contemporary topic that is undoubtedly an important algorithm for solving many modeling problems. Thus the parallelization of multigrid methods has broad applicability. The multigrid method in the context of parallel implementations is reviewed and implementation details relative to current programming methods are considered. This chapter is "must" reading for anyone interested in multigrid methods.

This article originally appeared in *SIAM News*, Vol. 25, No. 3, May 1992. It was updated during the summer/fall of 1995.

Parallel multilevel methods are shown to be the natural precursors to standard multilevel methods based on the personnel computing era of earlier this century. They are also the natural successors to standard multilevel methods in the age of computers. What makes six parallel multilevel methods practical and impractical is discussed in the context of the three algorithms that encapsulate them.

17.1. Preliminaries

Multigrid methods originated earlier this century, in the *personnel* computing era. Someone who needed to compute an approximation to the solution of a partial differential equation during that era would fill a room with people. After using very simple mechanical calculators to compute parts of the approximation, these people would pass their parts to the other people in the room who needed them. Except for the very different timescales and approximate solution accuracy, this process is similar to computing on today's distributed-memory parallel computers.

The most basic model problem for elliptic boundary value problems in multigrid has always been, effectively, the two-dimensional Poisson equation

on a square, namely,

(17.1) $-\Delta u = f$ in $\Omega = (0,1)^2$; $u = 0$ on $\partial\Omega$.

An $N \times N$ uniform mesh with mesh spacing $h = 1/(N-1)$ is placed over Ω as shown in Figure 17.1.

The grid points are x_{ij}, where $1 \le i, j \le N$. A central difference discretization is applied to (17.1) to get a set of linear equations

(17.2) $AU = F,$

where $F_{(i-1)N+j} = h^2 f(x_{ij})$ and $A = [-I, T, -I]$ is the block tridiagonal matrix defined using the $N \times N$ matrices I, the identity matrix, and $T = [-1, 4, -1]$, a tridiagonal matrix.

In the early part of the twentieth century, partial differential equations were approximately solved by relaxation techniques. (In fact, many types of equations are still solved in this way today.) Solving (17.2) by a Gauss–Seidel method could take the rest of the personnel computers' lives if N was large enough and the required accuracy was strict enough. Hence, new tricks were needed to reduce the computation time.

FIG. 17.1.   *Uniformly meshed domain.*

During the 1920s, and probably several decades earlier, two-level schemes were used by engineers. An auxiliary grid Ω_1 was used to generate an initial guess to the solution to (17.2) on Ω_2.

Let $GS(j, n) = n$ Gauss–Seidel sweeps on level j. Algorithm 1, a *one-way two-level* algorithm is defined as follows.

ALGORITHM 1. *One-way Two-Level*
(a)   *Solve the problem on Ω_1 by any means.*
(b)   *Interpolate U_1 onto Ω_2.*
(c)   *Do $GS(2, \cdot)$ until convergence.*

Bilinear interpolation was typically used in step b. Let N_i be the number of grid points on grid Ω_i and define

(17.3) $\sigma_i \approx \dfrac{N_i}{N_{i-1}}, \quad i > 1.$

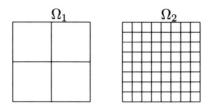

FIG. 17.2. *Meshes for the two-level schemes.*

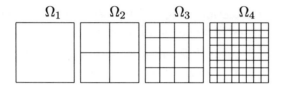

FIG. 17.3. *Multiple grid meshes.*

Before digital computers were commonplace, σ_2 was typically between 5 and 10 (see [30]).

During the 1960s, Soviets developed what is now considered true multigrid (see [3, 4, 19, 20]). In this case, the auxiliary grid Ω_1 was used to correct approximations to the solution to (17.2) on Ω_2; see Figure 17.2. Algorithm 2, a *two-level correction* algorithm is defined as follows.

ALGORITHM 2. *Two-Level Correction*
(a) $U_2 \leftarrow GS(2,n)$.
(b) $F_1 \leftarrow \sigma^2(F_2 - A_2U_2)$ *where* $\Omega_1 \cap \Omega_2$.
(c) $C_1 \leftarrow GS(1,m)$.
(d) $U_2 \leftarrow U_2 + Interpolate(C_1)$.
(e)   *Repeat (a)–(d) until convergence.*

Projection of the style of step b, known as *injection*, is in disfavor today; once again, bilinear interpolation has become typical. In this case, σ_2 is typically 4. (Note that this changes Ω_1 in the diagram.)

Algorithm 2 is a notoriously slow algorithm for solving (17.2). The best trick for speeding it up is to use more than two grids. Define Ω_j, $j = 1, \ldots, k$ according to Figure 17.3 and place uniform meshes Ω_j over Ω, with $h_j = \sigma^{k-j}h_k$, for some $\sigma \in \mathbb{R}$. Then solve

$$A_kU_k = F_k$$

using the U_j, $j < k$ in some prescribed manner.

Other tricks include

- using different orders of interpolation (e.g., bicubics instead of bilinears) or projection to transfer information between meshes,

- using better discretization methods (e.g., finite-elements or volumes) for the partial differential equation,

- solving the level-1 problem by a direct method,

- using a more sophisticated solver, such as conjugate residuals or BiCGSTAB,

- starting the computation on level 1 instead of level k. (This is commonly referred to as a *nested iteration* in multigrid lingo.)

Thus, for a general linear differential equation on some domain Ω that has been discretized into a sequence of problems $j = 1, \ldots, k$, we have

$$(17.4) \qquad A_j U_j = F_j, \quad U_j, F_j \in \mathcal{M}_j, \; A_j \in \mathcal{L}(\mathcal{M}_k).$$

The solution spaces $\mathcal{M}_j$ are typically function or vector spaces based on the discretization method (typically, finite differences, finite elements, or finite volumes), and $\mathcal{L}$ stipulates that A_j is a linear operator on $\mathcal{M}_j$. We assume that there exist mappings between the neighboring spaces:

$$R_j : \mathcal{M}_j \to \mathcal{M}_{j-1} \quad \text{and} \quad P_{j-1} : \mathcal{M}_{j-1} \to \mathcal{M}_j.$$

While not necessary (see [15]), we also assume that there are mappings $Q_j : \mathcal{M}_j \to \mathcal{M}_{j-1}$ such that

$$A_{j-1} = Q_j A_j P_{j-1}.$$

This leads to Algorithm 3, a truly abstract correction algorithm of the following form.

ALGORITHM 3. $MG(j, \{\mu_\ell\}_{\ell=1}^{j}, x_j, F_j)$
    (1) *If $j = 1$, then solve $A_1 x_1 = F_1$*
    (2) *If $j > 1$, then repeat $i = 1, \ldots, \mu_j$:*
        (2a) *Solving:* $x_j \leftarrow PreSolve_j^{(i)}(x_j, F_j)$
        (2b) *Residual Correction:*
            $x_k \leftarrow x_j + P_{j-1} \, MG(j - 1, \{\mu_\ell\}_{\ell=1}^{j-1}, 0, R_j(F_j - A_j x_j))$
        (2c) *Solving:* $x_j \leftarrow PostSolve_j^{(i)}(x_j, F_j)$
    (3) *Return x_j*

This definition assumes that $\mu_1 = 1$. Common values for μ_j, $2 < j \leq k$, are 1 (*V cycle*) and 2 (*W cycle*). It is quite common for either the PreSolver or the PostSolver (but not both) to be the identity operator. Common solvers are relaxation methods and conjugate gradient-like methods. On the coarsest grid, a direct solver may be used.

The purpose of this article is not to analyze Algorithm 3 (see [14] and [15] for extensive analysis) but to discuss various methods for parallelizing it. Further, no mention will be made of nonlinear and/or nested iteration versions of Algorithm 3—the commentary applies to these cases verbatim.

There are two major themes in parallel multigrid today; telescoping and nontelescoping methods. Telescoping methods ($\sigma > 1$) include domain decomposition methods and a method that computes on all levels simultaneously. Nontelescoping methods use multiple coarse subspaces in which the sum of the unknowns on any level equals the sum on any other level.

17.2. Telescoping Parallelizations

Domain decomposition techniques [5] offer a relatively easy path to parallelizing Algorithm 3. Each processor computes on the block of data per level that is assigned to it.

Because almost no work is required to get good convergence bounds, this is certainly the most attractive parallelizing technique from a theoretical standpoint. The method is merely a block iterative method used for the approximate solver combined with standard multigrid. Hence, many standard multigrid convergence theorems apply verbatim.

Some communications and processor scheduling issues arise with this method. The amount of data communication involved and the location of the processors in a network will determine whether or not the problem is converted into an input–output problem rather than a computational one.

The factor σ_j in (17.3) determines the telescoping of unknowns and thus plays a key part in processor scheduling. When the number of unknowns assigned to each processor falls below some threshold (which is a function of the problem, processor speed, and communications bandwidth and latency), some of the processors may be idle some of the time. To avoid this, an agglomeration of unknowns is typically performed, and a certain number of processors become completely idle or compute the same thing.

Having idle processors may seem like a waste of computer time on a parallel processor, but it actually is not necessarily the case. If the problem can be solved faster (as timed by a stopwatch) with some processors out of the computation some of time, then it is clearly a good approach, even if it is a bit wasteful. The principal reason for actually solving problems on parallel computers is that the results are perceived to be needed much sooner than just running many problems, one per processor.

As an aside, more and more parallel computing seems to be done on clusters of workstations (i.e., *distributed computing*) rather than on explicitly parallel machines. The workstation approach means that when a processor is out of the computation, its task scheduler can assign it to work on something else; thus, it may not actually be idle. This also occurs naturally on a parallel machine that has a multiuser or multitasking operating system on each node of the machine.

During the approximate solve steps, information must be passed between processors along neighboring data regions. If a solver like conjugate gradients is used, then information from dot products and matrix–vector multiplies will also clutter the communications mechanism.

An asynchronous relaxation method would seem to be ideal in this environment. With this approach, each processor does its local relaxation method and uses the information obtained most recently from other processors. This is fast, and if the standard relaxation procedure converges, so does the asynchronous version, just not quite as quickly [8]. Unfortunately, asynchronous relaxation methods have never caught on in the parallel multigrid community, mostly because the small number of smoothing iterations (2–6 iterations) makes it impossible to ignore changes in neighboring data.

Another approach to telescoping multigrid involves massively parallel computers (at least as many processors as unknowns on all of the meshes). Gannon and Van Rosendale [22] proposed what is referred to as a *concurrent multigrid* method. Typically, this means there should be $N \log_d N$ processors for a d-dimensional problem.

The concept is that all operations should be performed simultaneously on all unknowns on all levels. An initial approximation of zero is assumed for the solution. Two sets of vectors, q_j and d_j, are used to hold information about right-hand sides and data on the spectrum of levels in the sense that

$$F_k \approx \sum_{j=1}^{k} q_j + d_j.$$

A third set of vectors, x_j, contains the approximations to the solutions to each problem on each level. This information percolates to the finest level, k, to finally provide the approximate solution to the real problem. Algorithm 4, $CMG(k, \mu, F_k)$, is as follows.

ALGORITHM 4. $CMG(k, \mu, F_k)$
    (1) *Initialize in parallel:*
        $x_j = q_j = 0, \ 1 \leq j < k; \ q_k = F_k, \ x_k = 0.$
    (2) *Repeat* $i = 1, \ldots, \mu$:
        (2a) *Smoothing in parallel*
            $x_j \leftarrow PreSolve_j^{(i)}(x_j, q_j), \ 1 \leq j \leq k.$
        (2b) *Compute data corresponding to* x_j *in parallel:*
            $d_j \leftarrow A_j x_j, \ 1 \leq j \leq k.$
        (2c) *Compute residuals in parallel:*
            $q_j \leftarrow q_j - d_j, \ 1 \leq j \leq k.$
        (2d) *Project q onto coarser levels in parallel:*
            $q_1 \leftarrow q_1 + R_2 q_2;$
            $q_k \leftarrow (I - P_{k-1}R_k)q_k;$
            $q_j \leftarrow (I - P_{j-1}R_j)q_j + R_j q_{j+1}, \ 1 < j < k.$

(2e) *Inject x into finer levels in parallel:*
$$x_1 \leftarrow 0; \ x_k \leftarrow x_k + P_{k-1}x_{k-1};$$
$$x_j \leftarrow P_{j-1}x_{j-1}, \ 1 < j < k.$$
(2f) *Inject d into finer levels in parallel:*
$$d_1 \leftarrow 0; \ d_k \leftarrow d_k + P_{k-1}d_{k-1};$$
$$d_j \leftarrow P_{j-1}d_{j-1}, \ 1 < j < k.$$
(2g) *Put all the data back into q in parallel*
$$q_j \leftarrow q_j + d_j, \ 1 \le j \le k.$$
(3) *Return* x_k

This definition assumes that μ evenly divides twice the number of levels.

Algorithm 4 has a number of noteworthy aspects. First, communication between adjacent levels is twice the amount that would be expected. This is absolutely required in order for the algorithm to be consistent (without which it diverges).

Second, this algorithm limits what types of iterative methods qualify as approximate solvers. The cost of one iteration on any level must be the same as the cost on any other level, assuming one processor per unknown. If the matrices A_j are similar enough to each other, Jacobi and conjugate gradients are fine, but Gauss–Seidel and symmetric successive overrelaxation (SSOR) are not. The latter two methods assume that the data is traversed in a particular order rather than all at once. Hence, the length of time to complete each iteration is dependent on the number of unknowns, violating the requirement of identical time per level.

Third, the number of levels that information must traverse to move from the finest level to the coarsest one and back is twice the usual number. Hence, all work estimates will be $O(2 \log N)$ (with the 2 being part of the constant) asymptotically in the number of levels, the usual multigrid method of estimating the problem complexity. This may seem high, but it is actually equivalent to the complexity of a standard V cycle.

Finally, there is a similar, but quite different, approach to parallel multigrid: apply a domain decomposition method to the finest grid and then use a serial computer multigrid method on each of the subdomains. Studies have shown (e.g., [23]) that this is not as fast an approach as using domain decomposition methods on each level of the multigrid algorithm.

17.3. Nontelescoping Parallelizations

Several competing methods have the same number of unknowns on each level. Each method has advantages and disadvantages.

The concept of using multiple subspaces to solve a problem whose solution lies in a particular space is hardly new. In fact, no one from the era in which it was invented is alive today. We will never know who really invented it, but we can be certain that it was introduced no later than in 1869 [29]. Further, it has been an active area of research in the field of symmetry groups (see [18]) for many years.

Assume a rooted tree of problems (see (17.4)) that are arbitrarily numbered. For a given problem k, it either has a set C_k of coarse space correction problems or it has none at all (i.e., $C_k = \emptyset$). When $C_k \neq \emptyset$, there are restriction and prolongation operators for each coarse space problem $\ell \in C_k$ such that

$$R_\ell : \mathcal{M}_k \mapsto \mathcal{M}_\ell \quad \text{and} \quad P_\ell : \mathcal{M}_\ell \mapsto \mathcal{M}_k.$$

We also assume that there are mappings

$$Q_\ell : \mathcal{M}_k \to \mathcal{M}_\ell \quad \text{such that} \quad A_\ell = Q_\ell A_k P_\ell.$$

These mappings are defined much as in the serial case.

A multiple coarse space correction multigrid scheme is defined by algorithm 5, $MCSMG(j, \mu_j, C_j, x_j, F_j)$, as follows.

> ALGORITHM 5. $MCSMG(j, \mu_j, C_j, x_j, F_j)$
> (1) If $C_j = \emptyset$, then solve $A_j x_j = F_j$
> (2) If $C_j \neq \emptyset$, then repeat $i = 1, \ldots, \mu_j$:
>   (2a) Smoothing: $x_j \leftarrow PreSolve_j^{(i)}(x_j, F_j)$
>   (2b) Residual Correction:
> $$x_j \leftarrow x_j + \sum_{\ell \in C_j} P_\ell \ MCSMG(\ell, \mu_\ell, C_\ell, 0, R_\ell(F_j - A_j x_j))$$
>   (2c) Smoothing: $x_j \leftarrow PostSolve_j^{(i)}(x_j, F_j)$
> (3) Return x_j

The performance of any variant of Algorithm 5 is dependent mainly on how small δ_j is, where

$$(17.5) \qquad \left\| \left(I - \sum_{\ell \in C_j} Q_\ell^{-1} R_\ell \right) u \right\| \leq \delta_j |u|, \quad u \in \mathcal{M}_j.$$

This is a measure of how many of the error components in $\mathcal{M}_j$ are not completely represented in the subspaces.

Ta'asan [31] introduced this method to the multigrid community when standard multigrid failed to converge for a class of problems with highly oscillatory solutions. He uses standard interpolation and projection methods and a Kaczmarz relaxation method in his examples.

Using a different set of interpolation and projection methods Hackbusch [26] developed a variant of Ta'asan's method using standard smoothers. Ta'asan's and Hackbusch's methods are both referred to as *robust multigrid*, which adds confusion (and heated discussions) to the field. More recently, Hackbusch's method has been referred to as *frequency decomposition multigrid* [27].

Frederickson and McBryan [21], using an approach different from that of Gannon and Van Rosendale, also investigated ways to keep all of the processors busy on a massively parallel single-instruction multiple-data (SIMD) machine. They used standard interpolation and projection methods and an elaborate smoother on each level. Unless great care is taken, this method computes the correction in one of the correction spaces while the corrections in the remaining spaces add up (pointwise) to zero. Their method is referred to as *parallel superconvergent multigrid*. A comparison of this method with that of Gannon and Van Rosendale might make an interesting student exercise.

The methods of Ta'asan, Hackbusch, and Frederickson–McBryan all use interleaved grids. For a problem on a square, this translates into the following, where the numbers refer to which subproblem the unknowns belong:

$$
\begin{array}{cccccccc}
3 & 4 & 3 & 4 & 3 & 4 & 3 & 4 \\
1 & 2 & 1 & 2 & 1 & 2 & 1 & 2 \\
3 & 4 & 3 & 4 & 3 & 4 & 3 & 4 \\
1 & 2 & 1 & 2 & 1 & 2 & 1 & 2 \\
3 & 4 & 3 & 4 & 3 & 4 & 3 & 4 \\
1 & 2 & 1 & 2 & 1 & 2 & 1 & 2 \\
\end{array}
$$

(The motivation is similar to that for multicolored orderings [1] for standard iterative methods.) Each of these methods requires that all of the matrices associated with the spaces be generated, except in trivial cases, thus doubling the memory requirements expected for solving boundary value problems. In addition, the coarse space operators are more difficult to compute using Hackbusch's variant than with either of the other two methods. In general, these are all space-wasteful methods.

A fourth, very different, approach that my colleagues and I developed, is referred to as either *constructive interference* [17] or, more recently, *domain reduction* (many references can be found in [9] and [16]). Our original motivation for using a multiple coarse space parallel multigrid algorithm was to eliminate the approximate solve step from standard multigrid algorithms. The solver takes most of the computational time but contributes almost nothing to the convergence rate, whereas coarse grid corrections take little time and reduce the error substantially. A general theory and simple examples were developed for multiple coarse space methods using no solver on the finest grid and mutually orthogonal subspaces that covered all of the error components of the original space (thus, $\delta_j = 0$ in (17.5)). This leads to very efficient direct methods rather than the expected iterative ones.

A side benefit of this theory is that the fine grid problem and, if a trick is used, most of the coarse space matrices do not need to be generated [9]. This method can use substantially less memory than a standard iterative or multigrid algorithm.

An additional note about domain reduction is that it leads naturally to more than 2^d subspaces for a d-dimensional problem. An eight-way decomposition of a problem on a square can be constructed, leading to problems defined on squares, rectangles, and triangles [6]. Both 60- and 64-way decompositions of a problem on a cube can be constructed with moderate difficulty [9]. In theory, a 192-way decomposition of a problem on a cube is possible. The entire problem would be solved 2660 times faster, if each of the 192 subproblems were solved by sparse Gaussian elimination, than if the original problem is solved by the same method (the latter is not advised, however).

To see how each of the variants operates, consider various projection operators in one dimension (where $x_0 = c_{N+1} = 0$):

- Linear projection: for $1 \leq i \leq N$,

$$y_i = x_{i-1} + 2x_i + x_{i+1}.$$

Let one subspace consist of the odd numbered y_i's and another subspace for the even numbered y_i's. This is used by both Ta'asan and Frederickson–McBryan.

- Linear–linear orthogonal complement projection: for $1 \leq i \leq N$,

$$y_i = \begin{cases} x_{i-1} + 2x_i + x_{i+1} & i \text{ even}, \\ -x_{i-1} + 2x_i - x_{i+1} & i \text{ odd}. \end{cases}$$

The subspaces are defined as in the linear projection case. The values at odd numbered y_i's correspond to the orthogonal complement of the traditional space defined by linear projection. This is used by Hackbusch and is actually a set of prewavelets.

- Symmetric–antisymmetric projection: for $1 \leq i \leq N/2$,

$$y_i = \frac{x_i + x_{N-i+1}}{2} \quad \text{and} \quad \hat{y}_i = \frac{x_i - x_{N-i+1}}{2}.$$

One subspace consists of y_i's and the other of $\hat{y}_i$'s. This defines a domain folding (or reduction) where even and odd functions are annihilated in exactly one subspace and exactly reproduced in the other. This is the method that my colleagues and I used.

The two-dimensional definitions of the above are defined in the obvious manner using tensor products.

Unfortunately, these methods have not been directly compared on a nontrivial example problem. The closest is a simple problem from the literature:

$$-10^5 u_{xx} - 10^{-5} u_{yy} = f \text{ in } (0,1)^2; \quad u = 0 \text{ on } \partial(0,1)^2.$$

To make things comparable with known results, a uniform mesh is used, a standard (simple) discretization, the energy norm, one Jacobi iteration in the analysis for multigrid (MG) and robust MG, and a direct solve on the coarsest level(s). The contraction factors are the following:

Method	Contraction factor	
MG	0.97	(17×17 grid)
Robust MG	0.33	(h independent)
Domain reduction (iterative)	ϵ	(h independent)
Domain reduction (direct)	0.00	(h independent)
Parallel superconvergent	"small"	(h independent)

The solver in the domain reduction method is either iterative (solving each problem to an accuracy of ϵ) or direct. If the smoother called for in the parallel superconvergent method can be constructed, then the contraction factor missing from the table will be very small, on the order of 0.05. Note that a line relaxation method, instead of point Jacobi, would make multigrid work well.

17.4. Implementing Parallel Multigrid

17.4.1. Shared Virtual Memory.
I have been distributing a public domain multilevel, aggregation–disaggregation code, Madpack (see [10] and [13]) for some years. Madpack is really a linear algebra package, rather than a package designed specifically for partial differential equations. The user specifies the domain or differential operator to Madpack indirectly, not directly. Interpolations and projections are computed as matrix–vector multiplies. Several sparse matrix formats are allowed including a stencil format that is very efficient on regular meshes.

Madpack2, first distributed in early 1986, has a sparse matrix–vector (and matrix transpose–vector) multiplier, a sparse direct solver, and three smoothers, namely, Gauss–Seidel, conjugate gradients, Orthomin(1). The latter two are preconditioned by SSOR. It is quite compact.

As an experiment, I parallelized parts of it in 1989 in the spirit of finding out how painlessly it could be done. Since I wanted to run my code on distributed-memory (Intel iPSC2) and shared-memory (Sequent Symmetry) parallel processors, as well as a network of workstations with a minimal amount of code changes, I started from the C version of Madpack and adapted it to the Linda system [7]. (I rejected proprietary message-passing systems at the time on the grounds that I refuse to program in any computer's assembly language, so why would I voluntarily program my communications in exactly that sort of nonportable environment?)

First, I realized that the obvious approach of storing matrices by column strips and vectors by row strips meant that the sparse matrix–vector multiplies were trivial to implement in parallel. I then replaced the three smoothers by two—diagonally preconditioned conjugate gradient and Orthomin(1)—which added a parallel dot product routine. Then I load balanced the operating processors by maintaining the same number of unknowns per processor independent of the level (effectively agglomerating at each level). Finally, I made absolutely sure that if I was doing a direct solve on the coarsest level, that I had only one processor involved. (The point of this exercise was to learn something about implementing something substantial in parallel and get something running quickly, not to produce a product quality code.)

The good news is that it was reasonably efficient in all three machine environments. The first implementation (on the Sequent) took a long time; the second (on the Intel iPSC/2), which unfortunately required porting to the distributed-memory environments, took an afternoon. With twenty minutes more work, I also had a simple three-dimensional domain reduction example running (with 99% parallel efficiency) that produced publishable results. Assuming there was really enough data associated with each active processor to keep them all active computationally, the parallel efficiency could be kept in the 75–99% range for most problems tried, even accounting for idle processors.

The bad news is that the differential equation front end to Madpack had to determine how to break natural data objects like vectors and sparse matrices into strips. This was not nearly as painful as had been expected but introduced many subtle bugs into the process that had to be found and fixed.

17.4.2. Explicit Message Passing. Over the years, Madpack has evolved into an object-oriented code based on a combination of C and Fortran [13]. More solvers have been added, temporary memory is dynamically allocated and freed, and the extremely complicated data structures have been hidden from the user.

The latest version, Madpack5, has been parallelized [28] using several message-passing systems (IBM's MPL, Intel's NX/2, and MPI). Information concerning the global problem must be entered before the package can determine which processor has which part of the global data. A fairly simple approach was taken once again. The advantage is that the user interface remains almost identical to the serial case: each processor calls the same routines in the same way as in the serial case, but one more routine is called to register local and global information.

By using a small collection of routines for the parallel communications that is not based on any specific message-passing system, it is now fairly painless to add interfaces to message-passing systems not already addressed. Due to the large number of ports of systems like PVM [24] and MPI [25], this is likely to be the direction that almost everyone in the scientific computing community takes.

17.4.3. Cache Awareness. Many parallel computers of the middle 1990s use RISC-based processors. As is commonly known, RISC processors require very careful use of their memory caches in order to operate at a high percentage of their peak speeds. (A cache is a very fast memory system that is tightly coupled to the processor that duplicates small parts of the main memory.) Most multigrid codes, for parallel or serial computers, do not attempt to optimize for caches. This is a serious flaw since it can be done quite easily, though not completely portably.

For example, consider a red-black ordering of Gauss–Seidel on a grid like that in Figure 17.1. The red-black ordering specifies that the red points are relaxed first, followed by the black ones. The usual implementation does this. Hence, all of the data (matrix, unknowns, and right-hand side) passes through the cache twice. Instead, the following algorithm can be performed.

ALGORITHM 6. *Cache Aware Red-Black Gauss–Seidel*
(a)   *Update all of the red points in row 1.*
(b)   *Do j = 2, N*
    (b1)   *Update all of the red points in row j.*
    (b2)   *Update all of the black points in row j − 1.*
    (b3)   *End Do*
(c)   *Update all of the black points in row N.*

When four grid rows of data along with the information from the corresponding rows of the matrix can be stored in cache simultaneously, this is a cache-based algorithm (a domain decomposition approach can be used to do this when four rows do not fit). The advantage is that all of the data and the matrix pass through cache only once (instead of the usual twice) per iteration of Gauss–Seidel.

On many machines, this improvement alone speeds up a multigrid solver by 25–40% alone. Other steps, like residual computation, projection of residuals, and interpolation can be drawn into a complex, cache-aware multigrid algorithm [12]. The speedup can be quite startling (e.g., 600% on some machines with particularly badly designed memory systems).

17.5. Conclusions

Parallel multilevel methods, which originated earlier this century in the personnel computing era, were the natural precursors to standard multilevel methods on single processors. Parallel multilevel methods are also the natural successors, in the age of advancing computers, to standard multigrid methods. What makes the six parallel multilevel methods different, practical, and impractical was discussed in the context of the three algorithms that encapsulate them. Because no one has carefully compared all of these methods, on a collection of common problems, on a variety of machine architectures, it is difficult to determine the conditions in which a particular method is really the right or wrong choice.

Using high-level programming tools makes implementing these methods much easier, although nontrivial. The use of low-level tools, such as explicit message-passing methods, should have been dismissed by the scientific computing community as a waste of human time. However, the community has flocked to message-passing systems, and this choice has been considered here. In fact, the choice is not as bad as it once was now that a standard exists for message-passing systems (i.e., MPI).

Whether or not high-level tools are used, parallel multilevel methods scale well assuming there is enough data to make using a parallel computer quite worthwhile. By paying close attention to the cache, dramatic improvements in speed can also be achieved.

There are a number of parallel multigrid codes in existence. In particular, the SUPRENUM project produced a collection of interesting codes [2]. There are several parallel multigrid codes available by anonymous ftp from MGNet.

17.6. MGNet

MGNet (see [10] and [11]) is an Internet repository that contains numerous preprints, conference proceedings, codes, a large bibliography, and general information about multigrid methods. It can be accessed through either the World Wide Web or anonymous ftp.

Location	Anonymous ftp	WWW URL
USA	na.cs.yale.edu	http://na.cs.yale.edu/mgnet/www/mgnet.html
Europe	ftp.cerfacs.fr	http://www.cerfacs.fr

A monthly electronic newsletter can be subscribed to by sending a request by e-mail to mgnet-requests@cs.yale.edu.

References

[1] L.M. ADAMS AND H.F. JORDAN, *Is SOR color-blind?*, SIAM J. Sci. Statist. Comput., 7(1986), pp. 490–506.

[2] M. ALEF, *Implementation of a multigrid algorithm on SUPRENUM and other systems*, Parallel Comput., 20(1994), pp. 1547–1557.

[3] G.P. ASTRAKHANTSEV, *An iterative method of solving elliptic net problems*, Zh. Vychisl. Mat. i Mat. Fiz., 11(1971), pp. 439–448.

[4] N. S. BAKHVALOV, *On the convergence of a relaxation method under natural constraints on an elliptic operator*, Zh. Vychisl. Mat. i. Mat. Fiz., 6(1966), pp. 861–883.

[5] A. BRANDT, *Multigrid solvers on parallel computers*, in Elliptic Problem Solvers, M.H. Schultz, ed., Academic Press, New York, 1981, pp. 39–83.

[6] F. BREZZI, C.C. DOUGLAS, AND L.D. MARINI, *A parallel domain reduction method*, Numer. Methods Partial Differential Equations, 5(1989), pp. 195–202.

[7] N. CARRIERO AND D. GELERNTER, *Linda in context*, Comm. ACM, 32(1989), pp. 444–458.

[8] D. CHAZAN AND W.L. MIRANKER, *Chaotic relaxation*, J. Linear Algebra Appl., 2(1969), pp. 199–222.

[9] C.C. DOUGLAS, *A tupleware approach to domain decomposition methods*, Appl. Numer. Math., 8(1991), pp. 353–373.

[10] C.C. DOUGLAS, *MGNet Digests and Code Repository*. Monthly digests subscribed to by sending a message to `mgnet-requests@cs.yale.edu` and an anonymous ftp site (`casper.cs.yale.edu`) for codes and papers on multigrid and related topics.

[11] ———, *MGNet: A Multigrid and Domain Decomposition Network*, ACM SIGNUM Newsletter, 27(1992), pp. 2–8.

[12] ———, *Caching in with Multigrid Algorithms: Problems in Two Dimensions*, Technical Report 20091, IBM Research Division, Yorktown Heights, NY, 1995.

[13] ———, *Madpack: A family of abstract multigrid or multilevel solvers*, Comput. Appl. Math., 14(1995), pp. 3–20.

[14] C.C. DOUGLAS AND J. DOUGLAS, *A unified convergence theory for abstract multigrid or multilevel algorithms, serial and parallel*, SIAM J. Numer. Anal., 30(1993), pp. 136–158.

[15] C.C. DOUGLAS, J. DOUGLAS, AND D.E. FYFE, *A multigrid unified theory for non-nested grids and/or quadrature*, E.W.J. Numer. Math., 2(1994), pp. 285–294.

[16] C.C. DOUGLAS AND J. MANDEL, *A group theoretic approach to the domain reduction method*, Computing, 48(1992), pp. 73–96.

[17] C.C. DOUGLAS AND W.L. MIRANKER, *Constructive interference in parallel algorithms*, SIAM J. Numer. Anal., 25(1988), pp. 376–398.

[18] A. FÄSSLER AND E. STIEFEL, *Group Theoretical Methods and Their Applications*, Birkhäuser, Berlin, 1992.

[19] R.P. FEDORENKO, *A relaxation method for solving elliptic difference equations*, Zh. Vychisl. Mat. i Mat. Fiz., 1(1961), pp. 922–927. Also in U.S.S.R. Comput. Math. and Math. Phys., 1(1962), pp. 1092–1096.

[20] ———, *The speed of convergence of one iteration process*, Zh. Vychisl. Mat. i Mat. Fiz., 4(1964), pp. 559–563. Also in U.S.S.R. Comput. Math. and Math. Phys., 4(1964), pp. 227–235.

[21] P.O. FREDERICKSON AND O.A. MCBRYAN, *Parallel superconvergent multigrid*, in Multigrid Methods: Theory, Applications, and Supercomputing, S.F. McCormick, ed., Lecture Notes in Pure and Appl. Math., 110(1988), pp. 195–210.

[22] D.B. GANNON AND J.R. VAN ROSENDALE, *On the structure of parallelism in a highly concurrent pde solver*, J. Parallel Distrib. Comput., 3(1986), pp. 106–135.

[23] U. GÄRTEL AND K. RESSEL, *Parallel multigrid: Grid partitioning versus domain decomposition*, in Proc. 10th International Conference on Computing Methods in Applied Sciences and Engineering, R. Glowinski, ed., New York, 1992, Nova Science Publishers, Commack, NY, pp. 559–568.

[24] A. GEIST, A. BEGUELIN, J.J. DONGARRA, W. JIANG, R. MANCHEK, AND V. SUNDERAM, *PVM Parallel Virtual Machine*, MIT Press, Cambridge, MA, 1994.

[25] W.D. GROPP, E. LUSK, AND A. SKJELLUM, *Using MPI: Portable Parallel Programming with the Message-Passing Interface*, Scientific and Engineering Computation, MIT Press, Cambridge, MA, 1994.

[26] W. HACKBUSCH, *A new approach to robust multi-grid solvers*, in ICIAM '87: Proc. First International Conference on Industrial and Applied Mathematics, A.H.P. Burgh and R.M.M. Mattheij, eds., SIAM, Philadelphia, 1988, pp. 114–126.

[27] ———, *The frequency decomposition multigrid method, part I: Application to anisotropic equations*, Numer. Math., 56 (1989), pp. 229–245.

[28] B. MOON, G. PATNAIK, R. BENNETT, D. FYFE, A. SUSSMAN, C.C. DOUGLAS, J. SALTZ, AND K. KAILASANTH, *Runtime support and dynamic load balancing strategies for structured adaptive applications*, in Proc. Seventh SIAM Conference on Parallel Processing for Scientific Computing, D.H. Bailey, P.E. Bjorstad, J.R.

Gilbert, M.D. Mascagni, R. Schrieber, H.D. Simon, V.J. Torczon, and L.T. Watson, eds., SIAM, Philadelphia, 1995, pp. 575–580.

[29] H.A. SCHWARZ, *Über einige Abbildungsaufgaben*, Ges. Math. Abh., 11(1869), pp. 65–83.

[30] R.V. SOUTHWELL, *Relaxation Methods in Engineering Science*, Oxford University Press, Oxford, 1940.

[31] S. TA'ASAN, *Multigrid Methods for Highly Oscillatory Problems*, PhD thesis, Weizmann Institute of Science, Rehovot, Israel, 1984.

Chapter 18

Shared-Memory Emulation Enables Billion-Atom Molecular Dynamics Simulation

Eduardo F. D'Azevedo
Charles H. Romine
David W. Walker

Editorial preface

Exploiting parallel computing technology in molecular dynamics (MD) is a logical approach. There can never be enough atoms in the simulation, since the more atoms there are the closer the simulation is to providing macroscopic understanding. Parallel computing is the only viable path that can provide the requisite computing power to continuously increase the number of atoms in MD simulations.

This chapter details the simulation of one billion atoms on a distributed-memory machine using a software interface that provides a logical shared-memory view of the machine.

This article originally appeared in *SIAM News*, Vol. 28, No. 5, May/June 1995. It was updated during the summer/fall of 1995.

Very large scale molecular dynamics (MD) simulations, involving hundreds of millions of atoms, allow the study of "macroscale" material properties and are thus receiving a lot of attention from researchers. To simplify parallel programming for MD applications, we developed shared-memory emulations of large-scale (more than one billion atoms) MD codes, which we ran on Oak Ridge National Laboratory's Intel Paragon, a distributed-memory computer.

We began with a parallel MD code for simulating Lennard–Jones fluids that had been written originally for an explicit message-passing environment. Using shared-memory primitives, we rewrote the code; the resulting code has no explicit message primitives and resembles a serial code. The shared-memory code can perform dynamic load balancing, and its performance is competitive with that of other explicit message-passing MD codes using spatial decomposition. This code was then applied to a *billion*-atom simulation.

It was the inherent difficulty of programming for a message-passing

environment that motivated us to develop the Distributed Object Library
(DOLIB) [3, 5], which emulates a shared-memory environment. During the
design of DOLIB, we were guided by several questions: (1) What programming
environment will allow the rapid porting of a serial code to a massively parallel
machine? (2) How can particle tracking be done effectively in a situation
in which the velocity field can cause the particles to cross interprocessor
boundaries? (3) How can dynamic load balancing be supported in a natural
way on distributed-memory MPPs?

We have used DOLIB in the rapid parallelization of several large-scale serial
application codes, including groundwater and global atmospheric models, in
addition to the large-scale MD simulations described here. This article may
be of interest to scientists or programmers considering the use of shared-
memory emulation for more sophisticated MD simulations or for particle-in-
cell methods. The new parallel MD prototype code can perform dynamic load
balancing and achieves performance competitive with codes that use spatial
decomposition and explicit message-passing.

18.1. DOLIB

Distributed-memory multiprocessors have proved to be scalable and to offer
good performance. Because of the limited memory on each processor node,
however, programmers generally need to perform their own data decomposition, carefully moving needed data among nodes by explicit message-passing.
Writing parallel application code in this way can be difficult. Shared-memory
emulation enables a programmer to make full use of the aggregate (gigabyte)
memory resources of a system while avoiding the difficulties of message-passing.

Achieving good performance with strongly coherent, emulated shared
memory on a distributed-memory system requires an effective caching strategy.
Thus, much of the research into shared virtual memory, such as Li [6], Li and
Hudak [7], and Stumm and Zhou [16], concerns intricate network protocols
that maintain cache coherency in the presence of multiple concurrent updates.
Shiva [8] is a shared virtual memory system for the Intel iPSC/2 hypercube
multiprocessor; it uses the Memory Management Unit (MMU) page fault
mechanism on each Intel i386 node to generate memory requests for remote
pages. The CHAOS library [15] is an attempt to provide support for the
parallel solution of irregular problems, i.e., problems whose communication
patterns are not easily predictable. CHAOS is a runtime library that can
analyze the pattern of indirect addressing of arrays and automatically devise
an optimized schedule of communication. The Global Array (GA) library [11],
developed at Pacific Northwest Laboratory, supports asynchronous access
to logical blocks in physically distributed matrices for use in computational
chemistry. MetaMP [12] offers a programming environment with C++ classes
and preprocessors that supports lightweight sharing through *weakly coherent*
mechanisms.

DOLIB is similar in many ways to GA and MetaMP in that it provides a set of

Fortran and C callable routines to emulate weakly coherent shared memory in distributed-memory environments, such as Intel multiprocessors and clusters of workstations. DOLIB supports runtime dynamic creation and destruction of one-dimensional global arrays. Index calculations can be used to support higher-dimensional arrays. Explicit gather and scatter operations provide access to array elements. DOLIB is portable in that no language extension is introduced and no preprocessor, compiler, or operating system support is required.

A global array in DOLIB is stored as fixed-size pages in a block wrapped fashion across all processors. DOLIB translates requests for remote data (gather) or updates (scatter) into the appropriate message sent to the "owner" processor of that data page. These DOLIB requests are then serviced by the IPX [10] message system, which is based on the concept of active messages. We have also used global shared memory in DONIO [4] as a large disk cache to enhance I/O performance on the Intel Paragon.

18.2. The Message-Passing Code

To demonstrate that logically shared memory on distributed-memory computers is useful in large-scale computations and performs at least as well as traditional message-passing codes, we implemented a DOLIB version of a three-dimensional MD code for simulating a Lennard–Jones fluid. We created the DOLIB version by modifying a message-passing MD code that was ported to the Intel Paragon and enhanced by David Walker of ORNL. The message-passing code was originally written at the University of Southampton, England, and subsequently adapted to run on the Intel iPSC/2 hypercube at Daresbury Laboratory.[11]

The message-passing code uses a link-cell (geometric hashing) algorithm in which all particles are hashed into a three-dimensional mesh of $N_b \times N_b \times N_b$ cells to model short-range atomic interactions. The minimum cell size is the cut-off distance (r_c) used in the short-range force evaluation, so that each particle interacts only with particles in the same cell or in the neighboring cells. Exploiting the symmetry of Newton's third law, the code requires the examination of atoms in only 13 (instead of 26) neighboring cells. The code assumes an $N_c \times N_c \times N_c$ face-centered cubic (FCC) periodic lattice with a total of $N = 4N_c^3$ atoms. Use of a "shifted-force" [14] Lennard–Jones 6-12 potential ensures that the potential and its first derivative are continuous at the cut-off distance. A simple Verlet [17] leapfrog scheme is used to update particle positions at each timestep. Details of the MD algorithms used are described in [1].

The MD code distributes the cells in blocks over a three-dimensional mesh of processors in such a way that each process is responsible for the particles in a rectangular subdomain. Particle information for cells lying at the boundaries

---

[11]The source is available from the CCP5 archive at ftp.dl.ac.uk:/ccp5/SOTON_PAR.

of a process must be communicated to one or more neighboring processes, since these particles may interact with particles in neighboring processes.

The code uses the standard approach of creating "ghost" cells around the boundary of each process. The communication can then be performed in a series of six shift operations (one for each face of the rectangular subdomain).

Particles can migrate from the subdomain of one process to that of another. The induced communication can again be performed in a series of shift operations. In the message-passing code, communications of boundary data and particle migration are combined to reduce the frequency (and hence the overhead) of message-passing.

This communication scheme requires that the code developer explicitly manage the packing of particles into message buffers, the communication of the message buffers, and the unpacking of the message buffers at the destination process. The coding of these operations is tedious, error prone, and as we show in this article, unnecessary.

18.3. Overview of Parallelization with DOLIB

We used DOLIB to develop an emulated shared-memory version of the MD code. The two codes are similar in overall structure; however, we believe that the new code, without the complexities of explicit message-passing, is much easier to write and understand.

The two most time-consuming kernels in the MD codes are those for (1) geometric hashing and the migration of atoms and (2) the force and potential evaluations. Hashing the atoms, which can be parallelized easily, requires only $O(N)$ work. The link-cell method is memory efficient, requiring only $O(N_b^3)$ storage for the three-dimensional bins, nine real vectors of length N (x, y, z, v_x, v_y, v_z, f_x, f_y, f_z, for the positions, velocities, and forces, respectively), and an integer vector of length N for maintaining the link cells. The total storage requirements, then, are $40N + O(N_b^3)$ bytes for single precision or $52N + O(N_b^3)$ bytes if the vectors (f_x, f_y, f_z) are in double precision.

Another common technique is to construct and maintain, for *each* atom, a list of neighboring atoms [17] that is updated every few timesteps. These lists can grow quite long, however, resulting in prohibitive memory costs for very large scale MD calculations.

We implement the link-cell method by performing a *global reordering* or renumbering, so that all atoms in a bin are contiguously numbered; if each bin has, say, 10 atoms, then after the reordering, the first bin will contain atoms 1 to 10, the second bin will contain atoms 11 to 20, and so forth.

Two passes are required for the geometric hashing and reordering. The first pass performs the geometric hashing and stores the result of the mapping. For each bin, we compute the number of atoms to be assigned.

In Figure 18.1, element i in the `particle-to-bin` array holds the bin number of particle i. The `particles-in-bin` array gives the total number of particles in each bin. Both are DOLIB global arrays. The DOLIB atomic

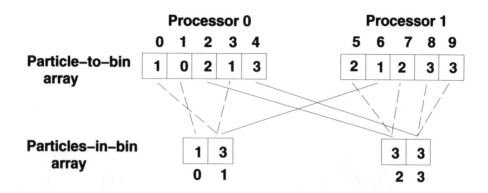

FIG. 18.1. *Hashing of particles into bins. The solid lines indicate interprocessor communications.*

accumulate operation is used to construct the `particles-in-bin` array.

We then use a pointer array to set up and allocate storage for each bin. The second pass performs the actual reordering and data movement. The vectors (f_x, f_y, f_z) are used as temporary storage and are cleared for the force computation.

We have found that after the first iteration, most of the particles will be hashed into the same bins (and processors) to which they belonged in the previous timestep, thus requiring very little data movement.

18.3.1. Force evaluation. For simplicity, we consider a two-dimensional partition of the computation. The $N_b \times N_b$ columns are *block* partitioned and assigned to individual processors. The code uses DOLIB's efficient contiguous block gather and update operations, since all atoms in each bin are contiguously ordered. Moreover, we exploit data reuse by selecting the 13 neighboring cells from five neighboring columns (see Figure 18.2). Position data from the shaded region in the left half of Figure 18.2 are required for processing column (i, j). In processing the next column, $(i + 1, j)$, position data from three of the five columns are reused; data must be brought in for only two new columns, $(i+2, j)$ and $(i + 2, j + 1)$. This organization reduces the amount of communication required for gather and update operations. No further communication is required once all position data for the two new columns have been gathered, which simplifies automatic thread parallelization by the Intel Paragon MP node compiler.

To avoid loss of accuracy by catastrophic cancellation, we have also modified the energy and potential computation to have the positive and negative contributions summed separately.

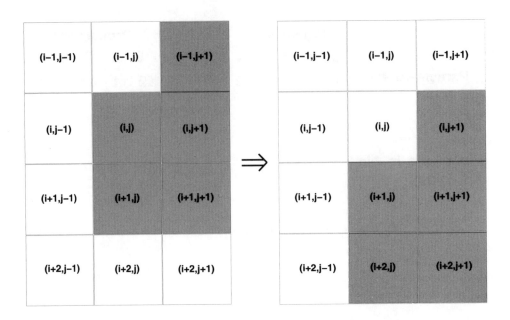

FIG. 18.2. *Top view of columns used in processing column* (i, j). *Only two new columns are needed for processing column* $(i + 1, j)$.

18.3.2. Dynamic Load Balancing.

Since a highly nonuniform distribution of particles can result in a serious load imbalance, we include an option for dynamic load balancing. We compute a work measure for each column and use this estimate to distribute columns to processors (see Figure 18.3). One simple measure of workload is the total number of atoms in the column.

For a sufficiently uniform distribution of atoms in the domain, this technique generally attains good load balance. For nonuniform distributions, however, a more reasonable work measure is to count, for each bin, the total number of possible atom–atom interactions for all other particles in the 13 neighboring bins. Both of these work estimates are provided as options to the load-balancing routine.

The load-balancing strategy then computes the overall total and average amounts of work for each processor, satisfying this average by using a greedy algorithm to assign columns to processors. In the future, we will provide a more sophisticated bin-packing algorithm. For the simple Lennard–Jones fluid simulations, however, each bin has approximately the same number of atoms, and there is almost no load imbalance.

18.4. Parallel Performance

Our parallel code was tested on an Intel MP node Paragon system. Each MP node contains three CPUs, one of which is configured as a dedicated message coprocessor, and at least 64 Mbytes of memory in a *local* shared-

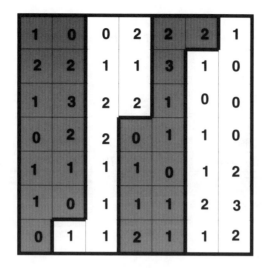

FIG. 18.3. *Simple block assignment of work (columns) to processors. The numbers are a measure of the work associated with the column.*

memory configuration. All our MD computations were done in single precision with two threads. One CPU runs the main computational thread, and the other is utilized for automatic thread parallelization by the Paragon MP Fortran compiler. We also implemented a double-precision option for force computations that ran about 15% slower, although we found essentially no difference in the numerical results.

We tested our code with a benchmark problem described in Plimpton [13, p. 23]: a Lennard–Jones 6-12 potential with reduced density ($\rho = 0.8442$) and reduced temperature ($T = 0.72$). The system is initialized with an *FCC* lattice, and randomized velocities are chosen from a Boltzmann distribution. The integration timestep is 0.00462 in reduced units, and the cut-off distance is $r_c = 2.5\sigma$. We estimate that each atom requires about 2200 flops per timestep for both the force and the potential evaluation.

Table 18.1 and Figure 18.4 show averaged runtimes per timestep for three problems: $N_c = 159$ (16,078,716 particles), $N_c = 200$ (32,000,000 particles), and $N_c = 250$ (62,500,000 particles). The total initialization and setup times were approximately 10 minutes for $N_c = 250$ on 64 processors. We achieved slightly faster runtimes without the overhead of computing work measure and dynamic load balancing, because the problem was already well balanced. We relied solely on the Intel f77 optimizing compiler (with -O3 -Mvect options when appropriate). Approximately 0.25 millisecond/atom/processor was required for each timestep.

For a 500,000-atom problem on a single processor (with parallel threads), our code achieved a speed of about 0.17 millisecond/atom/timestep. For the

TABLE 18.1

Time for one timestep (in seconds) on the Intel MP Paragon. Cutoff $r_c = 2.5\sigma$.

Particles	Processors					
	16	32	64	128	256	512
16,078,716	174	91	49	27	15	9
32,000,000	–	180	95	51	28	16
62,500,000	–	–	179	98	53	30

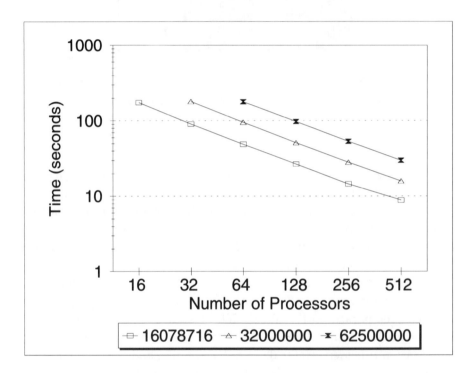

FIG. 18.4. *Time per timestep (in seconds) versus number of processors.*

one-processor run, the same "parallel" code was used on a single processor. No messages were generated in the DOLIB gather/scatter, which on a single processor are translated into memory copies.

Table 18.2 shows the performance of the code on very large simulations. About 75% of the time was spent in the force evaluation and 20% in hashing/reordering. Because the computation is based on a shared-memory programming paradigm, however, there is some overlap of message-passing and servicing of remote memory requests even within the force and potential computations.

TABLE 18.2

Time for one timestep (in seconds) on the Intel MP Paragon. Cutoff $r_c = 2.5\sigma$.

Particles	Processors		
	256	512	1024
256,000,000	182	103	63
500,000,000	–	194	120
1,000,188,000	–	–	233

Although we used single precision in our codes, our runtimes compare quite favorably with the times of 0.24 millisecond/atom/processor achieved on an Intel Delta [13] and about 0.26 millisecond/atom/processor [2, 9] (performed in double precision) on a 1024-processor CM-5 both using a spatial decomposition and linked-cell method.[12]

18.5. Acknowledgments

The authors take this opportunity to express appreciation to Bob Marr, Ron Peierls, and Joe Pasciak at Brookhaven National Laboratory for the IPX package, which simplified the development of DOLIB. We also thank Bill Shelton for his insight and advice on MD simulations and Al Geist for his helpful suggestions.

This work was supported by the U.S. Department of Energy.

References

[1] M.P. ALLEN AND D. TILDESLEY, *Computer Simulation of Liquids*, Claredon Press, Oxford, 1987.

[2] D.M. BEAZLEY AND P.S. LOMDAHL, *Message passing multi-cell molecular dynamics on the Connection Machine 5*, Parallel. Comput., 20(1994), pp. 173–196.

[3] E.F. D'AZEVEDO AND C.H. ROMINE, *DOLIB: Distributed Object Library*, Technical Report ORNL/TM-12744, Oak Ridge National Laboratory, Oak Ridge, TN, 1994.

[4] ——, *DONIO: Distributed Object Network I/O library*, Technical Report ORNL/TM-12743, Oak Ridge National Laboratory, Oak Ridge, TN, 1994.

[5] ——, *DOLIB: Distributed object library*, Proc. Seventh SIAM Conference on Parallel Processing for Scientific Computing, SIAM, Philadelphia, 1995, pp. 750–755.

[6] K. LI, *Shared Virtual Memory on Loosely Coupled Multiprocessor*, PhD thesis, Yale University, 1986.

[7] K. LI AND P. HUDAK, *Memory coherence in shared virtual memory systems*, ACM Trans. Comput. Systems, 7(1989), pp. 321–359.

[8] K. LI AND R. SCHAEFER, *Shared virtual memory for a hypercube multiprocessor*, Proc. Fourth Conference on Hypercubes, Concurrent Computers and Applications,

---

[12]The code used on the CM-5 computes about 7000 flops/atom.

March 1989, Monterey California, Golden Gate Enterprises, Los Altos, CA, 1989, pp. 371–378.

[9] P.S. LOMDAHL, P. TAMAYO, N. GRONBECH-JENSEN, AND D.M. BEAZLEY, 50 *GFlops molecular dynamics on the Connection Machine* 5, in Proc. Supercomputing '93, November 15–19, Portland, OR, Association for Computing Machinery and The Institute of Electrical and Electronics Engineers, Inc., 1993, pp. 520–527.

[10] B. MARR, R. PEIERLS, AND J. PASCIAK, *IPX—Preemptive Remote Procedure Execution for Concurrent Applications*, Technical Report, Brookhaven National Laboratory, Brookhaven, NY, 1994.

[11] J. NIEPLOCHA, R.J. HARRISON, AND R.J. LITTLEFIELD, *Global arrays: A portable "shared-memory" programming model for distributed memory computers*, in Proc. Supercomputing '94, IEEE Press, Piscataway, NJ, 1994, pp. 340–349.

[12] S.W. OTTO, *Parallel array classes and lightweight sharing mechanisms*, Scientific Programming, 2(1993), pp. 203–216.

[13] S. PLIMPTON, *Fast parallel algorithms for short-range molecular dynamics*, Technical Report SAND91-1144, UC-705, Sandia National Laboratories, Albuquerque, NM, May 1993.

[14] J.G. POWLES, W.A.B. EVANS, AND N. QUIRKE, *Non-destructive molecular dynamics simulation of the chemical potential of a fluid*, Molecular Phys., 46(1982), pp. 1347–1370.

[15] S. SHARMA, R. PONNUSAMY, B. MOON, Y.-S. HWANG, R. DAS, AND J. SALTZ, *Run-time and compile-time support for adaptive irregular problems*, in Proc. Supercomputing '94, IEEE Press, Piscataway, NJ, 1994, pp. 97–106.

[16] M. STUMM AND S. ZHOU, *Algorithms implementing distributed shared memory*, IEEE Computer, May 1990, pp. 54–64.

[17] L. VERLET, *Computer experiments on classical fluids* I. *Thermodynamical properties of Lennard–Jones molecules*, Phys. Rev., 159(1967), pp. 98–103.

Chapter 19

Probing the Playability of Violins by Supercomputer

Robert T. Schumacher
Jim Woodhouse

Editorial preface
_____

A journey into the culturally refined arena of a classic musical instrument, the violin, is the theme of this chapter. This chapter shows the inherent difficulties present in attempting to model the underlying behavior of a musical instrument. The computational advantage gained from a parallel implementation on the CM-2 is a significant help; however, much more remains to be explored.

This article originally appeared in *SIAM News*, Vol. 25, No. 5, September 1992. It was updated during the summer/fall of 1995.

The violins made by Antonio Stradivari and a few other luthiers from the 16th, 17th, and 18th centuries are surrounded with an air of mystery, and they command staggeringly high prices. It may seem surprising that this should continue to be the case, given the advances in both theoretical and experimental techniques for studying sound and vibration. But researchers who make a serious effort to apply these techniques find time and again that they are pushing against the limits of what can be achieved.

There are two related reasons for this situation. First, the assessment of a violin as "good" is based on subjective impressions of players and listeners, and before any useful physics can be done, it is necessary to try to pin down physical correlates of those impressions. Second, the violin, in common with any other successful musical instrument, has evolved to take the best possible advantage of human abilities: it allows motor actions, up to the limit of what we can achieve, to be turned into a range of sounds that we can process most acutely. The result is that the all-important nuances distinguishing a great violin from a moderate one may stem from rather small and subtle physical differences.

The most reliable judgments of violins, with respect to physical correlates of "quality," are obtained not from listeners but from players. A good player may be able to make compensations that mask the inadequacies of a poor instrument to a considerable extent, so that a listener is hardly aware of them. But the player, being inside the feedback loop of those compensations, will be quite well aware that the instrument is a poor one, precisely because it calls for such compensation. This suggests that physical correlates be sought for differences of "playability" between instruments and between notes on the same instrument. Such differences certainly exist, and they point to the possibility that the "stick-slip" oscillations produced by bowing a string are somehow influenced in their details by the acoustical behavior of the wooden box, which is the violin body, to which the string is attached.

This is a matter suitable for investigation by the physicist. Theoretical and experimental study of the bowed string has a long history, and the physical basis for much of the observed behavior is believed to be fairly well understood. A bowed string is a self-sustained oscillator, in which a complicated linear system (the string, with attached violin body, which is in turn weakly coupled to the acoustics of the auditorium) is driven by the friction force from the bow. The dependence of this friction force on the string motion under the bow is strongly nonlinear. Because this description is generically similar to that of various other nonlinear systems that have been much studied in the last 15 years or so, complicated behavior involving the possibility of many periodic and nonperiodic ("chaotic") regimes might be anticipated. This expectation is in qualitative agreement with the wide range of unmusical noises that can be elicited from a violin, especially in the hands of a novice.

Out of that multiplicity, the violinist is almost always trying to achieve one particular regime, which was first described by Helmholtz in the 19th century and is thus known as Helmholtz motion. It is a periodic, or at least approximately periodic, regime in which the string sticks to the bow for most of the time, slipping rapidly backward relative to the bow motion just once per vibration period. Many issues of playability therefore depend on how readily Helmholtz motion can be initiated and maintained, by means of various bowing techniques that the player wants to be free to use for musical reasons. Perhaps an instrument considered "easy to play" or to "speak easily" is one that readily yields a Helmholtz motion, with an acceptably short starting transient, under a wide range of bowing conditions.

To investigate how the vibration behavior of the violin body might influence this capability, we must resort to simulation. Certain knowledge, especially about the regimes of self-sustained oscillation that are possible under given conditions, can be gained by analytic calculation. Because of the strongly nonlinear character of the system, however, it is very hard to make progress on the questions of starting transients and of the choice among the possible regimes from a given bowing transient.

An efficient simulation scheme, based on the simplest physical model of a bowed string that seems to allow the main observed effects, has existed for some time [5, 6]. This scheme has been used to explore various questions of bowed-string behavior, and has also penetrated the commercial world, where musical synthesizers based on this technique have been reported to be under development for replacing or augmenting the standard FM-synthesis paradigm that has long dominated synthesizers and computer music. (No commercial products had been produced when this article was written, but the expected products were implicit as well as explicit at the International Conference on Physical Modeling, Grenoble, France, September 1990.)

Early implementations of the simulation scheme took the form of interactive programs, in which the playing parameters could be varied during a run so that the program could be "played" roughly like the real string. This yielded many valuable insights into the bowing process and the strengths and weaknesses of the particular model used. The parameter space explored in this way is so large, however, that it is extremely difficult to discern any structure in the overall behavior by watching individual interactive runs of the program. A more organized use of simulation is needed.

The aim of the project described here is to use simulations to map out some part of the player's parameter space and then represent the results in diagrammatic form so that any interesting structure can be readily discerned. Study of a two-dimensional subspace is suggested as it is hard to convey results in more than two dimensions. The choice of a suitable subspace for a preliminary investigation requires some care, since the computational cost is quite high and there are many possibilities to consider.

Previous discussions of regimes of bowed-string motion focused mainly on the influence of the "bow force," the normal force with which the bow is pressed against the string. Players know very well that this force must lie within certain limits for a Helmholtz motion to be obtained: too little force produces a "surface sound," which involves more than one slip per cycle of oscillation, while too much force can produce a raucous "crunch," in which the motion is not periodic at all. It therefore seems natural that the steady bow force at the end of the transient should be one variable to consider.

A second variable is then wanted, one that can be used to specify a range of different starting transients, all of which share the same eventual bow force. It was decided to use a second force-like variable. Starting transients can be simulated in which the initial bow force is different from the asymptotic value, with the offset decaying exponentially with time. In this way a range of somewhat plausible bowing transients can be simulated.

If the force starts from zero and increases to the final level, the result is a simple representation of a string-crossing transient in which the bow alights on the string and the force takes a finite time to build up. If the initial force is the same as the final value, a switch-on transient is produced. This probably does not represent anything done in normal playing, but it was a favorite condition for previous simulations and is useful for comparison. Finally, if the force starts high and decreases to the final value, we obtain at least a crude representation of a martelé transient. (Martelé bowing, a common way to give an emphatic start to a note, involves "digging in" to the string with the bow at the initial instant.) This transient will be less accurate than the string-crossing transient, because the bow speed will also vary significantly during a martelé. Despite this reservation, the two-dimensional space of transients in which the initial and final forces are varied (keeping the exponential timescale and the rest of the model constant) seems to be a promising candidate for a first study.

The calculation to be done is now analogous to the computation of the famous Mandelbrot set: for each point in the parameter plane, a nonlinear process is simulated with the coordinates of the point used as input data. The simulation is continued long enough to indicate the eventual outcome (in our case, whether a periodic Helmholtz motion is or is not produced); that point can be colored in some way to represent this outcome, and the calculation then moves on to the next point. When a reasonable area has been covered in this way, a picture will have been built up of the region of the parameter subspace in which Helmholtz motion actually occurs from a starting transient.

This computation lends itself very easily to implementation on a parallel computer. Values of the asymptotic force and the initial force may be assigned to different processors, which can simulate the string motion independently of one another. This suits the architecture of the Connection Machine-2 (CM-2) perfectly: in fact, it is the simplest possible application for such a machine. (The CM-2 used for the work described here is operated by the Pittsburgh Supercomputer Center.) We are interested, typically, in the performance of 16,384 processors running simulations that do not interact with one another.

The basic algorithm is described in [5, 6]. Two quantities are involved, the string velocity at the bowed point, $v(t)$, and the force applied to the string by the bow, $f(t)$. These quantities are connected in two different ways; the combination gives the governing equation for the process.

First, when the string and attached violin body are considered as a linear system, then

$$(19.1) \qquad v(t) = \frac{Y_0}{2} f(t) + \int_{-\infty}^{t} g(t - \tau) f(\tau) d\tau,$$

where Y_0 is the characteristic admittance of the string. The first term on the right-hand side of this equation represents the instantaneous response to the force (as would be found on an infinite string); the second represents the combined effect of all reflected velocity waves arriving back at the bowed point at time t. The latter is expressed as a convolution integral involving the impulse response function of the system, $g(t)$. For computational efficiency, this integral is best evaluated via two much shorter convolutions with the "reflection functions" of the two sections of string on either side of the bow, the impulse responses $h_1(t)$ and $h_2(t)$, which would apply if one or the other section of string were replaced by a semi-infinite string. It is readily shown that

$$(19.2) \quad g(t) = \frac{Y_0}{2} \left[h_1 + h_2 + 2h_1 \star h_2 + h_1 \star h_2 \star h_1 + h_2 \star h_1 \star h_2 + \cdots \right],$$

where $\star$ denotes the operation of convolution.

The second relation between $v(t)$ and $f(t)$ is a nonlinear function, $f = F(v)$, giving the velocity dependence of the frictional force between bowhair and string. The adequacy of such a function to characterize the physics of rosin friction is by no means obvious, and exploration of a more complete characterization is in the forefront of modern tribological (frictional) research. Here we simply assume a function $F(v)$ with a plausible and mathematically tractable form loosely based on experiment, which is known to give fairly realistic results when applied to the bowed-string problem [5, 6]. The function is plotted as the heavy curve in Figure 19.1. If when the convolution integral from (19.1) is denoted by $v_h(t)$, it is plain that the values of $f(t)$ and $v(t)$ are found at the intersection of this curve with a straight line with slope $2/Y_0$ and intercept $v_h(t)$, as shown.

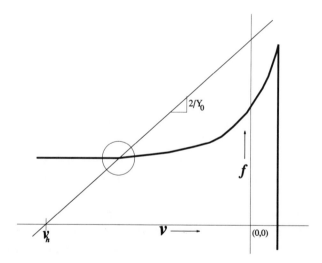

FIG. 19.1.   *Friction force as a function of string velocity at the bowed point (heavy curve). The vertical portion represents sticking, which occurs when the string velocity equals that of the bow. The curved portion represents sliding. The sloping line and ringed intersection illustrate the solution procedure for the governing equations at a given moment t.*

The two parts of the $F(v)$ curve require different calculations to solve for the intersection point, depending on whether the string is slipping or sticking at the given timestep in the calculation. Because the CM-2 is a SIMD (single-instruction multiple-data) machine, every processor is executing the same instruction at a given time. Because each processor has a different set of bowing force parameters, as described earlier, it is clear that not all processors are slipping or sticking at the same time. In fact, because of the possibility of multiple intersections of $F(v)$ and the straight line, an ambiguity whose correct resolution is described in [5], there are four separate branches in the computation. On a serial machine, they are handled straightforwardly with Fortran IF THEN statements. On a CM-2, they are dealt with by means of the WHERE statement. Each processor executes every instruction, whether or not it is appropriate to the particular state of the processor at that time, and the processors for which the instruction is inappropriate for their current states throw away the results of that step of the calculation.

Thus, the price paid for the parallelism is a certain inefficiency; each branch of the code written for a serial machine is executed on every processor. The advantage of the large number of processors far outweighs this inefficiency, however. A single period of oscillation takes about 0.4 second to execute. A comparison of speed with that of a serial machine running in "real time" (1/440 second for a period of oscillation, at a sampling rate corresponding to commercial CD recording, 44,100 samples per second) can be made by dividing the per period execution time by the number of processors. When we use half of the processors of the Pittsburgh Supercomputer Center's CM-2, 16K processors, the arithmetic shows a speedup of approximately 100 over the same mapping of the two-dimensional subspace of parameter space by a serial machine running in real time. That factor spells the difference between an investigation that is doable and one that would never be attempted.

Perhaps the most interesting technical part of the problem lies in the interpretation and classification of the output. The player is interested in how quickly an acceptable oscillation is established. An acceptable oscillation is that originally described by Helmholtz, with a single slip and recapture to the sticking state per period of oscillation. A great many other forms of periodic oscillation are possible, as is the nonperiodic noise that sometimes seems to be the most common result of a beginner's efforts. From this "zoo" of oscillations, periodic and nonperiodic, an experienced listener has no trouble distinguishing the Helmholtz motion from other regimes. Similarly, when watching a single simulation run on the screen, the human eye has no trouble classifying regimes into "Helmholtz" versus others. But it is not so easy to find a robust algorithm that allows each processor to recognize the characteristic Helmholtz pattern, so as to tell the experimenter which processors are oscillating properly.

For the first trials, we developed a detection criterion based on several simple tests. After a given number of iterations, we examine five nominal periods of the oscillation for periodicity by computing the autocorrelation function. We also keep track of the number of stick-slip transitions during that time, and of the number of "kinks," or velocity discontinuities, traveling on the string. The criteria for a successful Helmholtz oscillation are that the motion be periodic and that the number of stick-slip transitions and the number of kinks per period be equal to one. These tests must be fine-tuned to work adequately. Fine tuning is not easy since the main objective of the simulations is to cover a wide range of parameter values, to insure the Helmholtz motion itself varies quite widely. Both the waveform and frequency vary with bow force within a given model. There is no doubt that further effort will be needed in this area if a more sophisticated "expert system" is to be developed for regime classification.

From the tests just described, we construct a 128×128 pixel plane in which white space is Helmholtz motion and black space is not, for one reason or another. By saving the equivalent, color-mapped planes for autocorrelation, number of slips, and number of kinks, we can also make informed guesses about the nature of the oscillations in the non-Helmholtz regions. Some of the interesting ones can then be examined directly by reading out those portions of the simulation from selected processors in a separate run.

Another useful technique is to examine the output at periodic intervals, looking at the last five nominal periods at, say, 30-period intervals. We thus produce a "movie" of a few frames in which we can see the development of the Helmholtz region as a function of time since the onset of bowing. In that way the length of the transient can be assessed in various regions of the parameter plane. An example of the output is shown in Figure 19.2.

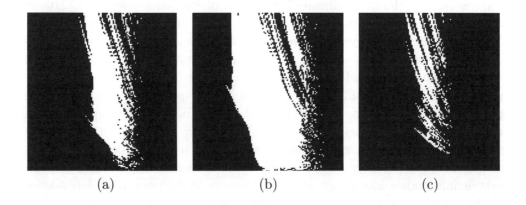

      (a)   (b)   (c)

FIG. 19.2. *Region of the parameter space in which a periodic Helmholtz motion has arisen (white space) after (a) 50 periods; (b) 60 periods; (c) 100 periods. The horizontal axis represents the asymptotic vertical force of bow on string, in the logarithm of natural units. The vertical axis represents the initial vertical bow force, from zero at the bottom to twice the asymptotic force of each column. The decay to the asymptotic force has an e-fold time of eight nominal periods of the oscillation. The picture is 128×128 pixels, where each pixel corresponds to a separate processor of the Connection Machine.*

In each diagram in Figure 19.2 the horizontal axis shows the asymptotic bowing force, on a logarithmic scale. The vertical axis shows the initial force

as a multiple of the asymptotic force, on a linear scale. The force begins as
the specified multiple of the asymptotic force and decays to the asymptotic
force with an e-fold time of about eight nominal periods. A very simple model,
developed in [1], is used for the rest of the system. It allows for the frequency-
dependent energy dissipation of a string on a real violin, but it does not attempt
to model the reactive, resonant nature of the response of the violin body. By
separating these two effects, this model makes a very useful benchmark case
against which the changes produced by more sophisticated models may be
judged.

For the example illustrated here, there is no "white space" in the diagrams
until about 35 period lengths have passed. After 50 period lengths, as shown in
Figure 19.2a, some small areas of the parameter plane have settled into periodic
Helmholtz motion. By 60 period lengths (Figure 19.2b), the white region has
broadened significantly. After 100 period lengths (Figure 19.2c), the diagram
shows a quite large area within which Helmholtz motion has been produced.
The three regions of different non-Helmholtz behavior that can be distinguished
in this final picture correspond well, qualitatively at least, to what happens
in real playing. At low bow forces, the Helmholtz motion gives way to other
periodic regimes with more than one slip per period, known to players as
"surface sound." At very high bow forces, the figure shows a rather fuzzy
vertical stripe. Here, the destabilizing effect of "negative resistance" [10] at
the bow during slipping prevents periodic motion of any kind being established.

In between these regimes is a large area of white space traversed by streaks
of black. These streaks indicate a quite different periodic oscillation regime,
which we have christened "multiple-flyback motion." That motion has more
than one slip per period, but instead of being spread through the whole
period, as in the surface sound obtained at low bow force, they appear in
a tight cluster. This regime has been observed on real violin strings and is
recognized by players as an undesirable sound, but it has not been the subject
of any detailed investigation. The diagrams shown in Figure 19.2 are the
first definite indication of any coherent structure in the parameter dependence
of this regime, and they provide encouragement for continuing investigations
along the lines described here.

It is very easy to think of further models and cases about which much
might be learned by applying this technique, and we intent to explore at least
some of them in the near future. Examples of model enhancements include the
following:

- incorporation of flexural stiffness in the string model,
- more correct allowance for torsional motion of the string (a major energy-
  loss mechanism in the bowed string),
- use of measured or predicted data on the reflection behavior of a player's
  finger, and
- inclusion of one or more resonances of the violin body.

This last possibility is of the most obvious interest: knowledge of the influence of instrument-body behavior on playability would immediately open the way to an investigation of how playability is influenced by specific constructional details (such as choice of wood, arching profile, and thickness distribution for the front and back plates of the violin). This would relate to physical measurements that can be made on old instruments and also to the concerns of present-day instrument makers.

19.1. Recent Developments

Since the original publication of this article, in September 1992, a number of developments anticipated in the article have occurred, along with some that were not anticipated.

It was anticipated that commercial applications would soon appear. That has indeed happened, and it is well documented by three articles in *Keyboard Magazine* [4, 7, 8]. The first, in February 1994, announces the forthcoming keyboard synthesizer by Yamaha; the second, in June, is a review of the Yamaha VL1; and the third, in September, is a review of the technical aspects of "physical modeling," as the technique used in this article is known in the commercial world.

The penultimate paragraph in the previous section anticipates some further developments of our own work. Work describing the incorporation of torsional motion and flexural stiffness into the model is described in [12, 13]. In addition, partly stimulated by some experimental work by one of us [11], we have recently focused on using our modeling methods to investigate maximum bow force for acceptable Helmholtz motion as a function of the string model.

The latter requires some explanation. The simulations that allow for torsional motion of the string require values for two parameters, conveniently expressed as ratios: the ratio of the torsional wave impedance to the transverse wave impedance (the reciprocal of the wave admittance Y_0 in (19.1) and (19.2)) and the ratio of the torsional wave velocity to the transverse wave velocity. However, these ratios are not independent for a given string, but are proportional, with a proportionality constant that is essentially the second moment of the radial mass distribution of the string. (Most modern stringed instrument strings are wrapped with one or more layers of metal, usually aluminum or silver, on a core that can itself be metal, the traditional sheep gut, or a thin bundle of very fine strands of nylon or similar plastic thread. The exception is that most violin E strings—the highest pitch string—are solid steel and hence are homogeneous.) That allows us to define an axis for a two-dimensional representation, for which the variable is the velocity ratio, with the constraint that the wave impedance ratio is constrained by a particular radial mass distribution. That this is useful not just for the class of homogeneous strings is shown in an illustration of a review article [13]. The maximum bow force work is intended to classify regimes in which the breakdown of Helmholtz motion is either from a transition to a different periodic regime or to an aperiodic regime, depending on where on the string the bow is placed.

This investigation is still in progress at this writing.

An unanticipated development was stimulated by a review [9] of a concert by violinist Mari Kimura in New York, in which she demonstrated the musical usefulness of a class of tones that have been dubbed anomalously low frequencies (ALF). These periodic oscillations of the bowed string at frequencies well below the usual fundamental frequency of the plucked string were in the process of being investigated experimentally by Hanson, Schneider, and Halgedahl and, theoretically, using computer physical modeling methods similar to ours, but on a PC-type computer, by Knut Guettler. Their work was published in companion papers a few months after the concert [3, 2].

Finally, it has to be admitted that although the original stimulation for our method of systematically classifying and exploring the very large parameter space of the act of bowing a string on a violin was ready access to a CM-2 massively parallel processor machine, the current work is done on a "von Neumann"-type machine—a single processor of the DEC Alpha cluster at the Pittsburgh Supercomputer Center (PSC). The change of machine was forced by the removal of the CM-2 from the PSC, but we have made a virtue of necessity by considering cases that the CM-2 architecture would not allow. The difficulty with the CM-2, as a SIMD machine, is that axes in a parameter plane that require changes in the iteration loop limits are not allowed. Among such axes is the bowing position: where on the string relative to the bridge is the bow placed? The DEC Alpha processor has allowed more flexibility in programming, at the expense of, typically, about a factor of six in CPU time for comparable jobs.

References

[1] L. CREMER, *The Physics of the Violin*, MIT Press, Cambridge, MA, 1983, Figure 2.5, p. 27.

[2] K. GUETTLER, *Wave analysis of a string bowed to anomalous low frequencies*, Catgut Acoustical Society Journal (Series II), 2(1994), pp. 8–14.

[3] R.J. HANSON, A.J. SCHNEIDER, AND F.W. HALGEDAHL, *Anomalous low pitched tones*, Catgut Acoustical Society Journal (Series II), 2(1994), pp. 1–7.

[4] M. MARANS, *The Next Big Thing*, Keyboard Magazine, February 1994, p. 100.

[5] M.E. MCINTYRE AND J. WOODHOUSE, *On the fundamentals of bowed string dynamics*, Acustica, 43(1979), pp. 93–108.

[6] M.E. MCINTYRE, R.T. SCHUMACHER, AND J. WOODHOUSE, *On the oscillations of musical instruments*, Journal Acoustical Society of America, 74(1983), pp. 1325–1345.

[7] E. RIDEOUT, *Yamaha VL1 virtual acoustic synthesizer. Physical modeling synthesizer*, Keyboard Magazine, June 1994, p. 104.

[8] C. ROADS, *Physical modeling: The history of digital simulations of acoustic instruments*, Keyboard Magazine, September 1994, p. 89.

[9] E. ROTHSTEIN, *A violinist tests limits in the music of her time*, The New York Times, April 21, 1994, p. B3.

[10] J.C. SCHELLENG, *The bowed string and the player*, Journal Acoustical Society of America, 53(1973), pp. 26–41.

[11] R.T. SCHUMACHER, *Measurements of some parameters of bowing*, Journal Acoustical Society of America, 96(1994), pp. 1985–1998.

[12] R.T. SCHUMACHER AND J. WOODHOUSE, *The transient behavior of bowed string motion*, Chaos, 5(1995), pp. 509–523.

[13] R.T. SCHUMACHER AND J. WOODHOUSE, *Computer modeling of violin playing*, Contemp. Phys., 36(1995), pp. 79–92.

Chapter 20

Ray Tracing with Network Linda

Rob Bjornson
Craig Kolb
Andrew H. Sherman

Editorial preface

Ray tracing, which is part of the overall field of visualization, is a computationally demanding task and thus a logical candidate for parallelization. A network of workstations is employed concurrently to solve the ray tracing problem. Using the programming language Linda, which offers an abstract view of parallelism different from normal message-passing, speedups of 30 on 40 workstations are achieved.

This article originally appeared in *SIAM News*, Vol. 24, No. 1, January 1991. It appears in its original form.

The potential of high-performance network computing has long been recognized. Recently, we used the Network Linda System[13] to achieve better than a 30-fold speedup on a network of 40 workstations at Yale University running a compute-intensive image rendering computation. The application was a parallelized version of the fractal image rendering program, Rayshade, which was developed by Craig Kolb while a student at Princeton University and a research assistant to Benoit B. Mandelbrot at Yale University. Rayshade is an example of the type of compute-intensive program now in everyday use to apply ray tracing algorithms to the rendering of color images.

The version of Rayshade discussed here was parallelized using the Network Linda System to exploit the substantial number of idle cycles available on a large network of Sun SPARCstations in the Department of Computer Science at Yale. It is obvious, of course, that most workstations will be idle at night or

---

[13] For additional information about the Network Linda System or other commercial Linda products, contact Scientific Computing Associates, Inc. at 246 Church Street, Suite 307, New Haven, CT 06510, (203)-777-7442, E-mail: `linda@sca.com`.

on weekends or holidays. However, even during normal working hours, most workstations in such networks are only lightly used, since common tasks such as editing, reading mail, and debugging programs generally consume only a small fraction of each workstation's available CPU power. In modern RISC-based workstations such as those at Yale, each CPU is capable of completing several million operations per second, so the idle cycles in a moderately large network can amount to a substantial computing resource in the aggregate. The Network Linda System provides the tools required to harness these idle cycles for useful work.

The Network Linda System is the network implementation of the Linda parallel programming language [1, 2]. Other versions of Linda are available for more traditional parallel computers, including Intel hypercubes and shared-memory computers from Silicon Graphics, Encore, and Sequent. More properly, Linda is a *coordination* language, comprised of a few simple statements that can be added to any base language to produce a parallel dialect of that language. Linda leaves to the base language the chores for which such languages were designed—I/O, arithmetic, loop control, procedure calling, etc.—and concerns itself solely with interprocess communication and control. Programmers continue to use their own familiar idioms for the bulk of their coding.

Linda operates on tuples and tuple spaces. A *tuple* is the fundamental Linda data object, consisting of an ordered list of typed values (called *actuals*) and typed wild-card placeholders (called *formals*). Tuples exist in *tuple space*, an unordered, logically shared, associative tuple memory.

Linda adds four new operations to the base language. Out produces a tuple and dumps it into tuple space. An in or rd operation is used to retrieve tuple data from tuple space. Either one specifies a template that is used to search tuple space for matching tuples. A tuple and template match provided that both have the same number of fields, the corresponding fields have the same type, and corresponding actual values are equal. A placeholder matches any value of the correct type and, as a side effect, is bound to the corresponding actual value in the matched tuple. The difference between the in and rd operations is that in removes the tuple from tuple space (guaranteeing unique access), while rd leaves it in tuple space for other processes to use. If in or rd fails to find a matching tuple, the invoking process blocks until an appropriate tuple is available. Finally, eval, which is similar to out, provides a method of process creation in Linda by dumping an unevaluated tuple into tuple space and creating new processes to evaluate each field of the tuple.

Although Linda may bear a passing resemblance to message-passing primitives commonly used on multiprocessors, it is fundamentally more powerful. message-passing requires that the sender know the identity of the recipient; data does not exist independently of computing processes. In contrast, data produced by a Linda program, in the form of tuples, is completely autonomous. Processes can create data without knowing or caring what process will eventually use it, or when this use will occur.

This persistence of tuples, independent of the processes that create them, substantially decouples processes from one another, making Linda programs much easier to write and understand than their message-passing counterparts.

The Network Linda System includes the C-Linda precompiler, analyzer, and runtime library. In outline, the precompiler assembles information about all Linda operations in the source and generates pure C code that is compiled by the native C compiler. At link time, the analyzer divides all the Linda operations into classes according to the style of tuple space access and maps them onto appropriate calls to the runtime library. Linda's in and rd operations support general associative lookup, but the goal at runtime is to use a tuple space access routine that performs efficiently by restricting the generality to that actually needed. Accordingly, each basic Linda operation is represented not by a single library routine, but by a family of routines; the system implements each Linda operation in the user's source by selecting the most appropriate member of the family.

The Network Linda System runs on networks of UNIX workstations, including HP-Apollo, IBM, Silicon Graphics, and Sun computers, among others. The user logs onto one node of the network, compiles the Linda program, and invokes TSNET (the Network Linda control program) to run the executable. The number of processors on which to run the application is a command-line input to TSNET, which selects the least-loaded workstations from a list of permissible candidates and starts the code as a separate process on each one. The individual processes compute independently, periodically coordinating through tuple space. The responsibility for managing tuple space is divided among the workstations to even out the Linda system overhead across the network and to prevent any single machine from becoming a bottleneck. Network Linda uses a reliable protocol built on UDP (Internet's User Datagram Protocol). Messages are point-to-point; no expensive broadcasts are necessary to transmit tuples from node to node. After the computation is complete, TSNET removes the processes and executables from the participating workstations.

We present performance data from running the Linda version of Rayshade on up to 40 Sun SPARCstation 1 workstations connected on a standard Ethernet network. Rayshade contains approximately 10,000 lines of C, lex, and YACC code (YACC and lex are standard UNIX tools for writing parsers and lexers, respectively) and is designed to run effectively across a wide variety of computing platforms [3]. It uses ray tracing to render color images composed of a number of primitive objects, currently including discs, planes, polygons, spheres, boxes, cylinders, cones, tori, and height fields. Arbitrary linear transformations, including translation, scaling, rotation, and skewing, can be applied to these primitives. Through the use of surfaces and textures, it is possible to make objects appear to be made out of plastic, wood, glass, or a wide variety of other materials.

Object appearance in Rayshade is also controlled by the number, type, color, and position of light sources throughout the scene. Supported light sources types include point, extended, directional (infinite), quadrilateral, and spotlights. Additional features include atmospheric effects (e.g., fog or mist), depth of field adjustment, support for stereo rendering, and options for control of overall image quality.

Ray tracing is an extremely compute-intensive floating-point computation. Rendering an image requires finding, for each sample of the image plane, the primitive closest to the viewer's eye. Determining the closest primitive requires that every object in the scene be tested, in one way or another, for intersection with the ray. While there are a variety of techniques for reducing the computational complexity of ray tracing (through, for example, partitioning of the model space into discrete regions), there are nevertheless typically dozens of ray/object intersection tests to perform per ray. On top of this, it is necessary to compute lighting functions, to apply complex procedural textures, and to trace additional rays from the surface of the object to each light source in order to determine shadowing. As a result, extremely complex images may require days of computation on the most powerful workstations.

FIG. 20.1. *Rayshade-rendered scene.*

Rayshade was parallelized using a standard Linda paradigm, the master–worker model. The master process runs on the user's workstation. Using eval, the master starts one worker process for each of the other participating workstations. It then uses out to create a task tuple for each computational task (rendering one scan line of the final image). Each worker looks for task tuples and grabs one with `in`. After finishing the computation required by the task, the worker uses `out` to dump a tuple containing the completed scan line into tuple space and looks for more task tuples. Meanwhile, the master collects the output scan lines and saves them in a file. When the master has collected all the scan lines, it causes all the workers to terminate by creating a special "poison-pill" task tuple.

Figure 20.1 shows a scene rendered by the parallel Rayshade program on the network at Yale. The image shown is a fractal "sphereflake," consisting of a mirrored ball, floating in space, surrounded by other, smaller mirrored balls, themselves surrounded by still smaller balls, ad infinitum.[14] The mirroring causes many rays to be reflected during the rendering process, perhaps several times. The image resolution is 512×512 pixels, and it was rendered using nine samples per pixel. On a single Sun SPARCstation 1, the computation requires almost four hours.

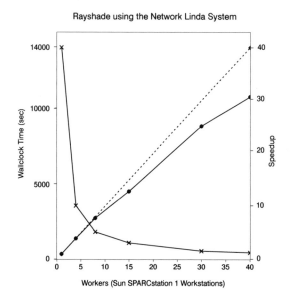

FIG. 20.2. *Rayshade using the Network Linda System: rendering times for different numbers of workstations.*

Figure 20.2 shows the speedup realized for various numbers of workers; the 45-degree line represents perfect speedup based on the number of workers. With 40 workstations, the rendering time is reduced to just over seven minutes (433 seconds). The master is excluded from the machine count in the plot, since for this application the demands on it are quite modest. The times

---

[14]We are indebted to Eric Haines from 3D/Eye of Ithaca, New York for providing the necessary data, which is part of his Standard Procedural Database.

shown are wallclock times and include all time involved in initialization and shutdown of Network Linda (which amounts to roughly 60 seconds, or 14% of the 40-processor time, for this image) and demonstrate the high-performance computing capability latent in large networks of workstations. For comparison, the same code run on a single processor of a Cray-2 (with no effort at vectorization) required over 66 minutes (3972 seconds) of CPU time.

A wide variety of other Linda applications have been developed, including genetic sequence comparison, erosion simulation, LU factorization, electromagnetic calculation, two-dimensional fast Fourier transform computation, parameter sensitivity analysis, and data fusion. Applications written in Linda are portable to any of a wide range of machines that support Linda, including most shared- and distributed-memory multiprocessors, as well as networks. Experience to date indicates that applications parallelized using Linda usually run essentially as efficiently as versions coded with nonportable native parallel programming constructions (such as explicitly shared memory or explicit messages), even when the Linda versions are transported without change from one type of parallel processor to another.

References

[1] N. CARRIERO AND D. GELERNTER, *Linda in Context*, Communications of the ACM, 32(1989), pp. 444–458.

[2] N. CARRIERO AND D. GELERNTER, *How to Write Parallel Programs: A First Course*, MIT Press, Cambridge, MA, 1990.

[3] Rayshade is available freely via anonymous ftp from `weedeater.math.yale.edu` (130.132.23.17) in `pub/rayshade.3.0`. For more information, contact Craig Kolb (`craig@weedeater.math.yale.edu`).

Chapter 21

A SIMD Algorithm for Intersecting Three-Dimensional Polyhedra

David Strip
Michael Karasick

Editorial preface

We seldom make the connection between high-performance computing and geometric modeling. However, geometric modeling can be a rather computationally expensive task and thus is a natural candidate for parallel machines. The intersections of polyhedra on a parallel architecture are determined, and using a unique combination of data structure and careful attention to the host architecture, the authors describe an implementation of geometric modeling on a single-instruction multiple-data (SIMD) machine.

This article originally appeared in *SIAM News*, Vol. 27, No. 3, March 1994. It was updated during the summer/fall of 1995.

Solids are most often represented using boundary representations, which describe a solid by enumerating the zero-, one-, and two-dimensional boundary sets. Descriptions of solid representations, systems, and algorithms are found in [6, 7]. Set-operation algorithms for solids are computationally intensive because they need to deal with many special cases. In addition, the computation time is a function of the output size, and interesting models tend to be large.

Theoretical results have addressed parallel algorithms for some of the simpler geometric problems in solid modeling (see, for example, [1]). Singular configurations are frequently ignored in these treatments, although they must be addressed in practical applications. (In a singular configuration, two solids intersect in such a way that small perturbations in location change the topology of the boundary of their intersection.)

Special-purpose, multiple-instruction multiple-data (MIMD) computers have used boundary representations to compute the intersection of polyhedra, with coarse-grain load balancing [4, 5]. In this article we describe a polyhedral representation for SIMD architectures, sketch an intersection algorithm, and show how some of the singular configurations are handled. We also present performance results comparing our Connection Machine implementation with a serial solid modeler.

21.1. Defining and Representing Solids

A boundary representation is an organized enumeration of the point sets on the surface of a solid. Face f of solid S is a regular two-dimensional set on the surface of S. The interior of S, $Interior(S)$, is below f. More precisely, if f is contained in plane $P(x, y, z) = ax + by + cz + d = 0$, then $Below$ is the half-space defined by $\{(x, y, z) : P(x, y, z) \leq 0\}$. For every neighborhood, $Nbhd$, of every point of f, $Nbhd \cap Below \cap Interior(S)$ is nonempty. Plane P is the oriented plane of f, and vector $N = (a, b, c)$ is the (outward) normal to f. If the plane Q is coincident with but oriented opposite to P, then the face contained in Q is distinct from f. The edges of a solid are the line segments defined by the intersection of two or more faces, and the vertices of a solid are the points defined by the intersection of three or more faces.

The Star-Edge data structure [3] is a typical serial boundary representation. An edge e with initial vertex u and terminal vertex v can bound many faces. Directed edges describe these incidences. If N is the outward normal of f, then d-edge (rhymes with hedge) e_f^+ exists if $(v - u) \times N$ points from e into the interior of f, and e_f^- exists if $(u - v) \times N$ points into the interior of f. D-edge e_f^+ has $InitialVertex$ u and $TerminalVertex$ v. D-edge e_f^- has $InitialVertex$ v and $TerminalVertex$ u. The $Tangent$ vector of a d-edge is $(TerminalVertex - InitialVertex)$. The $FaceDirection$ vector of a d-edge is $Tangent \times N$. A d-edge with unspecified orientation is denoted by $e_f^\bullet$; the subscript is omitted when the face is unspecified. Face f and edge e are referred to as the underlying face and edge of $e_f^\bullet$, respectively.

The d-edges of an edge e are ordered by radially ordering their face direction vectors on a plane perpendicular to e (see Figure 21.1a). The interior of the solid will lie between e_f^+ (face f at e) and its radial successor, e_g^- (face g at e). This d-edge pair is called a volume-enclosing pair. Similarly, the d-edges incident to a vertex and contained in a face f are ordered by radially ordering their tangents in the plane of f, and certain adjacent d-edge pairs are area-enclosing (see Figure 21.1b).

SIMD architectures optimize performance on independent, identical computations. This uniformity is attained by having a homogeneous data representation. Rather than describing vertices, edges, and faces with linked data structures, we use a single distributed data structure, called a parallel d-edge, with the following slots:

InitialVertex	TerminalVertex	Edge	Face
Label	Label	Label	Label
Point	Point	Successor	Normal
Successor	Successor	Orientation	Distance

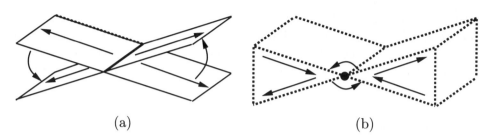

FIG. 21.1. (a) *Radial ordering of faces around edges;* (b) *radial ordering of edges around a vertex.*

For example, if the faces of the solid are labeled $0, \ldots, F$, then *FaceLabel* is simply the number assigned to the face of this d-edge. In the previous section we said that d-edges are radially ordered around vertices and edges. If the d-edges of a solid are labeled $0, \ldots, D$ and $e^{\bullet}$ is a d-edge, then *EdgeSuccessor* is the label of the radially counterclockwise d-edge around its edge e, and *Orientation* is a bit that is true if and only if the d-edge is oriented with e. *InitialVertexSuccessor* and *TerminalVertexSuccessor* are the labels of the d-edges of f that are radially counterclockwise around the initial and terminal vertices of the d-edge, respectively. The equation of the plane P of f is represented by *FaceNormal*, the unit outward normal, and *FaceDistance*, the signed distance from the origin to P. Each d-edge of a solid is represented as a parallel d-edge, one per processor, with the label of a d-edge corresponding to the name of the processor containing the d-edge. For example, 24 parallel d-edges (and hence 24 processors) are used to describe a cube, which is defined by 12 edges and 24 d-edges.

This data structure provides a natural assignment of work among processors for certain tasks. For example, solid modeling computations often require computing information, such as d-edge tangents, face-direction vectors, and so on. Parallel d-edges allow such quantities to be computed directly:

$$Tangent \equiv e^{\bullet}.TerminalVertex.Point - e^{\bullet}.InitialVertex.Point.$$

The calculation for the serial representation, by contrast, uses a total of four instances of three different data structures:

$e = e^{\bullet}.Edge$
if $e^{\bullet}.Orientation$ **then**
    $Tangent \equiv e.TerminalVertex.Point - e.InitialVertex.Point$
else
    $Tangent \equiv e.InitialVertex.Point - e.TerminalVertex.Point$

Parallel solid modeling algorithms and serial algorithms manipulate boundary representations in different ways. Rather than iterations through several data structures, computation on a massively parallel processor is done by selecting a set of processors for a calculation and then performing the calculation on that set of *active* processors. For the example of the d-edges of a face, selection is accomplished by activating those processors whose *FaceLabel* matches the given face. Associative selection is an important property of parallel d-edges.

21.2. Labeling Boundary Elements for Efficient Computation

The efficiency of algorithms that use the representation is determined by the particular assignment of parallel d-edges to processors. For example, the processors can be viewed as a spatial mesh—each processor represents a volume, and each vertex is assigned to the processor for the containing volume. This assignment works well only if the spatial distribution of a solid's vertices is uniform.

Alternatively, because each boundary element is labeled by an integer, we can simply assign the vertex, edge, and face labeled i to processor i. This assignment works well if processor i does not need to operate on more than one of its boundary elements simultaneously. For example, if we wish to enumerate the boundary elements containing point p, this assignment works well if at most one of the boundary elements labeled i contains p.

Each parallel d-edge contains the coordinates of two vertices and the plane of a face. Unless the corresponding edge labeled i is described by the parallel d-edge assigned to processor i, another data structure on processor i is required. (Similarly, the vertex labeled j and the face labeled k should be described by the parallel d-edges assigned to processors j and k, respectively.) Given an assignment of d-edges to processors, we need to label vertices, edges, and faces in such a way that additional data structures are not needed. Furthermore, we would like to compute this boundary element labeling efficiently. The intersection algorithm that motivates the parallel d-edge data structure requires that a boundary element labeling has the following properties:

> **Uniqueness Property:** No two distinct boundary elements of the same type have the same label. (This is the minimal requirement for a labeling.)
>
> **Incidence Property:** Each boundary element is labeled by an incident d-edge.
>
> **Intersection Property:** Given A, a labeled solid, and Ω, a point, line, or plane, no two distinct boundary elements of A with the same label have point intersections with Ω. (For example, a plane cannot transversely intersect an edge and a vertex with the same label.)

The importance of the uniqueness property is self-evident. The incidence property ensures that the data describing a boundary element labeled i are present in the parallel d-edge contained in processor i. The intersection property allows significant reduction in the complexity and communication costs of parallel solid modeling procedures.

Consider, for example, the problem of labeling the vertices of a solid C formed by intersecting solid A with solid B. Some of the vertices of C are vertices of A or B. The others arise from boundary intersections. Suppose we need to label a vertex that is the point intersection of an edge of A with an edge of B. If the edge of A has m adjacent faces and the edge of B has n, then at least $n + m$ processors contain parallel d-edges incident to the point. Each of these processors can label its copy of the point identically, using the label (α, β), where α is the label of the edge of A and β is the label of the edge of B. Any point intersection of a boundary element of A with a boundary element of B can be labeled analogously with the labels of the intersecting boundary elements. This computation can be done efficiently, with no interprocessor communication. The incidence property of the labeling guarantees that any other object of A labeled α is incident to edge α. A case analysis using the intersection property shows that only one point common to both solids is labeled (α, β).

There is a simple algorithm for computing the labeling from an assignment of parallel d-edges to processors: label each parallel d-edge by the processor to which it is assigned; label face f with the largest label of a d-edge of f; label edge e with the largest label of a d-edge of e; and label vertex v with the largest label of a d-edge with $TerminalVertex$ v ($InitialVertex$ could also be chosen). Figure 21.2 contains an example. The uniqueness, incidence, and intersection properties follow from this algorithm.

InitialVertex.Label	15
InitialVertex.Point	$(1, 0, 0)$
InitialVertex.Successor	12
TerminalVertex.Label	9
TerminalVertex.Successor	8
TerminalVertex.Point	$(0, 0, 0)$
Edge.Label	15
Edge.RadialSuccessor	15
Orientation	false
Face.Label	12
Face.Normal	$(0, 0, 1)$
Face.Distance	0
SolidLabel	6

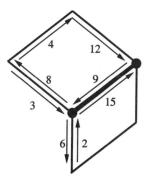

FIG. 21.2. *Labels for boundary elements referenced in d-edge* 9.

21.3. Communication Protocols Used by the Intersection Algorithm

The intersection algorithm uses a variety of primitive communication protocols that are supported by the Connection Machine architecture. The simplest is *nearest-neighbor* communication, in which a processor sends (receives) a message to (from) one of its nearest neighbors along an arc of the hypercube. We treat the hypercube as a linear array of processors and can thus describe communications using shorthand like "send data to the right (or left) processor," denoted $a' := a[\pm 1]$.

Although most communication is nearest-neighbor, there is some communication between arbitrary processors in the machine, most often just a permutation of data among (a subset) of the processors. The algorithm also uses a small number of *one-to-many broadcasts*, where each of n processors sends data to m (nonoverlapping) processors (such communications are very efficiently handled on the Connection Machine). Finally, we use a very few *many-to-one reductions*, in which, say, mn processors send their data to one of n processors, and results at each of the n processors are combined logically or arithmetically. Details can be found in [2].

In addition to primitive communications, the intersection algorithm uses the (primitive) *rank* function, denoted $r := rank(a)$, which assigns to each processor's *parallel variable (pvar)* r its order in a sort of the values of pvar a. (A parallel variable p is the logical aggregation of the values of the variable p on each processor.) Finally, we use a *scan*, denoted $a' := scan^{\pm}(a, t, \odot)$, operation, which scans the value of a into the a' of neighboring processors, applying operation $\odot$ cumulatively and terminating when one of these processors has (boolean) t set true ($\odot$ defaults to copy a).

21.4. Relocating Data to Reduce Communication Costs

For the exchange of a sequence of general messages, it is beneficial to reassign the data to be exchanged to physically adjacent processors so that nearest-neighbor protocols can be used. For example, consider the radial ordering of faces around an edge described with the *Edge.RadialSuccessor* field of a parallel d-edge. Suppose that each processor p containing d-edge $e_f^{\bullet}$ needs to communicate with processor q containing processor $e_g^{\bullet} = e_f^{\bullet}.Edge.RadialSuccessor$. Even though p and q are not physically adjacent, two ranks and a single permutation-send can be used to relocate the interchange data, D, to processors that are adjacent:

(1) Construct a local coordinate system with z-axis $Tangent(Edge.Label)$.

(2) Compute the pseudo-angle ϕ from $+x$ to $FaceDirection \equiv Tangent \times Face.Normal$. For unit vectors $\vec{x}$, $\vec{y}$, and $\vec{v}$, $pseudoAngle(\vec{v}) \equiv sign(\vec{v} \cdot \vec{y})(1 + \vec{v} \cdot \vec{x})$.

(3) Construct the (integer) sort key given by $K \equiv (Edge.Label, rank(\phi))$. (The sort key has $2d$ bits, where d is the dimension of the hypercube.)

(4) Each processor sends D and $Edge.Label$ to the processor $rank(K)$.

The radial orderings are now linearized, so that D can be exchanged using nearest-neighbor communication.

As an example of the use of this reassignment technique, suppose that each processor wishes to classify a pvar-vector $\vec{v}$ that intersects d-edge $e^\bullet$. Each processor classifies $\vec{v}$ as follows: $\vec{v}$ is parallel to $Tangent(e^\bullet)$ and oriented with or opposite to $Tangent(e^\bullet)$; $\vec{v}$ points from $e^\bullet$ into the interior of face $e^\bullet.Face.Label$, or into the interior (alt. exterior) of the solid between faces $e^\bullet.Face.Label$ and $(e^\bullet.Edge.Successor).Face.Label$. These first two cases are easily determined by each processor. The remaining cases are distinguished by the above assignment procedure, for which the relocated data are

$$(pseudoAngle(\vec{v}), Orientation).$$

21.5. Sketch of the Intersection Algorithm

Figure 21.3 shows the intersection of two solids and the result of the intersection. The edges of the new solid can be divided into three groups. Figure 21.3a shows the edges that lie on the boundaries of both input solids (i.e., on the "intersection curve" of the two input solids); Figure 21.3b shows the edges that meet the intersection curve at one or two points; and Figure 21.3c shows the edges that lie on the boundary of one of the input solids and are completely contained in the interior of the other.

The three phases of the algorithm correspond to these three classes of edges. For each phase there are numerous subcases, depending on how the boundaries of the input solids meet. The challenge is to arrange the data and computations efficiently on a SIMD architecture. (See [2, 8] for details.)

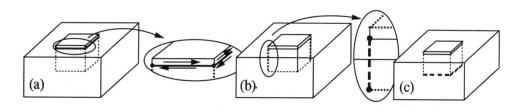

FIG. 21.3.   (a) *Phase* I: *Construction of the d-edges of the intersection curve.* (b) *Phase* II: *Construction of d-edges incident to the intersection curve. D-edges that are incident to the intersection curve are constructed by testing to see whether either segment of the d-edge incident to the intersection point lies in the interior of the other solid.* (c) *Phase* III: *D-edges of one solid that are contained in the interior of the other are identified by ray-casting and added to the intersection-solid.*

21.5.1. Preprocessing. Before starting to identify the d-edges of the output solid, we carry out a preprocessing step that calculates how the d-edges of each solid intersect the planes of the faces of the other. Our algorithm does this by assigning each d-edge/face pair to a (virtual[15]) processor and then having that processor calculate the intersection of the d-edge with the face. This step can be viewed as the creation of two matrices of processors. Each matrix is indexed by d-edges of one solid ("rows") and faces of the other solid ("columns"); two one-to-many broadcasts, two scans, and a permutation are used to construct the matrices. A d-edge can intersect or be parallel to the plane of the face, or a projection (ray) of the d-edge can intersect the plane of the face. Eventually, we will be interested in whether the d-edges that intersect the plane actually intersect the interior of the face or the boundary of the face, or do not intersect the face at all.

We construct axis-aligned bounding boxes for each solid and use these boxes to eliminate intersections that are outside the bounding boxes. We classify the intersections of the d-edges with the face planes and then create a sparse-matrix representation by reassigning the d-edge/face pairs of interest. Typically, the remaining set is much smaller than the complete d-edge/face matrix. It is also more sensitive to output size than to input size.

21.5.2. Phase I—Constructing D-Edges of the Intersection Curve. The first group of d-edges of the new solid that we identify are those that lie on the boundaries of both input solids (see Figure 21.3a). These d-edges are formed by the intersecting faces of the two input solids. Recall that each processor is associated with a d-edge/face pair. If f is a face of solid A and g a face of B, an intersection of d-edge $e_f^{\bullet}$ with the plane of g is denoted as $e_f^{\bullet}/g$. The intersection points $e_f^{\bullet}/g$ and $e_g^{\bullet}/f$ lie along the line defined by the intersection of the planes of f and g. We reassign these linearly ranked intersection points to physically adjacent processors. Using scans, we identify those regions of the intersection line that are on both faces, and we then create the new d-edges that lie on the intersection curve. Many cases must be analyzed to determine the d-edges that are built, which depends on whether the faces intersect in their interiors or on their boundaries. While this step is conceptually similar to that used in a serial algorithm, there are major implementational differences, especially in the use of the ordering of points along the intersection line to make the computation efficient.

When the first phase of the algorithm is complete, we have identified precisely the d-edges that intersect the faces (as opposed to the planes) of the other solid. We can then further refine the number of intersection points that we carry around. This in turn reduces the (virtual) processor requirement

---

[15]Throughout the algorithm we assume that the machine has enough processors. Many processors, i.e., millions, are often required. We assume that the SIMD machine provides a means for simulating more processors than are physically available. In the remainder of the paper we drop the modifier "virtual."

to a number more tightly bounded by output size and hence reduces the time required for future computations.

21.5.3. Phase II—Constructing D-Edges Incident to the Intersection Curve.
The second group of d-edges to be identified for the new solid are those that are incident to the intersection curve at one or two points (see Figure 21.3b). These output d-edges are segments of d-edges of one of the input solids that intersect boundary elements of the other. In Phase I of the intersection algorithm, we determined which of the d-edge/plane intersections actually intersect the solid boundary. What remains to be determined is which portions of the d-edge are in the interior of the other solid, and which are not.

The most difficult case, which occurs when a d-edge intersects a vertex, is refined to an equivalent d-edge/edge or d-edge/face test. The d-edge/face incidence test is simple (just a dot product). (The d-edge/edge test was described in the section on the relocation of data. While conceptually similar to algorithms used at this stage on serial machines, the parallel implementations of the d-edge/edge and d-edge/vertex incidence tests are necessarily quite different.)

21.5.4. Phase III—Constructing D-Edges Interior to an Input Solid.
The final set of d-edges we identify for the new solid are those d-edges of one input solid contained in the interior of the other, as shown in Figure 21.3c. These d-edges, which are built in a manner completely unlike that typically used on serial machines, can be divided into two groups: those contained and those not contained in lines that intersect the other solid. For the first group, we use the results of the incidence tests computed in Phase II to determine whether the d-edge is interior or exterior to the solid. For the second group, it is sufficient to know whether the other input solid is unbounded.

This ray-casting approach contrasts with the methods typically used in serial algorithms. Since the incidence tests of Phase II of the algorithm are comparatively expensive, they are not, in general, applied to rays. In a serial algorithm, a transitive closure is used instead for Phase III. It is possible that isolated regions of the input solid boundaries cannot be classified by transitive closure. In that case, the serial algorithm must use a ray-casting procedure to classify some point on that isolated portion of boundary. Another transitive closure step is then needed to classify the rest of that portion. Transitive closure is very inefficient on a parallel machine—its execution time is proportional to the distance (in a graph sense) of a d-edge from the closest d-edge intersecting the boundary.

21.6. Performance Measurements
We implemented the parallel solid modeling algorithm described in this paper on a Thinking Machine CM-2. (Since these tests were made (in 1990), a faster model of the CM-2 has appeared.) In order to evaluate the performance of

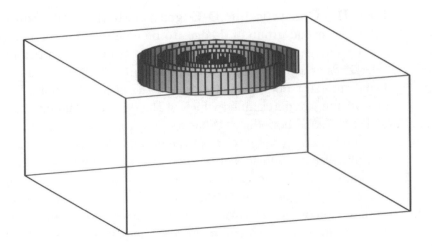

FIG. 21.4. *Toolpath test problem—block-spiral.*

our algorithms, we compared our modeler with GDP, a production modeler developed internally at IBM Research. GDP is executed on an IBM 3090 600S. (This is the only supercomputer for which we have been able to identify a general-purpose polyhedral solid modeler.)

This comparison is difficult, as the CM-2 has a large number of very simple (bit-serial) interconnected processors, and the IBM 3090 is a classic "big-iron" supercomputer that achieves its speed by such means as very high speed components, pipelines, and vector processors. The two intersection algorithms are also quite different. With the SIMD algorithm, the number of virtual processors required can grow asymptotically with the product of the number of faces in the two input solids. Runtime is in turn a function of this number, except that the code will skip (large and time-consuming) portions if there are no intersections of certain types, such as edge–vertex intersections. We have generated two examples at roughly opposite ends of the range.

In the first example (Figure 21.4), all the faces of one solid intersect a single face of the other. Thus, "complexity" grows linearly with the problem size. In addition, all the intersections are in "general position"; i.e., they are all edge–face intersections and, therefore, easy to analyze. Figure 21.5a compares the SIMD algorithm with that of GDP, showing speedups of up to 35. When the performance of GDP is extrapolated to larger problems, speedups of 50 or better are predicted. For problems that fully occupy the machine, the results as well as data for 32K processors (which are not shown here) show an almost perfectly linear speedup in the number of processors.

The second example balances the sizes of the two solids, so that the number of face–face pairs grows like n^2. In addition, all the intersections are edge–edge intersections, which require execution of many of the time-consuming procedures. This pathological example case takes a faceted approximation to a cylinder, duplicates it, rotates the duplicate by half a facet, and then intersects the two solids, obtaining a faceted approximation to a cylinder with twice the number of facets. As the number of singular intersection points increases, we expect the CM-2 intersection time to degrade relative to GDP because of increased interprocessor communication. As shown in Figure 21.5b, the CM-2 intersection algorithm is still faster, but the disparity is smaller. The discontinuities are induced by the virtual processor requirement, which is a step function. (The spike in the Connection Machine data is the result of a known router utilization bug for $n = 512$ facets.)

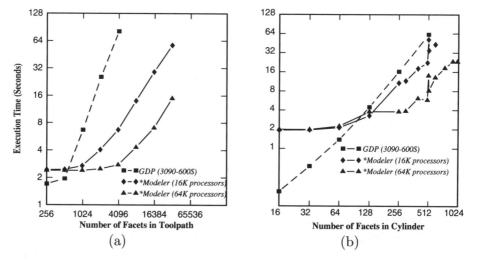

FIG. 21.5.   (a) "Milling" a path on a cube with an n-faceted approximation to a spiral. (b) Making a 2n-faceted approximation to a cylinder by intersecting two n-faceted cylinders.

21.7. Conclusions

We believe the work described in this article has produced three main contributions. The first is the parallel d-edge data structure. We have shown how the rich combinatorial structure of a boundary representation can be embedded in a distributed, uniform structure appropriate to a SIMD processor. Second, although not described here, we have dealt with all singularities that arise in an intersection algorithm. This is difficult even for serial algorithms, and we have efficiently incorporated the treatment of singular cases into a

SIMD algorithm. Third, our avoidance of transitive closure procedures in the intersection algorithm allows us to have an algorithm whose execution time is independent of input solid topology.

Because the best solid modeling algorithms are primarily combinatorial, standard measures like megaflop throughput achieved are not effective. Furthermore, our algorithm is structurally different from serial analogues, substituting a larger, homogeneous data-structure for serial computations. Hence, the only meaningful comparison of performance between these two approaches requires the use of benchmarking.

We have shown in our tests that our parallel intersection outperforms a serial one by a factor of 35 or better. As the models become bigger, the disparity increases. Furthermore, the nature of our intersection algorithm and the architecture of the Connection Machine allow scaling to essentially arbitrarily large arrays of processors with linear scaling in algorithm performance.

References

[1] M.T. GOODRICH, *Intersecting line segments in parallel with an Outt-sensitive number of processors*, in Proc. 1989 ACM Symposium on Parallel Algorithms and Architectures, 1989, Santa Fe, NM, pp. 127–136.

[2] M.S. KARASICK AND D.R. STRIP, *Intersecting solids on a massively parallel processor*, ACM Trans. Graphics, 14(1995), pp. 21–58.

[3] M. KARASICK, *On the Representation and Manipulation of Rigid Solids*, McGill University, Montreal, Quebec. Report 89-976, Dept. of Computer Science, Cornell University, Ithaca, NY, 1988.

[4] Y. NAKASHIMA, H. NIIMI, K. SHIBAYAMA, AND H. HAGIWARA, *A Parallel Processing Technique for Set Operations Using Three-Dimensional Solid Modeling*, Trans. Info. Proc. Soc. Japan, 30(1989), pp. 1298–1308.

[5] C. NARAYANASWAMI, *Parallel Processing For Geometric Applications*, Rensselaer Polytechnic Institute, Troy, NY, 1990.

[6] A. REQUICHA AND H. VOELCKER, *Boolean operations in solid modeling: Boundary evaluation and merging algorithms*, in Proc. IEEE, 73(1985), pp. 30–44.

[7] M. SEGAL AND C. SÉQUIN, *Partitioning polyhedral objects into nonintersecting parts*, IEEE Computer Graphics and Applications, 8(1988), pp. 53–67.

[8] D.R. STRIP AND M.S. KARASICK, *Solid modeling on a massively parallel processor*, Internat. J. Supercomputer Appl., 6(1992), pp. 175–191.

Chapter 22

Numerical Simulation of Laminar Diffusion Flames

Craig C. Douglas
Alexandre Ern
Mitchell D. Smooke

Editorial preface

Combustion problems, which include detailed chemistry, are difficult to solve. Using the convergence of a number of sophisticated algorithms and a parallel architecture, the authors formulate a solution scheme for these types of problems. The parallelism is achieved in a portable way, permitting the same code to run on WAN-based clusters of RS/6000 workstations or the IBM SP1 or SP2 machines. Although space constraints in this article don't permit a more complete coverage of the algorithms, there is sufficient reference to published literature to complete the picture.

This article originally appeared in *SIAM News*, Vol. 27, No. 9, November 1994. It was updated during the summer/fall of 1995.

Not too long ago, anyone wanting to solve large science or engineering problems had to have access to a supercomputer costing millions of dollars. Quite recently, a new breed of relatively inexpensive workstations has become widely available, making it possible to solve such problems on machines individuals can afford to own. The scalar peak speeds of these machines are 30–275 megaflops, with 400 megaflops or more on the horizon (which compares rather favorably with the speeds of vector supercomputers of not so long ago). While these rates have been attained only for simple problems, like dense matrix–matrix multiplication, the rates seen for many other problems are also quite high.

The class of problems described in this article—the numerical simulation of combustion systems—is an example of the large problems being solved on the new workstations. Of course, the problem for the user of a single one of these machines is that the options of either connecting a collection of machines or buying a parallel version of the workstation become more and more tantalizing.

In fact, over the course of two years, we did all of the above. We started on a single machine with a peak rate of 100 megaflops (an IBM RISC System/6000 model 560 computer) and then used a farm of these machines. We then moved to an IBM SP1 and, finally, to an IBM SP2. Due to a nice feature of the communications library we used (MPL), the executables worked on the Ethernet at Yale and on the fast switches in the SP1/SP2s without either recompiling or relinking.

22.1. Problem Formulation

Improvements in computational algorithms and computer capabilities have provided new, extremely powerful tools for investigations of chemically reacting systems that were computationally infeasible only a few years ago. Diffusion flames are one example of such systems. These flames are important in studies of the interaction of heat and mass transfer with chemical reactions in commercial burners, gas turbines, and ram jets. The ability to predict the coupled effects of chemical reactions and complex transport phenomena is critical in modeling turbulent reacting flows, improving engine efficiency, and investigating the processes by which pollutants are formed.

Practical combustion systems require multidimensional studies. In the past, the modeling was done along two independent lines, with either the chemistry or the fluid dynamics effects given priority. For many years, most of the detailed chemistry computational studies therefore involved strictly one-dimensional configurations: freely propagating or burner-stabilized premixed flames and counterflow premixed or diffusion flames. Recently, two-dimensional configurations have become common. Three-dimensional models combining fluid dynamics effects with finite-rate chemistry will appear shortly, thanks in large part to fast RISC-based parallel machines with adequate memory per node (.125–2 gigabytes).

In this article we discuss the formulation and numerical solution of two-dimensional, axisymmetric laminar diffusion flame models, including detailed chemistry, in which a cylindrical fuel stream is surrounded by a coflowing oxidizer jet. We obtain a computationally feasible solution, while being able to study the interaction of the fluid flow and the chemical reactions in the diffusion flames. The full elliptic problem, including up to 50 chemical species in addition to the temperature and the fluid dynamics variables, is treated.

The fuel jet shown in Figure 22.1 discharges into a laminar air stream. The concentric tubes through which the fuel and oxidizer flow have radii R_I and R_O, respectively. The two gases make contact at the outlet of the inner tube. Once the flame forms, it resembles a candle unless it is lifted (as it is in Figure 22.1). Since the solution is axisymmetric, the computational domain is only the upper right quadrant of Figure 22.1. The right and bottom boundaries are the z and r axes, respectively, in Figure 22.1.

Our model consists of the full set of two-dimensional axisymmetric governing equations expressing the conservation of total mass, momentum, energy,

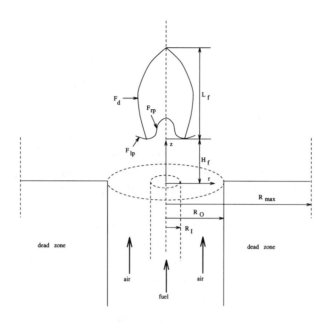

FIG. 22.1. *Physical configuration of the fuel jet (not in scale).*

and species mass in steady-state form. The dependent variables are the axial and radial velocity components, the pressure, the temperature, and as many chemical species as considered in the chemical reaction mechanism (typically 16–50). Boundary conditions are specified for the dependent unknowns at the axis of symmetry ($r = 0$), outer zone ($r = R_{max}$), inlet ($z = 0$), and exit ($z = L$). The formulations for the fluid dynamics processes that have been used to date are stream function–vorticity [8], primitive variables [10], and vorticity–velocity [3]. Each of these formulations has advantages and disadvantages.

In the stream function–vorticity formulation, pressure is eliminated as a dependent variable from the momentum equations, the number of equations to be solved is reduced by one, and continuity is explicitly satisfied locally. Despite these attractions, this formulation presents a severe difficulty in the specification of vorticity boundary conditions. A zero vorticity boundary condition at the inlet of the computational domain results in a rough approximation of the true solution. The specification of vorticity boundary values in terms of the stream function, however, requires the discretization of second-order derivatives, making it necessary to solve severely ill conditioned linear systems. In addition, the stream function–vorticity formulation is not extendible to three-dimensional configurations in a simple form.

The primitive variable formulation allows for accurate boundary conditions and can be used for three-dimensional unsteady problems, but it generally requires a staggered grid arrangement. Staggered grid schemes have drawbacks in complex geometric configurations where nonorthogonal curvilinear coordinates are used.

The vorticity–velocity formulation is a relatively new formulation for reacting flows. It eliminates the pressure variable while replacing the first-order continuity equation with additional second-order equations. Unlike stream function–vorticity, vorticity–velocity is easily extendible to three dimensions and allows more accurate formulation of boundary conditions in a numerically compact way. Off-diagonal convective terms in the linear systems that exert a strong influence in a stream function–vorticity formulation disappear. Another attractive feature of vorticity–velocity is that the governing equations can be discretized on a nonstaggered grid, thus allowing easy implementation of a multigrid algorithm.

Once the fluid dynamics model has been selected, the evaluation of the chemistry and the thermodynamics and transport properties of the mixture needs to be specified. This task is an important and often expensive part of the numerical calculation. A set of databases and general-purpose subroutines are available. Data in the literature are used for the chemical reaction mechanism, which can include up to 50 chemical species along with 100 elementary reactions. For example, a typical chemical reaction is of the form

$$CH_3 + H \rightleftharpoons CH_4$$

with some constants associated with how the reaction is accomplished. The complete reaction table we used, containing 46 reactions, is in [4]. On the other hand, a general theory for fast and accurate multicomponent transport evaluation has become recently available [5]. In the framework of this theory, the transport properties of the mixture are expanded into convergent series. Accurate and cost-effective approximations for the transport properties are then obtained by truncation. For a more mathematical discussion of the iterative algorithms involved, we also refer to [6].

22.2. Some Details

Our present laminar diffusion flame model consists of the full set of two-dimensional, axisymmetric equations expressing the conservation of total mass, momentum, energy, and species mass. Because of the advantages described earlier, we adopt the vorticity–velocity formulation of the Navier–Stokes equations. The vorticity (ω) is defined in terms of the radial and axial components of the velocity vector $v = (v_r, v_z)$ as follows:

$$(22.1) \qquad\qquad \omega = \frac{\partial v_r}{\partial z} - \frac{\partial v_z}{\partial r}.$$

The vorticity transport equation is formed by taking the curl of the momentum equations, which eliminates the partial derivatives of the pressure field. Due to the high temperature gradients present in the flame, no viscosity derivatives in the right-hand side of the vorticity transport equation are neglected. On the other hand, a Laplace-type equation is obtained for each velocity component by taking the gradient of (22.1) and using the continuity equation. A complete description of the governing equations for diffusion flames in vorticity–velocity form can be found in [3, 4]; these equations are simply restated here.

In the notation we use, ρ denotes the mass density of the mixture, μ its shear viscosity, and g the gravitational acceleration. For convenience, we define the components of $\overline{\nabla}\beta$ to be $(\frac{\partial}{\partial z}\beta, -\frac{\partial}{\partial r}\beta)$ for any scalar β. The cylindrical divergence of the velocity vector v is denoted by $\operatorname{div}(v) = \frac{1}{r}\frac{\partial}{\partial r}(rv_r) + \frac{\partial}{\partial z}v_z$. The unknowns are v_r, v_z, ω, the temperature, and the species mass fractions. The governing equations for the diffusion flame can then be written as follows:

Radial Velocity.

$$(22.2) \qquad \frac{\partial^2 v_r}{\partial r^2} + \frac{\partial^2 v_r}{\partial z^2} = \frac{\partial \omega}{\partial z} - \frac{1}{r}\frac{\partial v_r}{\partial r} + \frac{v_r}{r^2} - \frac{\partial}{\partial r}\left(\frac{v\cdot\nabla\rho}{\rho}\right).$$

Axial Velocity.

$$(22.3) \qquad \frac{\partial^2 v_z}{\partial r^2} + \frac{\partial^2 v_z}{\partial z^2} = -\frac{\partial \omega}{\partial r} - \frac{1}{r}\frac{\partial v_r}{\partial z} - \frac{\partial}{\partial z}\left(\frac{v\cdot\nabla\rho}{\rho}\right).$$

Vorticity.

$$\frac{\partial^2 \mu\omega}{\partial r^2} + \frac{\partial^2 \mu\omega}{\partial z^2} + \frac{\partial}{\partial r}\left(\frac{\mu\omega}{r}\right) = \rho v_r\frac{\partial \omega}{\partial r} + \rho v_z\frac{\partial \omega}{\partial z} - \frac{\rho v_r}{r}\omega + \overline{\nabla}\rho\cdot\nabla\frac{v^2}{2} - \overline{\nabla}\rho\cdot g$$

$$+ 2\left(\overline{\nabla}(\operatorname{div}(v))\cdot\nabla\mu - \nabla v_r\cdot\overline{\nabla}\frac{\partial \mu}{\partial r} - \nabla v_z\cdot\overline{\nabla}\frac{\partial \mu}{\partial z}\right).$$

$$(22.4)$$

Equations for the energy and species conservation can be found in any textbook on the area. The density is computed from the ideal gas law as a function of the pressure p, the mean molecular weight of the mixture $\overline{W}$, the universal gas constant R, and the absolute temperature T as follows:

Equation of State.

$$(22.5) \qquad \rho = \frac{p\overline{W}}{RT}.$$

Since we consider only low-Mach-number flames, we can use the constant outlet pressure in (22.5) to compute the density. The pressure field is then eliminated from the governing equations as a dependent unknown. Once a computed numerical solution of (22.2)–(22.4), plus the energy and species equations, is obtained, pressure can be recovered by solving a Laplace-type equation, which is derived by taking the divergence of the momentum equations.

22.3. Solution Methods

The computational method used is a combination of a steady-state and a time-dependent approach. The latter is used to find an approximate solution on a coarse grid, with a flame sheet used for the starting estimate.

A discrete solution of the governing equations is computed on the (tensor product) mesh M_2, whose initial nodes are at the intersection of the lines of the meshes

$$M_r = \{0 = r_0 < r_1 < \cdots < r_i < \cdots < r_{n_r} = R_{max}\}$$

and

$$M_z = \{0 = z_0 < z_1 < \cdots < z_j < \cdots < z_{n_z} = L\}.$$

Local mesh refinement techniques on M_r and M_z are used to refine the mesh M_2. A steady-state solution is attempted on the new tensor product mesh M_2.

22.3.1. Damped Newton Method.

A finite difference procedure is used to approximate the spatial operators in the governing equations. Central and upwind differences are used to approximate diffusion and convection terms, respectively. An approximation to the analytic solution is then found at each node of the mesh. With the difference equations written in residual form, we seek a solution U^* to the system of nonlinear equations

$$(22.6) \qquad\qquad F(U) = 0$$

starting from an initial guess U^0. If this guess is sufficiently close to U^*, then the damped Newton iteration

$$(22.7) \qquad J(U^n)(U^{n+1} - U^n) = -\lambda^n F(U^n), \quad n = 0, 1, \ldots$$

converges to the correct solution. Here, $J(U^n) = \partial F(U^n)/\partial U$ is the Jacobian matrix and λ^n, $0 < \lambda^n \leq 1$, is the nth damping parameter.

The Jacobians are effectively nine-point operators, but the points are dense square blocks equal in size to the number of components in the calculation. Hence, for a flame calculation with 50 components, the Jacobians have up to 450 nonzeros per row. The number of spatial points in the mesh determines whether the Jacobian is still considered a sparse matrix.

Once the Jacobian is formed, the Newton equations (22.7) are solved by either a preconditioned (Gauss–Seidel) BiCGSTAB or a generalized minimum residual (GMRES) procedure. Rather than working with dimensionless variables, we introduce a scale factor for each dependent variable. The scale factors are chosen so that the Newton corrections $U^{n+1} - U^n$ are similar for variables of equal importance [4]. The Newton iteration continues until the scaled norm of the discrete vector $U^{n+1} - U^n$ is reduced appropriately. An appropriate choice of the scale factors can yield significant savings in the execution time, on the order of a factor of 10.

22.3.2. Adaptive Mesh Multigrid Methodology. In the solution of the governing equations, the dependent variables in some regions in each coordinate direction exhibit extensive spatial activity (a steep front and sharp peaks). In some cases, the solution components can vary by three orders of magnitude between neighboring mesh points. These active regions must be refined adequately. Adaptive techniques that attempt to equidistribute positive weight functions have been used successfully. The equations for flames with more than 50 chemical species and 100 chemical reactions can be solved efficiently by this technique [8].

In the *nested iteration* multigrid method, (22.6) is first solved on a coarse mesh. This mesh is then refined, and the solution from the coarser mesh is interpolated (using linear or cubic interpolation) onto the new, finer mesh. Then (22.6) is solved on the new mesh. These steps can be repeated until there are k meshes, with the kth one sufficiently refined to resolve the flame.

The standard approach uses a *one-way multigrid* method. This means that each of the coarser meshes is used only to initialize the next finer mesh. By saving the last Jacobian used in the damped Newton iteration (22.7), *correction problems* can be calculated on the coarser meshes. This method can be used to accelerate the iterative solver used for (22.7) [1, 2].

22.3.3. Flame Sheet Starting Estimates. In order to converge, the solution procedure just described requires an adequate starting estimate. Determining an estimate that is *good enough* can be challenging. The difficulty lies in the exponential dependence of the chemistry terms on the temperature and in the nonlinear coupling of the hydrodynamic and the thermochemistry solution fields. One approach that has been used for more than a decade (cf. [8]) begins with the solution of a flame sheet problem.

In the flame sheet model, the fuel and oxidizer are assumed to undergo an infinitely fast and irreversible conversion reaction into stable products in the presence of an inert gas. Such an approach provides profiles for each of the major components of the flame, including temperature and major species, and is used to initialize diffusion flame problems with detailed chemistry. In the two-dimensional flame sheet models used in our calculations, the vorticity–velocity equations are coupled with a Shvab–Zeldovich equation of the form

$$\frac{1}{r}\frac{\partial}{\partial r}\left(r\rho D\frac{\partial S}{\partial r}\right) + \frac{\partial}{\partial z}\left(\rho D\frac{\partial S}{\partial z}\right) = \rho v_r \frac{\partial S}{\partial r} + \rho v_z \frac{\partial S}{\partial z}.$$

Here, S is a conserved scalar and D is a diffusion constant. The temperature and major species profiles are recovered from the conserved scalar S.

22.3.4. Time Relaxation. The adaptive mesh multigrid methodology and the flame sheet starting estimate help eliminate some of the convergence difficulties associated with the direct solution of the governing equations [8]. Nevertheless, neither the interpolated solution from one mesh to the next finer one nor the flame sheet starting estimate generally lies in the convergence domain of (22.7). Therefore, a scaled pseudotransient term $D_{scale}\partial U/\partial t$ is appended to the left-hand side of the conservation equations to produce a problem that is parabolic in time. The time derivative is replaced by a backward Euler approximation [8]. At each timestep, a damped Newton iteration is again used to solve a system of nonlinear equations. This new problem is quite similar to (22.6); in fact, the major difference is that the diagonal of the steady-state Jacobian is weighted by the reciprocal of the timestep. After an appropriate number of timesteps, a switch to the steady-state form of the equations becomes possible.

22.4. Parallel Computing Methodology

Reasonable solvers for this class of problems separate the chemistry parts of the code from the algebra parts. The complete rewrite of a few thousand lines of code that at first appeared to be required for parallelization is no longer necessary. This is due to the large amounts of memory per node on recently delivered parallel computers. Similar considerations hold for workstations used in cluster configurations for distributed computing.

The Jacobian matrices have a regular block sparse structure, which we exploited in our parallelization strategy. Additionally, we wanted the operation of our parallel code to be similar to that of our serial code.

The four principal sets of operations we needed to parallelize were the following:

1. Jacobian matrix construction,

2. matrix–vector multiplication,

3. nonlinear residual evaluation,

4. inner products and norms (i.e., simple level-1 BLAS).

The vast majority of the serial computer time is spent on operations 1–3.

The block structure of the Jacobians is shown in Figure 22.2. Each block has a similar block structure with three subblocks. The subblocks are $n_c \times n_c$ and dense, where n_c is the number of dependent unknowns. Hence, the Jacobians resemble nine-point operators but have $9n_c$ nonzeros per row on average. Assuming $n_c \in [16, 50]$, the Jacobians use a considerable amount of memory in comparison with the solution vector.

We decided to use a sparse matrix domain decomposition method; i.e., we considered the Jacobian matrix to be a two-dimensional domain, which we decomposed by rows of blocks. This corresponds to a strip domain decomposition method. For example, in Figure 22.2, we decomposed every four rows of blocks. Hence, for a matrix–vector multiply, only the $n_c \cdot n_r$ unknowns associated with end blocks must be transferred between processors.

We considered two other domain decomposition methods. Schur complement methods simply use too much extra memory in computing the comple-

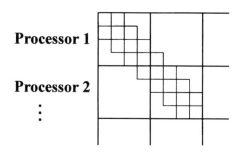

Processor 1

Processor 2

FIG. 22.2.   *Upper left corner of the Jacobian matrix block structure.*

ment of a matrix with $9n_c$, $16 \leq n_c \leq 50$ nonzeros per average row. Alternating methods on two-dimensional subdomains, additive or multiplicative, have two problems: first, the gradients of the solutions along the internal subdomain boundaries are quite large. Second, standard theory based on condition numbers indicates that a huge number of iterations of the alternating procedures would be required to solve our problems.

We have two matrix–vector multiplication routines; in one the preconditioner is prefactored into the operations, and in the other it is not. The parallelization of each routine follows the techniques just described. The preconditioner is local to each processor, not global. Hence, the total numbers of iterations for our parallel solvers are not always the same as the serial computer equivalents.

The nonlinear residual calculation uses the same vector decomposition as the matrix–vector and Jacobian routines. Hence, no data need to be transferred between processors to start this procedure.

One aspect of the large number of nonzeros in the Jacobians is that we can afford to store the complete solution vector on a single processor without using much extra memory. In our case, our parallel computers had from 128 megabytes to 2 gigabytes of local memory on each node. Hence, if some operation is quick and does not parallelize well (or at all), we can actually gather all the data, do the operation on a single node, and then scatter data back to the remaining processors. This was particularly useful in parallelizing the code, one major operation at a time, and for debugging purposes.

The general computing methodology used was that of a single processor directing all the processors to do various computing tasks. Tasks assigned to processors included carrying out their part of the parallel iterative procedures, computing a Jacobian, and evaluating a nonlinear residual. Large tasks are the typical operation, not small tasks, such as requiring all processors to do an inner product.

The communications library we used, MPL [9], allowed us to produce one executable for each of the following IBM computers: SP1, SP2, and clusters of

RS/6000s. The fast switches could be used with no relinking or recompiling. Hence, only one copy of the source files and one executable were maintained for all these environments.

Since we want the code to run on other machines, we are porting the code from MPL to MPI [7]. An excellent, free version of MPI can be retrieved by anonymous login from info.mcs.anl.gov. A number of proprietary versions of MPI will soon be available; the one from IBM still has the nice feature of one executable for all three platforms.

22.5. Numerical Results

The flame sheet model provides a test problem of moderate computational cost for probing the efficiency of solution algorithms that can then be used to tackle diffusion problems with detailed chemistry. Such a study is presented in [2, 3]. The flame sheet considered is adequately resolved on a grid with up to 2×10^4 nodes, with 20% of the nodes clustered in a region covering 0.1% of the computational domain. This requires up to 100 megabytes of work space. The adaptive mesh multigrid procedure outlined here is particularly efficient for solving flame sheet problems. In comparison with traditional solution procedures, the total execution times drop by a factor of 10. Speedups as high as 166 in the time relaxation phase have been seen on a single processor. For three-dimensional problems, speedups much greater than 10 should be obtained. At the beginning of this study, using an IBM RISC System/6000 model 560 workstation, we spent as much as 96 minutes on the flame sheet phase. Recently, using an eight-node SP2, we spent as little as 45 seconds.

The step from a flame sheet problem to a finite-rate chemistry flame problem is a challenging one, involving a dramatic increase in the difficulty of the problem. This is attributable mostly to the much larger number of dependent unknowns, the nonlinear fluid dynamics–thermochemistry coupling, and the disparate length scales that must be resolved in the computed solution. In particular, excellent resolution of a flame sheet problem requires up to 128 megabytes, while a reasonable solution of a diffusion flame with 50 chemical species can require up to a gigabyte. Traditional solvers for two-dimensional laminar diffusion flames used to require more than a hundred CPU hours on a supercomputer. Owing to recent improvements in computational algorithms and computer capabilities, this is no longer the case.

The flame configuration is a methane–air lifted laminar diffusion flame with a triple flame structure at its base, as illustrated in Figure 22.1. Although extremely difficult to compute numerically, this flame configuration was chosen in [4] because of the availability of experimental and numerical results for it [10]. The numerical solution that is considered first includes 16 chemical species engaged in a C1-chain reaction mechanism; i.e., only molecules with at most *one* carbon atom are considered. In addition to the 16 chemical species, four unknowns are associated with each mesh point. Hence, there are a total of 20 unknowns per mesh point. The flame is appropriately resolved with 5×10^3

mesh nodes, and very good agreement with previous experimental (2.5×10^5 data points) and numerical data is obtained. Consider a one-way nonlinear multigrid method with BiCGSTAB/Gauss–Seidel as the solver on all levels. On the finest level, the wallclock times for convergence on the 89×85 grid are given in Table 22.1.

TABLE 22.1

Fine-grid solution times for various machines. Peak speeds are 130 Mflops for the SP1 and RISC System/6000 model 580 and 266 Mflops for the SP2.

Machine	Processors	Wall-time	Speedup
RISC System/6000–580	1	602.59	1.00
SP1	4	137.21	4.39
	8	70.37	8.56
	12	58.89	10.23
	16	53.85	11.19
SP2	8	40.10	15.03
	16	20.96	28.75

The chemical mechanism just considered yields adequate resolution for the temperature and several chemical species. More detailed investigation of the flame structure requires additional chemical species in the model. For instance, the study of the formation of pollutants, such as nitric oxide, requires up to 50 chemical species. While the memory requirements for such problems are too large for current serial computers, these problems can be adequately solved on parallel computers with large amounts of memory per node. The diffusion flame just described can then be conveniently used to initialize diffusion flame calculations with more detailed chemistry submodels. Such work is in progress.

References

[1] C.C. DOUGLAS, *Implementing abstract multigrid or multilevel methods*, in Proc. Sixth Copper Mountain Conference on Multigrid Methods, Vol. CP 3224, N.D. Melson, T.A. Manteuffel, and S.F. McCormick, eds., NASA, Hampton, VA, 1993, pp. 127–141.

[2] C.C. DOUGLAS AND A. ERN, *Numerical solution of flame sheet problems with and without multigrid methods*, in Proc. Sixth Copper Mountain Conference on Multigrid Methods, Vol. CP 3224, N.D. Melson, T.A. Manteuffel, and S.F. McCormick, eds., NASA, Hampton, VA, 1993, pp. 143–157.

[3] A. ERN, *Vorticity-Velocity Modeling of Chemically Reacting Flows*, PhD thesis, Mechanical Engineering Department, Yale University, February 1994.

[4] A. ERN, C.C. DOUGLAS, AND M.D. SMOOKE, *Detailed chemistry modeling of laminar diffusion flames on parallel computers*, Internat. J. Supercomputer Appl., 9(1995), pp. 167–186.

[5] A. ERN AND V.GIOVANGIGLI, *Multicomponent Transport Algorithms*, Vol. m 24, Springer-Verlag, Heidelberg, 1994.

[6] A. ERN AND V.GIOVANGIGLI, *Projected iterative algorithms with application to multicomponent transport*, Lin. Alg. Appl., (1996).

[7] W.D. GROPP, E. LUSK, AND A. SKJELLUM, *Using MPI: Portable Parallel Programming with the Message-Passing Interface*, MIT Press, Cambridge, MA, 1994.

[8] M.D. SMOOKE, R.E. MITCHELL, AND D.E. KEYES, *Numerical solution of two-dimensional axisymmetric laminar diffusion flames*, Combust. Sci. and Tech., 67(1989), pp. 85–122.

[9] M. SNIR AND P. HOCHSCHILD, *The communication software and parallel environment of the IBM* SP2, IBM Systems Journal, 34(1995), pp. 205–221.

[10] Y. XU, M.D. SMOOKE, P. LIN, AND M.B. LONG, *Primitive variable modeling of multidimensional laminar flames*, Combust. Sci. and Tech., 90(1993), pp. 289–313.

Chapter 23

Applications of Algebraic Topology to Concurrent Computation

Maurice Herlihy
Nir Shavit

Editorial preface

All parallel programs require some amount of synchronization to coordinate their concurrency to achieve correct solutions. It is commonly known that synchronization can cause poor performance by burdening the program with excessive overhead. This chapter develops a connection between certain synchronization primitives and topology. This connection permits the theoretical study of concurrent computing with all the mathematical tools of algebraic and combinatorial topology.

This article originally appeared in *SIAM News*, Vol. 27, No. 10, December 1994. It was updated during the summer/fall of 1995.

Today, the computer industry is very good at making computers run faster: speeds double roughly every two years. Eventually, however (and perhaps as early as the turn of the century), fundamental limitations, such as the speed of light or heat dissipation, will make further speed improvements increasingly difficult. Beyond that point, the most promising way to make computers more effective is to have many processors working in parallel, the approach known as multiprocessing.

The hard part of multiprocessing is getting the individual computers to coordinate effectively with one another. As a typical coordination problem, if two computers, possibly far apart, both try to reserve the same airline seat, care must be taken that exactly one of them succeeds. Coordination problems arise at all scales in multiprocessor systems—at a very small scale, processors within a single supercomputer might need to allocate resources, and at a very large scale, a nationwide distributed system, such as an "information highway," might need to allocate communication paths over which large quantities of data will be transmitted.

Coordination is difficult because multiprocessor systems are inherently *asynchronous*: processors can be delayed without warning for a variety of reasons, including interrupts, preemption, cache misses, and communication delays. These delays can vary enormously in scale: a cache miss might delay a processor for fewer than ten instructions, a page fault for a few million instructions, and operating system preemption for hundreds of millions of instructions. Any coordination protocol that does not take such delays into account runs the risk that a sudden delay of one process in the middle of a coordination protocol may leave the others in a state where they are unable to make progress.

The need for effective coordination has long been recognized as a fundamental aspect of multiprocessor architectures. As a result, modern processors typically provide hardware mechanisms that facilitate coordination. Until recently, these mechanisms were chosen in an ad hoc fashion, but it is becoming increasingly clear that some kind of mathematical theory is needed if the implications of such fundamental design choices are to be understood.

In this article, we focus on some new mathematical techniques for analyzing and evaluating common hardware synchronization primitives. Aside from its inherent interest to the computer science community, we believe this work may be of interest to the mathematical research community because it establishes a (perhaps unexpected) connection between asynchronous computability and a number of well-known results in combinatorial topology.

In many multiprocessor systems, processors communicate by applying certain operations, called *synchronization primitives*, to variables in a shared-memory. These primitives may simply be reads and writes, or they may include more complex constructs, such as *test-and-set*, *fetch-and-add*, or *compare-and-swap*. The test-and-set operation atomically writes a 1 to a variable and returns the variable's previous contents. The fetch-and-add operation atomically adds a given quantity to a variable and returns the variable's previous contents. Finally, the compare-and-swap operation atomically tests whether a variable has a given value and, if so, replaces it with another given value.

Over the years, computer scientists have proposed and implemented a variety of different synchronization primitives, and their relative merits have been the subject of a lively debate. Most of this debate has focused on the ease of implementation and ease of use of the primitives. More recently, however, it has emerged that some synchronization primitives are inherently more powerful than others, in the sense that every synchronization problem that can be solved by primitive A can also be solved by primitive B, but not vice versa. This article describes the new conceptual tools that are making it possible to provide a rigorous analysis of the relative computational power of different synchronization primitives. This emerging theory could provide the designers of computer networks and multiprocessor architectures with mathematical tools for recognizing when problems are unsolvable, for evaluating alternative synchronization primitives, and for making explicit the assumptions needed to make a problem solvable.

Our discussion focuses on a simple but important class of coordination tasks called *decision problems*. At the start with such problems, processors are assigned private *input values* (perhaps transmitted from outside). The processors communicate by applying operations to a shared-memory, and eventually each process chooses a private *output value* and halts. The decision problem is characterized by (1) the set of legitimate input value assignments and (2) for each input value assignment, the set of legitimate output value assignments. For example, consider the following *renaming* problem: as input values, each processor is assigned a unique identifier taken from a large range (like a social security number). As output values, the processors must choose unique values taken from a much smaller range. (Renaming is an abstraction of certain resource allocation problems.)

To solve a decision problem, a processor executes a program called a *protocol*. Because processors are subject to sudden delays, and because halting one processor for an arbitrary duration should not prevent the others from making progress, we require that each processor finish its protocol in a fixed number of steps, regardless of how its steps are interleaved with those of other processors. Such a protocol is said to be *wait-free*, since it implies that no processor can wait for another to do anything.

23.1. Simplicial Complexes

A decision problem has a simple geometric representation. Assume we have $n + 1$ processes, each assigned a different color. A processor's state before starting a problem is represented as a point in a high-dimension Euclidian space. This point, called an *input vertex*, is labeled with a process color and an input value. Two input vertices are *compatible* if (1) they have distinct colors and (2) there exists a legitimate input value assignment that simultaneously assigns those values to those processes. For example, in the renaming problem described earlier, input values are required only to be distinct, so two input vertices are compatible if and only if they have distinct colors and distinct input values. We join any two compatible input vertices with a line segment, any three with a solid triangle, and any four with a solid tetrahedron. In general, any set of k compatible input vertices spans an *input k-simplex* in k-dimensional space. The set of all possible input simplexes forms a mathematical structure, called a *simplicial complex*. We call this structure the problem's *input complex*.

The notions of an *output vertex*, *output simplex*, and the problem's *output complex* are defined analogously, simply replacing input values with output values. The decision problem itself is defined by a relation Δ that carries each input n-simplex to a set of output n-simplexes. This relation has the following meaning: if S is an input simplex, T is an output simplex, and the processors start with their respective input values from S, then it is acceptable for them to halt with their respective output values from T.

For example, consider the instance of the renaming problem in which three processors are assigned unique input values in some large range and must coordinate to choose unique output values in the range 0 to 3. Here, an output simplex is a triangle whose vertices are labeled with distinct colors and distinct input values in the range 0 to 3. There are $4 \cdot 3 \cdot 2 = 24$ distinct output triangles, and it is not difficult to draw them on a sheet of paper. The result, shown in Figure 23.1, is topologically equivalent to a torus.

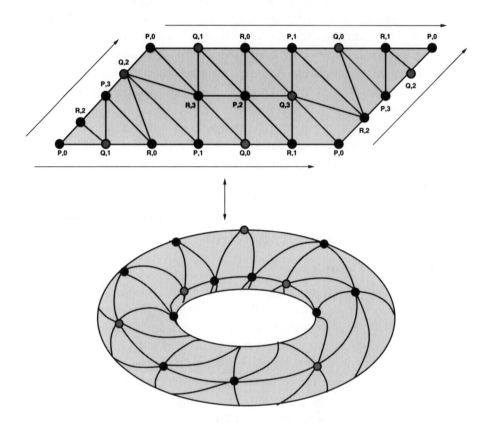

FIG. 23.1. *Three-process renaming with four names.*

Having shown how to specify a decision problem with a geometric model, we now do the same for the protocols that solve such problems. Recall that a protocol is a program: each processor starts out with its input value in a private register, applies a sequence of operations to variables in the shared-memory, and then chooses an output value based on the results of the computation. We can view any such protocol as accumulating a history of shared-memory operations—when the protocol has "seen enough," it computes its output value by applying a *decision map* to its history.

Any execution of a protocol generates a set of histories, one for each processor. The set of all possible executions also defines a simplicial complex: each vertex is labeled with a processor color and a history, and two vertices are compatible if they are labeled with distinct colors and if in some protocol execution, they see those two histories. We call this the *full-information complex* for the protocol. More precisely, for every input simplex S, any protocol induces a corresponding full-information complex $\mathcal{F}(S)$. The union of these complexes is the full-information complex for the protocol.

What does it mean for a protocol to solve a decision problem? Recall that a *decision map* δ carries each history h to the output value chosen by the protocol after observing h. The decision map induces a map from the full-information complex to the output complex: $\delta(\langle P, h \rangle) = \langle P, \delta(h) \rangle$. We are now ready to give a precise geometric statement of what it means for a protocol to solve a decision problem: given a decision problem with input complex $\mathcal{I}$, output complex $\mathcal{O}$, and relation Δ, a protocol solves a decision problem if and only if, for every input simplex $S \in \mathcal{I}$ and every full-information simplex $T \in \mathcal{F}(S)$, $\delta(T) \subset \Delta(T)$.

This definition is simply a formal way of stating that every execution of the protocol must yield an output value assignment permitted by the decision problem specification. Roundabout as this formulation of this property might seem, it has an important and useful advantage. We have moved from an operational notion of a decision problem, expressed in terms of computations unfolding in time, to a purely combinatorial description expressed in terms of relations among topological spaces. It is typically easier to reason about static mathematical relations than about ongoing computations, but, more importantly, this model allows us to exploit classical results from the rich literature on algebraic and combinatorial topology.

To prove that certain decision problems cannot be solved by certain classes of protocols, it is enough to show that no decision map exists. We can derive a number of impossibility results by exploiting basic properties that any decision map must have. In particular, any decision map is a *simplicial map*: it carries vertices to vertices, but it also carries simplexes to simplexes. Simplicial maps are also *continuous*: they preserve topological structure. If we can show that a class of protocols generates full-information complexes that are "topologically incompatible" with the problem's output complex, then we have established impossibility. Conversely, if we can prove that the decision map exists, then we have shown that a protocol exists.

A complex has *no holes* if any sphere embedded in the complex can be continuously deformed to a point. (More technically, the complex has trivial homotopy groups.) It has *no holes up to dimension d* if the same property holds for spheres of dimension d or less. (Notice that when d is zero, this condition means the complex is connected.) For example, a two-dimensional disk (e.g., a plate) has no holes, and a two-dimensional sphere (e.g., a basketball) has no holes up to dimension one, because any loop (e.g., a rubber band) on the sphere can be deformed to a point. By contrast, a torus has no holes only up to dimension zero—it is connected, but not every 1-sphere (loop) placed on the surface can be deformed to a point.

23.2. Read/Write Protocols

The simplest interesting synchronization primitives are atomic reads and writes to variables in shared-memory. We recently used this simplicial model to give a complete combinatorial characterization of the decision problems that can be solved by read/write protocols [8].

The full-information complexes for read/write protocols have a remarkable property: for any input simplex S, the full-information complex $\mathcal{F}(S)$ has no holes. This property holds for any read/write protocol, no matter how many variables it uses or how long it runs. This property is a powerful tool for proving impossibility results. A careful analysis of the renaming problem shows that if there are fewer than $2n+1$ possible output values, then the output complex has a hole. Moreover, any decision map must "wrap" a particular sphere in the full-information complex around that hole in such a way that the image of the sphere cannot be continuously deformed to a single point. Because the full-information complex has no holes, however, that sphere can be continuously deformed to a point in the full-information complex. Because the decision map is continuous, the image of that sphere can also be contracted to a point, and we have a contradiction. The same kind of analysis shows that a variety of fundamental synchronization problems have no wait-free solutions in read/write memory.

This topological model also yields a "universal" algorithm that can be used to solve any problem that can be solved by a wait-free read/write protocol. Any decision problem can be considered as a kind of "approximate agreement" problem in which each processor chooses a vertex in the output complex, and the processors negotiate among themselves to ensure that all processors choose vertices of a common simplex. This problem, which we call "simplex agreement," provides a simple normal form for any decision task protocol.

We can combine these two notions to give a complete characterization of the decision problems that can be solved by wait-free read/write protocols. Because the exact conditions require some technical definitions beyond the scope of this article, the focus here is on the underlying intuition. A decision problem has a wait-free read/write protocol if and only if the relation Δ can be "approximated" by a continuous map on its underlying point set, in the following sense. Given the input complex $\mathcal{I}$, construct a new complex, $\sigma(\mathcal{I})$, by subdividing each simplex in $\mathcal{I}$ into smaller simplexes. If v is a vertex in $\sigma(\mathcal{I})$, define $carrier(v)$ to be the smallest simplex in $\mathcal{I}$ that contains v. The decision problem is solvable in read/write memory if and only if there exists a subdivision $\sigma(\mathcal{I})$ and a simplicial map $\mu : \sigma(\mathcal{I}) \to \mathcal{O}$ such that for each vertex $v \in \sigma(\mathcal{I})$, $\mu(v) \in \Delta(carrier(v))$. Informally, this condition states that it must be possible to "stretch" and "fold" the input complex so that each input simplex can cover its corresponding output simplexes.

This condition is shown schematically in Figure 23.2. The top half of the figure illustrates the relation Δ for a generic decision problem, and the bottom half shows how Δ can be approximated by a simplicial (continuous) map μ.

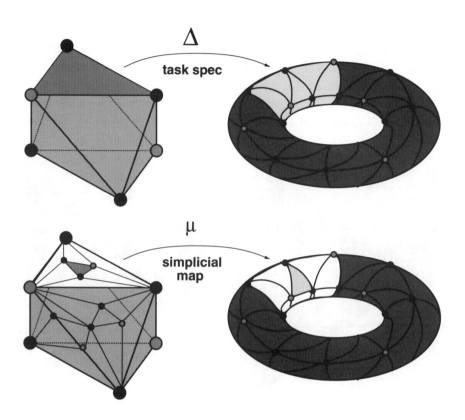

FIG. 23.2. *Existence condition for read/write protocols.*

23.3. Other Kinds of Protocols

Although read/write protocols have considerable theoretical interest, real multiprocessors typically provide more powerful synchronization primitives. The topology of full-information complexes for such protocols is more complicated. For example, Figure 23.3 shows the full-information complexes for two simple protocols in which processors communicate by applying test-and-set operations to shared variables. Casual inspection shows that these full-information complexes differ from their read/write counterparts in one fundamental respect: they have one-dimensional holes. Nevertheless, they do resemble them in another respect: they are connected. In general, any protocol in which $(n + 1)$ processors communicate by pairwise sharing of test-and-set variables has a full-information complex with no holes up to dimension $\lfloor n/2 \rfloor$.

In a recent paper, Herlihy and Rajsbaum [6] analyzed the topological properties of full-information complexes for a family of synchronization primitives called *k-consensus* objects, which encompasses many of the synchronization primitives in use today. The larger the value of k, the more powerful is the primitive. The full-information complex for any protocol in which processes communicate via k-consensus objects has no holes up to dimension $\lfloor n/k \rfloor$. So at one extreme, when $k = 1$, the complex has no holes at all, and at the other extreme, the complex becomes disconnected. As k ranges from 1 to $n + 1$, holes appear first in higher dimensions and then spread to lower dimensions. A surprising implication of this structure is that there exist simple synchronization primitives that are *incomparable*: it is impossible to construct a wait-free implementation of one from the other.

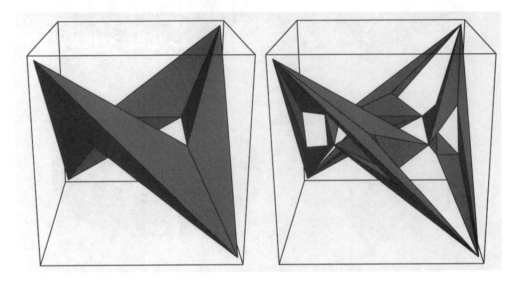

FIG. 23.3. *Full-information complexes for some test-and-set protocols.*

23.4. Related Work

The *consensus* problem is an idealized form of the transaction commitment protocols commonly used in distributed databases. In 1985, Fischer, Lynch, and Paterson [5] showed that if processors communicate by exchanging messages, then any consensus protocol has a "window of vulnerability" during which the failure or delay of a single processor will cause the protocol itself to fail or delay. This result showed that the notion of "asynchronous computability" differs in important ways from conventional notions of computability. Since then, a variety of research efforts have focused on characterizing the decision problems that can be solved by particular synchronization primitives in the presence of unpredictable failures and delays.

In 1988, Biran, Moran, and Zaks [1] gave a graph-theoretic characterization of decision problems that can be solved in the presence of a single failure in a message-passing system. This result was not substantially improved until 1993, when three independent research teams—Borowsky and Gafni [2, 3], Saks and Zaharoglou [9], and Herlihy and Shavit [7]—succeeded in applying combinatorial techniques to protocols that tolerate delays by more than one processor.

We provided the complete characterization of read/write solvability in two recent papers [7, 8]. The analysis of the topological properties of full-information complexes for protocols using more powerful primitives appears in a recent paper by Herlihy and Rajsbaum [6]. Recently, Chaudhuri et al. [4] were able to use similar topological techniques to derive the first lower bounds for a class of decision problems in a message-passing system in which processors execute in lockstep, but in which a processor can fail at any time by halting.

23.5. Conclusions

We believe this topological approach has a great deal of promise for the theory of distributed and concurrent computation, and that it merits further investigation. It has already produced a number of new and unexpected results and has illuminated an unexpected connection between the emerging theory of concurrent computation and the well-established theories of algebraic and combinatorial topology.

References

[1] O. BIRAN, S. MORAN, AND S. ZAKS, *A combinatorial characterization of the distributed tasks which are solvable in the presence of one faulty processor*, in Proc. 7th Annual ACM Symposium on Principles of Distributed Computing, August 1988, pp. 263–275.

[2] E. BOROWSKY AND E. GAFNI, *Generalized flp impossibility result for t-resilient asynchronous computations*, in Proc. 1993 ACM Symposium on Theory of Computing, May 1993.

[3] E. BOROWSKY AND E. GAFNI, *Immediate atomic snapshots and fast renaming*, in Proc. 12th Annual ACM Symposium on Principles of Distributed Computing, August 1993.

[4] S. CHAUDHURI, M. HERLIHY, N. LYNCH, AND M.R. TUTTLE, *A tight lower bound for k-set agreement*, in Proc. 34th IEEE Symposium on Foundations of Computer Science, October 1993.

[5] M. FISCHER, N.A. LYNCH, AND M.S. PATERSON, *Impossibility of distributed commit with one faulty process*, J. Assoc. Comput. Mach., 32(1985), pp. 374–382.

[6] M.P. HERLIHY AND S. RAJSBAUM, *Set consensus using arbitrary objects*, in Proc. 13th Annual ACM Symposium on Principles of Distributed Computing, August 1994.

[7] M.P. HERLIHY AND N. SHAVIT, *The asynchronous computability theorem for t-resilient tasks*, in Proc. 1993 ACM Symposium on Theory of Computing, May 1993.

[8] M.P. HERLIHY AND N. SHAVIT, *A simple constructive computability theorem for wait-free computation*, in Proc. 26th Annual Symposium on Theory of Computing, May 1994, pp. 243–252.

[9] M. SAKS AND F. ZAHAROGLOU, *Wait-free k-set agreement is impossible: The topology of public knowledge*, in Proc. 1993 ACM Symposium on Theory of Computing, May 1993.

Chapter 24

Parallel Computation of Economic Equilibria

Anna Nagurney

Editorial preface

The mathematical formulation of economic equilibria presented in this chapter involves an interesting blend of techniques. Variational methods and mathematical programming are combined, and a decomposition to induce parallelism is then applied. An interesting contrast is provided by the performance of the solution algorithm on two very different architectures, the CM-2 and the IBM 3090.

This article originally appeared in *SIAM News*, Vol. 25, No. 1, January 1992. It was updated during the summer/fall of 1995.

Problems in which agents compete for scarce resources until the system is driven to an equilibrium state are numerous in economics. Examples of economic equilibrium problems include oligopolistic market equilibrium problems, spatial price equilibrium problems, problems of human migration, and general financial equilibrium problems. The concept of equilibrium, hence, is central to economics, and the determination of equilibrium prices and quantities is central to computational economics. Here we explore the application of parallel architectures to this problem domain.

The methodology that we utilize—the theory of variational inequalities— was introduced initially for the study of partial differential equations, primarily those in mechanics. It has now been used (cf. [9]) to study a plethora of *economic* equilibrium problems, governed by such distinct equilibrium concepts as Nash equilibrium, spatial price equilibrium, and Walrasian price equilibrium. By way of explanation, Nash equilibrium denotes the state in which no agent can increase his utility by unilateral action, given that all other agents keep their actions fixed, whereas Walrasian price equilibrium denotes the state in which there is no excess demand of a commodity at a positive price. In addition to providing qualitative properties of existence, uniqueness, and sensitivity of equilibrium solutions, variational inequality theory helps us to develop mathematically correct algorithms for the computation of solutions to equilibrium problems.

Interestingly, besides the variational inequality formulation, many economic equilibrium problems also have an underlying network structure (see, e.g., [9]). A classical example is the spatial price equilibrium problem where the nodes represent the markets and the links the transportation routes. In this case the underlying network is bipartite. Recently, networks have been used to model migration equilibrium problems, pure exchange economies, and general financial equilibrium problems. Network models have also been used to assist in policy analyses, where the effects of regulatory instruments such as price controls, trade restrictions, tariffs, and taxes are to be determined. Such policy interventions are used by governments as part of agricultural and energy programs and may result in disequilibrium.

Finally, we note that the connection established recently in [3] (see also [11]) between the set of solutions to finite-dimensional variational inequality problems and the set of solutions to projected dynamical systems allows one to study a wide range of equilibrium problems not only at the equilibrium state but in the richer (and behavioral) framework of this new class of dynamical system.

The computational approach to equilibrium problems in economics has been primarily serial in nature. Here we illustrate how parallel computation can be applied to solve equilibrium problems that are very large. For additional background and applications, see [13].

24.1. Background

The finite-dimensional variational inequality problem, $\text{VI}(F, K)$, is to determine the vector x^* in a closed convex subset K of $\mathbb{R}^n$ such that

$$(24.1) \qquad F(x^*)^T (x - x^*) \geq 0 \quad \forall x \in K,$$

where $F(\cdot)$ is a known function from K to $\mathbb{R}^n$.

$\text{VI}(F, K)$ contains, as special cases, complementarity problems, min-max problems, as well as minimization problems, and is also related to fixed point problems; hence, it is a natural framework for the study of equilibrium problems. For example, the connection between variational inequalities and minimization problems is as follows. Let $f(\cdot)$ be a continuously differentiable scalar-valued function defined on some open neighborhood of K, and denote its gradient by $\nabla f(\cdot)$. If there exists an $x^* \in K$ such that

$$(24.2) \qquad f(x^*) = \min_{x \in K} f(x),$$

then x^* is a solution to the variational inequality

$$(24.3) \qquad \nabla f(x^*)^T (x - x^*) \geq 0 \quad \forall x \in K.$$

On the other hand, if $F(\cdot)$, again on an open neighborhood of K, is the gradient of a convex, continuously differentiable function $f(\cdot)$, then $\mathrm{VI}(F, K)$ and the minimization problem (24.2) are equivalent. Hence, an optimization form of a variational inequality exists only when the Jacobian matrix $[\partial F/\partial x]$ is symmetric, but $\mathrm{VI}(F, K)$ can also handle problems with asymmetric Jacobians for which no equivalent optimization formulation exists.

Now consider the dynamical system defined by the ordinary differential equation (ODE):

$$(24.4) \qquad \dot{x} = \Pi_K(x, -F(x)), \quad x(0) = x_0 \in K,$$

where K is a convex polyhedron and where given $x \in K$ and $v \in R^n$, the projection of the vector v at x is defined by

$$(24.5) \qquad \Pi_K(x, v) = \lim_{\delta \to 0} \frac{P_K((x + \delta v) - x)}{\delta},$$

and the orthogonal projection $P_K(x)$ with respect to the Euclidean norm is defined by

$$(24.6) \qquad P_K(x) = \arg\min_{z \in K} \|x - z\|.$$

ODE (24.4) has been termed a "projected" dynamical system (cf. [11]). As shown in [3], the stationary points of (24.4), i.e., those that satisfy $0 = \Pi_K(x, -F(x))$, coincide with the solutions of (24.1).

24.2. Parallel Decomposition

We consider variational inequality problems in which the feasible set K can be expressed as a Cartesian product of sets; that is,

$$(24.7) \qquad K = \prod_{\alpha=1}^{\zeta} K_\alpha,$$

where each $K_\alpha \subset \mathbb{R}^{n_\alpha}$, $\sum_{\alpha=1}^{\zeta} n_\alpha = n$, x_α denotes a vector in $\mathbb{R}^{n_\alpha}$, and $F_\alpha(x) : K \mapsto \mathbb{R}^{n_\alpha}$ for each α. In this case parallel variational inequality decomposition algorithms can be applied for the computation of the solution x^* to $\mathrm{VI}(F, K)$. The motivation is to resolve the variational inequality problem into simpler variational inequality or optimization problems, each of which can then be allocated to a distinct processor. Many equilibrium problems in economics can be defined over a feasible set K of the form (24.7). For example, in the case of multicommodity problems, each subset K_α would correspond to the constraints of a commodity α [6]; in the case of multiclass problems in human migration, each K_α would correspond to a distinct class α [5].

The linearized parallel decomposition algorithm, which we now describe, decomposes $\mathrm{VI}(F, K)$ into ζ simpler subproblems, which are quadratic programming problems [1]. Each of these problems can then be allocated to a distinct processor, and all can then be solved simultaneously. The statement of the algorithm is as follows.

Initialization:
    Start with an initial vector $x^0 \in K$. Set $\tau = 1$.
Step τ, $\tau=1,2, \ldots$:
    Construct the functions

$$(24.8) \qquad F_\alpha^\tau(x_\alpha) = D_\alpha(x^{\tau-1}) \cdot x_\alpha + (F_\alpha(x^{\tau-1}) - D_\alpha(x^{\tau-1})x_\alpha^{\tau-1})$$

for $\alpha = 1, \ldots, \zeta$, where $D_\alpha(x^{\tau-1})$ is the diagonal part of $\nabla_\alpha F_\alpha(x)$, the gradient of F_α with respect to x_α, and solve the ζ subproblems:

$$(24.9) \qquad\qquad F_\alpha^\tau(x_\alpha) \cdot (x_\alpha' - x_\alpha) \geq 0 \quad \forall x_\alpha' \in K_\alpha.$$

    Let the solution to (24.9) be x_α^τ; $\alpha = 1, \ldots, \zeta$.
If (equilibrium conditions are satisfied) **then**
        stop
Else
    Set $\tau = \tau + 1$, and go to (24.8).

Because variational inequality problems are usually solved iteratively as mathematical programming problems, the overall efficiency of a variational inequality algorithm depends upon the efficiency of the mathematical programming algorithm used at each iteration. Furthermore, in the case in which there is a special network structure to each subproblem (24.9), an even finer decomposition may be possible. We now present an algorithm for the solution of a network subproblem with special structure that arises in the realization of the linearized parallel variational inequality decomposition algorithm. The notable feature of this procedure is that it lends itself to a massively parallel implementation, as we will demonstrate later. This demand market exact equilibration procedure is described and theoretically analyzed in [2]. Additional applications, ranging from finance to transportation, in which the exact equilibration algorithm can be used as a subroutine, can be found in [9], [11], and [13].

In particular, we are interested in computing the flows $x_{1l}, \ldots, x_{ml}$ from the supply markets $1, \ldots, m$ to the demand market l that satisfy the following equilibrium conditions: the cost of the good from i to l, $g_i x_{il} + h_{il}$, is equal to the demand price, $-r_l \sum_i x_{il} + q_l$, at demand market l, if there is a positive flow x_{il}; if the cost exceeds the demand price, then there will be zero flow between the pair of markets. Mathematically, the conditions are

$$(24.10) \qquad g_i x_{il} + h_{il} \begin{cases} = -r_l \sum_i x_{il} + q_l & \text{if } x_{il} > 0, \\ \geq -r_l \sum_i x_{il} + q_l & \text{if } x_{il} = 0, \end{cases}$$

where the g_i, $i = 1, \ldots, m$, terms are all greater than zero, as are the terms r_l and q_l.

This problem may be viewed as a network problem with two nodes and l links connecting the origin node with the destination node. The x_{il}'s then may be interpreted as corresponding to the flows on the respective links where the associated cost on the ith link is given by $g_i x_{il} + h_{il}$. The function $-r_l \sum_i x_{il} + q_l$, in turn, may then be interpreted as the demand price associated with trade between the origin/destination pair.

The algorithm below computes the solution to the above system (24.10) in closed form.

SORT:
    Sort the h_{il}'s, $i = 1, \ldots, m$, in nondescending order and relabel the h_{il}'s accordingly. Define $h_{m+1,l} = \infty$.
TEST:
    If $(q_l \leq h_{1l})$ **then**
        $x_{il} = 0,\ i = 1, \ldots, m$,
        stop
    Else
        $v = 1$.
COMPUTE:

$$(24.11) \qquad \rho_l^v = \frac{\sum_{i=1}^{v} h_{il}/g_i + \frac{q_l}{r_l}}{\sum_{i=1}^{v} 1/g_i + \frac{1}{r_l}}.$$

    If $(h_{vl} < \rho_l^v \leq h_{v+1,l})$ **then**
        $s' = v$
        goto **SET:**
    Else
        $v = v + 1$
        goto **COMPUTE:**
SET:
    $x_{il} = \frac{\rho_l^{s'} - h_{il}}{g_i}, \quad i = 1, \ldots, s'$
    $x_{il} = 0, \quad i = s' + 1, \ldots, m.$

In the case where the demand $\sum_i x_{il}$ is known and fixed, the procedure that will equalize the costs for all of the positive trade flows can be obtained from the above scheme by replacing the q_l/r_l term in the numerator in (24.11) by the known demand and by deleting the second term in the denominator in (24.11). The **TEST** step is then unnecessary, provided that v is initialized to one.

Finally, we recall the general iterative scheme proposed in [3] for the computation of stationary points of the projected dynamical system (24.4), equivalently, the solutions to the finite-dimensional variational inequality problem (24.1). In the case where the feasible set K is the nonnegative orthant, as occurs in many economic equilibrium problems, which require, for example, that the commodity outputs and prices must be nonnegative, the equilibrium problem can be massively decomposed into subproblems that can, in most cases, be computed explicitly and in closed form. The general iterative scheme takes on the following form.

Initialization:
   Start with an initial vector $x^0 \in K$. Set $\tau = 1$.

Step τ, $\tau = 1, 2, \ldots$:
   Compute
(24.12) $$x^{\tau+1} = P_K(x^\tau - a_\tau F_\tau(x^\tau)),$$

where a_τ is a sequence of positive scalars and F_τ is a sequence of vector fields that "approximates" $F(\cdot)$ (cf. [3, 11]), with the simplest approximation being $F_\tau = F$, and yields an Euler method.

   Note that in the case that $K = R_+^n$, each term of the vector $x^{\tau+1}$ can be computed independently and simultaneously according to (24.12) as

(24.13) $$x_i^{\tau+1} = \max\{0, x_i^\tau - a_\tau (F_i)_\tau(x^\tau)\} \quad \text{for} \quad i = 1, \ldots, n.$$

24.3. Applications

We now illustrate the above ideas in the context of several applications. For example, the parallel variational inequality decomposition procedure applied to the multiclass migration equilibrium model developed in [5], with known populations of each class in the economy, would yield, per iteration, as many fixed demand network problems of the form described above as there are classes. In the case of the pure exchange general economic equilibrium problem (cf. [9]), the algorithm would yield a series of single network subproblems in which there are as many links as there are commodities.

   Next we expand upon the ideas in the context of other partial and general economic equilibrium problems. In a partial equilibrium model only segments of producers, consumers, and/or commodities are considered, whereas in general equilibrium models, all of the economic entities are treated.

24.3.1. Partial Economic Equilibrium Models.   In the case of static multicommodity spatial price equilibrium problems, the variational inequality decomposition algorithm would yield as many classical, single commodity spatial price equilibrium problems as there are commodities. In this application, each single commodity subproblem (24.9) could then be solved using a demand market equilibration algorithm [2]. This in turn resolves the subproblem into precisely the specially structured network problem of the form outlined above. Such a decomposition was applied to multicommodity market equilibrium problems in [6] and implemented on a coarse grain architecture, the IBM 3090/600E, at the Cornell National Supercomputer Facility (CNSF). The results of the parallel implementation for problems up to 50 supply markets, 50 demand markets, and 12 commodities, are presented in [6].

A dynamic market equilibrium model, formulated as a projected dynamical system, was introduced in [12] in which the optimal commodity production, consumption, and trade patterns are to be determined over space and time. The Euler method (cf. (24.12)) was then implemented on the Thinking Machine CM-2 Connection Machine and on problems with as many as 50 supply markets and 50 demand markets, that is, with 250,000 unknown commodity shipment variables solved using this parallel architecture. In [14], an alternative dynamic market model was introduced, in which both the commodity shipments and prices were to be computed. Therein is discussed the implementation of the Euler method, along with numerical results on both the Thinking Machines CM-2 and CM-5 architectures.

In many applications, subproblems of the form (24.8) that possess a network structure, such as the bipartite network problems, may be decomposed into even finer subproblems, each of which will then be of the special single origin/destination pair form and be amenable to solution on massively parallel architectures (see also, e.g., [7]). Toward the goal of exploiting precisely the simplicity of the exact equilibration procedures, we have developed a splitting equilibration algorithm (SEA) [7], which resolves a variety of bipartite-type network problems into series of single network subproblems, each of which can then be solved explicitly in closed form using the exact equilibration procedures.

24.3.2. General Economic Equilibrium Models. The SEA has been used not only to compute solutions to spatial price equilibrium problems, but also to estimate input/output tables of the U.S. economy, to estimate migration flows in the U.S., and to estimate the social/national accounts for the U.S. Department of Agriculture. For example, an input/output table is a matrix whose rows correspond to the origin sectors of the goods (the producers) and whose columns correspond to the destination sectors (the purchasers). The matrix elements in an input/output table are the flows of the products from each of the producing sectors to each of the consuming sectors. A social accounting matrix, on the other hand, is a general equilibrium data system that consists of the accounts in the economy of a nation. The rows represent the receipts of the accounts, the columns the expenditures of the accounts, and the individual matrix entries the transactions in the economy. Numerical results for the application of the algorithm to a spectrum of such problems in both a serial and in a parallel environment on the IBM 3090/600E can be found in [7].

Macromonetary policy is another area of application that holds particular promise for the above ideas. We have developed a network model of financial flow of funds accounting that explicitly incorporates feedback and can be used to calculate reconciled values of all outstanding financial instruments in an economy, as well as tangible assets and net worth. The model captures, as special cases, distinct data-specific problem scenarios and permits the estimation of sector holdings of both assets and liabilities, total outstanding financial instrument volumes, and total sector holdings. Examples of sectors are households, private businesses, commercial banks, etc.; examples of financial instruments are money market fund shares, corporate equities, life insurance reserves, etc.

To solve such problems, we have recently developed network decomposition algorithms based on the SEA and applied them to establish a consistent set of financial flow of funds accounts for the Federal Reserve Board (see [8]). This is the first stage in the development of a general equilibrium model of credit policy reform for the U.S. Department of Agriculture. We have also recently developed a multisector, multi-instrument general financial equilibrium model in which each sector uses the market value of its portfolio and its individual assessment of price risk as simultaneous choice criteria to maximize utility of its expected future portfolio. The variational inequality decomposition algorithm applied to this general equilibrium model results in series of network subproblems, which, again, can be solved simultaneously and in closed form (cf. [13] and [10]).

24.4. Implementation on the CM

The Connection Machine model CM-2 from the Thinking Machines Corporation (TMC) is a distributed-memory, SIMD (single-instruction multiple-data) massively parallel processing system with 64K processors in its full configuration. Each processing element is under the control of a microcontroller that sends instructions from a front-end computer to all of the elements for execution. The model of computation is data level parallelism; that is, all processors execute identical operations.

The language that we used for the implementation was CM Fortran version 1.0. It is a high-level language that compiles into Paris, the assembly-level language of the machine. It is a very compact language: the addition of two matrices, for example, is expressed in a single statement.

We now briefly describe some of the intrinsic functions of CM Fortran that make it very well suited for implementing the exact equilibration algorithms outlined in §24.3. For example, the intrinsic function cmf_order sorts elements of a matrix either rowwise or columnwise and returns the indices. The minval and maxval functions return the smallest and largest elements, respectively, in either a row or a column of an array. The transpose feature of a matrix is useful in minimizing the cost of communication between processors in which the data elements are located.

Since matrix operations must be conformable, i.e., the matrices operated on must be of the same dimensions, one may need to change a matrix into a vector or vice versa; the functions pack and unpack are very useful for such transformations. Also one may use the spread command to replicate a vector into a matrix.

Finally, we note the availability of logic statements, such as the where, `else`, `end` statements that check conditions on vector/matrix elements in parallel.

In order to take advantage of the data level parallelism a large number of processors are needed to operate on multiple copies of the data simultaneously. Note that in an input/output matrix consisting of 500 rows and 500 columns we would need 250,000 processors, which is greater than the number of physical processors available to us even in a fully configured CM-2. The CM-2, however, has the notable feature known as *virtual processors* (VP) that permits a processor to operate on multiple copies of the data. This feature is identical to having multiple physical processors operating on their own copy of the data. The VP ratio is defined as the ratio of the number of virtual processors to physical processors.

We considered the estimation of an input/output matrix for the U.S.; the matrix consisted of 485 row and 485 columns. We implemented the SEA (cf. [7, 4]), which resolves the general equilibrium system to be estimated into a series of supply market and demand market equilibrium problems of the form described in §24.3 (cf. (24.10)). In particular, there were 485 supply subproblems and 485 demand subproblems to be solved at each iteration, until convergence.

We now briefly highlight some important implementation issues. In particular, the solution of each of the 485 subproblems of the form (24.10), which consisted of 485 unknown x_{il} variables, was carried out by using 485 of the processors to first compute the ρ_l^v given in (24.11) for $v = 1, \ldots, 485$. A `shift` command was then utilized to bring the neighboring $h_{vl}, h_{v+1,l}$ values to the same location to minimize the communication. The $h_{vl} < \rho_l^v \leq h_{v+1,l}$ check condition was implemented using the `where`, `else`, `end` construct. All 485 demand problems were solved in the same fashion, simultaneously. The x_{il}'s for $i = 1, \ldots, 485$; $l = 1, \ldots, 485$ were then updated, also simultaneously.

We also implemented a parallel version of the SEA on the IBM 3090/600E. For this purpose we utilized as the base the serial Fortran code and added the Parallel Fortran (PF) constructs to handle the task allocation, that is, the assignment of each of the 485 demand/supply subproblems to the six CPUs. The conversion of the serial code to the parallel code was relatively straightforward in that only task origination statements, dispatch statements that allocated a demand/supply subproblem to the next available processor, a waiting statement for synchronization, and task termination statements had to be added to the original serial code.

Our experience with implementing the parallel version of the code on an IBM 3090/600E will now be compared and contrasted with our experience with implementing the code on the CM-2. As the discussion above reveals, a highlight of our experience was the ease with which the parallel implementation on the IBM 3090 was done. On the other hand, our serial Fortran code developed on the IBM 3090 was of limited value in the preparation of our CM Fortran code. Indeed, programming in CM Fortran on the CM-2 required

the use and application of entirely different concepts, and, consequently, even
the fundamental approach to the implementation of the exact equilibration
algorithm had to be rethought, as discussed above. Essentially, rather than
allocating a particular market subproblem to a processor, as was the philosophy
and realization on the 3090 implementation, all trade flows were, instead,
computed simultaneously on the CM-2; this was the finest level possible of
decomposition for the problem, both conceptually and architecturally.

Since we were able to eliminate most of the DO loops and to use CM Fortran
intrinsic functions, which are well-suited to our algorithm, we were able to
produce a very compact code vis-à-vis the 3090 PF code. The convenience
and elegance of having critical matrix operations represented in a single step
and the availability of useful intrinsic functions, however, was tempered by
the absence of certain features that are taken for granted in Fortran. For
example, in the version of CM Fortran available to us, only vectors and not
matrices could be used as indices into arrays. Even though vectors could be
used as indices into other vectors or matrices, however, no two elements of
such an index vector could have the same value, which would have resulted in
"collisions."

We now present the results of our implementations on the two architectures.
In Table 24.1 we present the results of the computations for the CM-2
system. The CM-2 used for these results was located at the Northeast Parallel
Architectures Center (NPAC). It consisted of 32K processors; each processor
contained a local memory of 8K bytes. A SUN workstation was used as the
front end. The problem was solved using 8K processors with a VP ratio of 32,
16K processors with a VP ratio of 16, and 32K processors with a VP ratio of 8.

TABLE 24.1

Results of SEA implemented on a CM-2.

No. of physical processors	Real time (sec.)	CM time (sec.)	Front-end virtual time	CM % utilization
8K	52.05	51.74	52.05	99%
16K	29.86	29.58	29.86	99%
32K	16.76	16.34	16.72	98%

Observe that the CM time decreases approximately linearly as the number
of processors is increased. The estimated CM time for 64K processors would be
approximately nine seconds. We note that the same problem was solved on an
IBM 3090/600E and required 438.35 CPU seconds for the serial Fortran code
(cf. [7]). In Table 24.2 we report the stand-alone results of SEA implemented
on an IBM 3090/600E for the same example. SEA required four iterations for
convergence on both architectures.

TABLE 24.2

Results of SEA implemented on an IBM 3090/600E.

Number of processors	Wallclock time (seconds)
1	444.18
2	229.85
4	118.76
6	86.32

These numerical results strongly suggest that our implementation on the CM-2 using CM Fortran is very promising. We believe that further enhancements to the language will make even more efficient implementations realizable.

24.5. Conclusions

A wealth of computational problems in economics are amenable to solution by algorithms developed for the exploitation of advanced architectures. Here we have attempted to interest the reader in economic equilibrium problems that can be formulated as variational inequality problems or as projected dynamical systems and, moreover, may have an underlying network structure. For additional background on parallel algorithms and applications to economics, we refer the reader to [13].

24.6. Acknowledgments

Support for this research was provided, in part, from the National Science Foundation under the Faculty Awards for Women Program, NSF grant DMS 902471, and the US Department of Agriculture and is gratefully acknowledged. This research was conducted using the computational resources of the NPAC at Syracuse University, the National Center for Supercomputer Applications at the University of Illinois, and the CNSF at Cornell University.

References

[1] D.P. BERTSEKAS AND J.N. TSITSIKLIS, *Parallel and Distributed Computation*, Prentice–Hall, Englewood Cliffs, NJ, 1989.

[2] A. EYDELAND AND A. NAGURNEY, *Progressive equilibration algorithms: The case of linear transaction costs*, Comput. Sci. Econom. Management, 2(1989), pp. 197–219.

[3] P. DUPUIS AND A. NAGURNEY, *Dynamical systems and variational inequalities*, Ann. Oper. Res., 44(1993), pp. 9–42.

[4] D.S. KIM AND A. NAGURNEY, *Massively parallel implementation of the splitting equilibration algorithm*, Comput. Econom., 6(1993), pp. 151–161.

[5] A. NAGURNEY, *Migration equilibrium and variational inequalities*, Econom. Lett., 31(1989), pp. 109–112.

[6] A. NAGURNEY AND D.S. KIM, *Parallel and serial variational inequality decomposition algorithms for multicommodity market equilibrium problems*, Internat. J. Supercomputer Appl., 3(1989), pp. 34–59.

[7] A. NAGURNEY AND A. EYDELAND, *A splitting equilibration algorithm for the computation of large-scale constrained matrix problems: Theoretical analysis and applications*, in Computational Economics and Econometrics, Advanced Studies in Theoretical and Applied Econometrics 22, H.M. Amman, D.A. Belsley, and L.F. Pau, eds., 1992, pp. 65–105.

[8] A. NAGURNEY AND M. HUGHES, *Financial flow of funds networks*, Networks, 22(1992), pp. 145–161.

[9] A. NAGURNEY, *Network Economics: A Variational Inequality Approach*, Kluwer Academic Publishers, Boston, MA, 1993.

[10] ——, *Variational inequalities in the analysis and computation of multi-sector, multi-instrument financial equilibria*, J. Econom. Dynamics Control, 18(1994), pp. 161–184.

[11] A. NAGURNEY AND D. ZHANG, *Projected Dynamical Systems and Variational Inequalities with Applications*, Kluwer Academic Publishers, Boston, MA, 1995.

[12] A. NAGURNEY, T. TAKAYAMA, AND D. ZHANG, *Massively parallel computation of spatial price equilibrium problems as dynamical systems*, J. Econom. Dynamics Control, 19(1995), pp. 3–37.

[13] A. NAGURNEY, *Parallel computation*, in Handbook of Computational Economics, H. M. Amman, D. Kendrick, and J. Rust, eds., Elsevier, North Holland, 1996, pp. 331–400.

[14] A. NAGURNEY, T. TAKAYAMA, AND D. ZHANG, *Projected dynamical systems modeling and computation of spatial network equilibria*, Networks, 26(1995), pp. 69–85.

Chapter 25

Solving Nonlinear Integer Programs with a Subgradient Approach on Parallel Computers

Robert Bixby
John Dennis
Zhijun Wu

Editorial preface

Mixed integer programming ranks among the most difficult computational problems to solve. It is further exacerbated by a nonlinear objective function, and the complexity of developing solution schemes for realistic problem sizes is significant. An intricate algorithm is described that, in the context of gas pipeline network optimization, offers parallelism in which good speedups are achieved on an nCube.

This article originally appeared in *SIAM News*, Vol. 25, No. 4, July 1992. It appears in its original form.

Many large, and hard, nonlinear integer programming problems—or, more generally, mixed integer nonlinear programming problems—arise in both theoretical study and practical applications. No general and efficient solution to these problems can be found with traditional computers. Thus, it becomes necessary to develop new algorithms for solving these problems on advanced architectures, such as parallel computers. This article describes a new algorithm for solving nonlinear integer programs, developed at the Center for Research on Parallel Computation at Rice University, and shows how the solution to a nonlinear integer program can be achieved on a parallel computer.

Nonlinear integer programming is concerned with solving the problem

$$(25.1) \qquad \min \qquad f(x)$$
$$x \in B^n = \{0,1\}^n$$

or its natural extension

(25.2)   **min** $f(x)$

                                $x \in R^n$ integral,

where $f : R^n \longrightarrow R$ is a general nonlinear function.

This class of problems contains many NP-hard problems and has both theoretical and practical applications. For example, consider the problem that for any norm $\| \cdot \|$,

(25.3)   **min** $\| b - Ax \|$

                                $x \in R^n$ integral,

where $b \in R^m$ and A is an $m \times n$ matrix with integer elements. This problem, called the closest vector problem in integer programming, has been proven to be NP-complete even for simple norms such as l_2 and l_∞.

Another example is related to the solution of a class of more general problems: mixed integer nonlinear programming problems. It turns out that under some circumstances, problems of this class can be reduced to general nonlinear integer programs. For instance, an unconstrained mixed integer nonlinear program,

(25.4)   **min** $g(x, y)$

(25.5) $y \in R^m$

                                $x \in R^n$ integral,

can be formulated, under some appropriate assumptions, as the following problem:

(25.6)   **min** $f(x)$

                                $x \in R^n$ integral,

with $f(x) = $ **min** $\{g(x, y) : y \in R^n\}$.

Mixed integer nonlinear programming has recently found an important application in the steady-state optimization of gas pipeline network operation. Percell [4] studied three model problems for pipeline networks with up to 12 compressor stations, each of which contains several compressor units. Given demands and resources for the network, the goal of the optimization is to find steady-state profiles of pressure, flow, temperature, and compressor station configurations (i.e., choice of compressor units to be run). These solutions are optimal for chosen objectives, such as minimizing the amount of fuel, maximizing the total flow, and maximizing gas inventory.

Mathematically, the problem is formulated as a minimization of a nonlinear objective function subject to nonlinear equality constraints. There are two groups of variables: continuous and discrete. The continuous variables correspond to pressures, flows, temperatures, speeds, powers, etc. The discrete variables correspond to compressor stations, and their values are the number of compressor units to be run.

There are several well-studied methods for nonlinear continuous optimization. The challenge of applying them to the pipeline optimization problem lies in finding an effective way to handle the discrete variables, i.e., in deciding how many compressors should be used and which should be turned on or off for the network operation. Simply enumerating all possible values for the discrete variables is infeasible as there are exponentially many combinations with respect to the number of discrete variables. Besides, the number of discrete variables tends to be large in practice. Anglard and David [1] considered a gas pipeline network with 196 compressor stations, each of which contained up to eight compressor units. A robust way to deal with the discrete variables is to solve a nonlinear integer program, as illustrated in (25.6).

Several approaches to the solution of a nonlinear integer program have been studied in the last 30 years (see [3] for a general review). The main ones are enumeration, algebraic, and linearization approaches. Most of them work for problems with special structures. But for problems with general objective functions, such as (25.6), the enumeration method is hardly efficient, and the other two approaches cannot be applied owing to their special requirements for the forms of the objective function.

25.1. The Subgradient Approach

Bixby, Dennis, and Wu [2] have proposed a subgradient approach to nonlinear integer programming problems with more general or complicated objective functions. With this approach, a nonlinear integer program in the form of (25.1) is considered as a nonsmooth problem over the set of 0–1 integer points. Notions of subgradients and supporting planes are then introduced for the objective function at integer points. By computing subgradients and supporting planes, a sequence of linear approximations to the objective function is constructed, and the optimal solution is found by successively solving the sequence of linear subproblems.

More specifically, the subgradient algorithm iteratively searches for the solution among integer points. At each iteration, it generates the next iterative point by solving the problem for a local piecewise linear model that is constructed from the supporting planes for the objective function at the set of iterative points already generated. The supporting planes are computed by using special continuous optimization techniques. The problem for the local piecewise linear model in each iteration is equivalent to a linear integer minimax problem, which can be solved with a standard method for linear integer programming.

In Algorithm 1, f^r, the restriction of f to B^n, where $B^n = \{0,1\}^n$, is called the discrete objective function and $\partial f^r(x^{(i)})$ is the subdifferential of f^r at $x^{(i)}$. Formally, the algorithm can be outlined as follows.

Algorithm 1 {*The Subgradient Algorithm*}
0 {*Initialization*}
   $T = \emptyset$, $H = \emptyset$, $i = 0$
   pick up $x^{(i)} \in B^n$
1 {*Iteration*}
   do while $i \leq m$
      1.1 {*Optimality testing*}
         if $x^{(i)} \in T$ or $0 \in \partial f^r(x^{(i)})$ is known
         then
            $x^{(i)}$ is the optimal solution, stop
      1.2 {*Generating supporting planes*}
         $T = T \cup \{x^{(i)}\}$
         $H = H \cup \{g_{x^{(i)}} : g_{x^{(i)}}(x) = f^r(x^{(i)}) + s_{x^{(i)}}^T(x - x^{(i)}), \quad s_{x^{(i)}} \in \partial f^r(x^{(i)})\}$
      1.3 {*Solving a linear integer minimax problem*}
         find a solution $x^{(*)}$ for
         $\min_{x \in B^n} \{p(x) = \max \{g(x) : g \in H\}\}$
      1.4 {*Updating*}
         $i = i + 1$
         $x^{(i)} = x^{(*)}$
   end do

There are three major steps at each iteration: optimality testing, generation of a supporting plane, and solution of a linear integer minimax problem. The essential work in the first step is to construct the optimality criteria. The challenge of the second step lies in finding a method for computing a subgradient such that a supporting plane can be generated. For the third step a special linear integer program needs to be solved.

The optimality criteria are based on the following facts (given in [2]):

Fact 1: A **necessary** and **sufficient** condition for $x^* \in B^n$ to be the minimizer of f^r (or f) over B^n is $0 \in \partial f^r(x^*)$.

Fact 2: For the sequence $T = \{x^{(j)} \in B^n, \ j = 1, \ldots, i\}$ generated by the algorithm at the ith iteration, if $\exists j < i$ such that $x^{(j)} = x^{(i)}$, then $x^{(i)}$ must be an optimal solution.

It follows from Fact 2 that the algorithm stops whenever an iterate is repeated. Because there are only finitely many distinct iterates, the algorithm is finite.

Consider the generation of a supporting plane for the function f^r at a given integer point $\bar{x}$. If g is the function for the supporting plane, then g is a linear function and

$$(25.7) \qquad g(x) = f^r(\bar{x}) + s^T(x - \bar{x}), \qquad s \in \partial f^r(\bar{x}).$$

To obtain this function, $f^r(\bar{x})$ can be computed easily, but the subgradient s must be chosen such that g bounds f^r from below as tightly as possible. In the case where f is convex and differentiable, it is easy to verify that $\nabla f(\bar{x})$, the gradient of f at $\bar{x}$, is a subgradient of f^r at $\bar{x}$. A trivial way to choose s, therefore, is to set s to $\nabla f(\bar{x})$. However, with this subgradient, g could be too "steep" to be a preferred supporting plane; in this case a subgradient other than $\nabla f(\bar{x})$ is required, such that g is as "flat" or "close" to f^r as possible.

Unfortunately, there are no simple methods for computing any subgradients for general nonlinear nonsmooth functions. In this algorithm subgradients are obtained by a process that can successively improve a given subgradient. The process starts with the subgradient $s = \nabla f(\bar{x})$ and then updates it such that the corresponding supporting plane g is "lifted," i.e., made "flatter" or "closer" to f^r. The updated s remains a subgradient as long as g still supports f^r at $\bar{x}$, i.e.,

$$(25.8) \qquad g(x) \leq f^r(x) \qquad \forall x \in B^n.$$

The lifting process continues until the best possible supporting plane is obtained. However, for every update, condition (25.8) must be verified. For a given subgradient s, if S is defined such that $x \in S$ if $f(x) \leq g(x)$, then condition (25.8) is equivalent to the following statement:

> A vector s is a subgradient of f^r if and only if the interior of S does not contain 0–1 integer points.

Figure 25.1 illustrates with a simple example how the lifting process is conducted and condition (25.8) is guaranteed. In this example, the lifting process is applied to find a subgradient of f^r at $\bar{x}$. First, s is set to $\nabla f(\bar{x})$. The supporting plane defined by this subgradient is $g_{(0)}$. Then s is updated to "lift" $g_{(0)}$ a little bit, and $g_{(1)}$ and $S_{(1)}$ are obtained. Let $A = \{x \in R^n : x_i \geq 0 \text{ if } \bar{x}_i = 1, \text{ and } x_i \leq 1 \text{ if } \bar{x}_i = 0, i = 1, \ldots, n\}$. Geometrically, A is a region that contains B^n, and its boundaries are formed by hyperplanes $x_i = 1 - \bar{x}_i$, $i = 1, \ldots, n$. Once it has been observed that the interior of A contains no points in B^n other than $\bar{x}$, condition (25.8) holds if $S_{(1)}$ is inside of A. In general,

> for each S obtained in the lifting process, the interior of S does not contain 0–1 integer points as long as S is contained in A.

To obtain better subgradients, s can be further updated until the corresponding S hits the boundary of A (see $g_{(2)}$ and $S_{(2)}$ in Figure 25.1).

Now consider the updated subgradient s and its corresponding S. Let d_i be the distance between S and the ith boundary of A. Then $d = (d_1, \ldots, d_n)$ is a function of s. It can be proven that the function is well defined under some assumptions. The lifting process can then be formulated mathematically as an

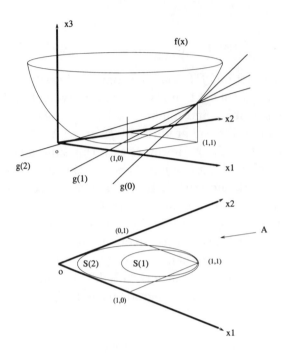

FIG. 25.1. *The lifting process for computing subgradients.*

optimization problem:

$$(25.9) \qquad\qquad \textbf{min} \quad \| d(s) \|$$
$$\textbf{st} \quad d_i(s) \geq 0, \qquad i = 1, \ldots, n.$$

The major computation for this optimization problem is the evaluation of the function $d(s)$ for each s. In terms of the lifting process, distances between the boundaries of S and A need to be calculated for each lifting step. If the distances are positive, S is inside A and hence condition (25.8) holds. In any case, $d_i(s)$ for any i can be calculated by first computing an extreme point of S along the x_i direction and then calculating the distance between the extreme point and the ith boundary of A. The extreme point of S can be found by solving a relatively simple continuous optimization problem (linear objective function with only one nonlinear constraint).

Finally, the third step of each iteration involves the solution of a linear integer minimax problem. Because the problem can be formulated as a special linear integer program, a branch-and-bound procedure can be applied. To compute the bound for every branching step, the standard simplex method is used to solve the dual problem of the linear relaxation problem.

25.2. Generating a Supporting Plane in Parallel

As described in the preceding section, the subgradient algorithm requires that a number of subproblems be solved at every iteration. These subproblems are difficult, and their solutions involve large amounts of computation. Reducing the time required to solve these subproblems appears to be a very important and challenging consideration. Bixby, Dennis, and Wu [2] propose the use of parallel computers to speed up the algorithm so that reasonably large problems can be solved. In fact, parallelism can indeed be exploited for the algorithm to achieve high performance.

The subgradient algorithm carries at the top level an iterative procedure that is sequential and cannot be done in parallel. Numerical experiments show that for most problems the algorithm can find an optimal solution in only a few, at most $\mathcal{O}(n)$, iterations. Therefore, the algorithm can be effectively parallelized if the computation at each iteration can be done in parallel.

For each iteration the major computational costs are those of generating a supporting plane and solving a linear integer minimax problem. To generate the supporting plane, the lifting process is conducted to achieve a solution to problem (25.9), where most of the work is in the evaluation of the function $d(s)$, as discussed previously. Computing each component of $d(s)$ involves the solution of a continuous optimization subproblem. A total of n subproblems, therefore, need to be solved to obtain all components of $d(s)$ for each given s. The computation here could be very expensive.

Each of the n subproblems is totally independent, however, and they all have almost the same structure and size. Thus, it is easy to introduce parallelism to do the function evaluation, and if up to n processors are used, subproblems can be distributed evenly over the processors and solved in parallel with little communication overhead. Parallelism of this type is suitable for such parallel systems as the Intel iPSC/860 hypercube, with up to 128 processors, and the nCube, with up to 8192 processors. For large problems ($n = 100 \sim 1000$), very high performance can be achieved as many, up to n, processors can be used.

25.3. Parallel Branch-and-Bound for the Linear Subproblem

The linear integer minimax problem induced at each iteration by the subgradient algorithm is solved by applying a branch-and-bound procedure, a popular scheme for solving linear integer programming problems. But for large problems—those with, say, 100 to 1000 variables—the method may still produce so many subproblems that the solution cannot be obtained in a reasonable time.

The branch-and-bound procedure can be represented by a tree, with nodes corresponding to subproblems and branches corresponding to relations among subproblems. The process can thus be parallelized by exploiting the tree structure, although it is not straightforward to do so. The tree structure is constructed dynamically as the procedure moves forward, and the parallelism among subproblems often is not known until the subproblems are generated.

Algorithm–2 is the parallel branch-and-bound procedure used within the subgradient algorithm for the linear subproblems. The algorithm is based on the general branch-and-bound method, but a multiple branching strategy is used instead of the more common binary branching. More precisely, if p processors are used, the algorithm always makes p branches at every branching step, producing p subproblems and solving them, one for each processor, in parallel. After solving the subproblems, the algorithm proceeds by making branches recursively for the new subproblems.

With the multiple branching strategy, processors can be scheduled in a systematic way: each time p subproblems are produced, they are assigned to the p processors, one for each processor. All groups of p subproblems produced in this way are almost the same, except for some variables set to different values. The load is thus balanced automatically in solving the subproblems. Moreover, because subproblems are generated regularly and correspond to processors, first subproblem for first processor, second subproblem for second processor, etc., processors do not need to trace a global subproblem stack to find the subproblems they need to solve. Instead, each processor has only a small local stack of its own subproblems.

Globally, the parallel branch-and-bound procedure conducts a depth-first search because at each step the new subproblems always are processed first. But after every group of p subproblems is solved, they can be sorted according to some priority. The branching can then be made for the subproblems in the sorted order (local best-first branch).

> **Algorithm 2** { *The Parallel Branch-and-Bound Algorithm* }
>   * { *Initial Procedure* }
>   initialize p, $\underline{z}_p$, z, and x (p represents the initial subproblem)
>   solve p
>   let $\underline{z}_p$ and x_p be the optimal value and solution
>   **if** x_p is integral **then**
> $z = \min\{z, \underline{z}_p\}$, $x = x_p$, stop
>   **push**(p, P) (P is a local subproblem stack)
>   **branch-and-bound**$(1, x, z)$
>   **pop**(P)
>   * { *End of Initial Procedure* }
>
>   * { *Recursive Procedure* }
>   **branch-and-bound**(i, x, z)
>   broadcast z_p from processor i
>   **if** $z_p \geq z$, return
>   **if** processor $\# = i$ **then**
>   select branching variables
>   broadcast branching variables from processor i
>   generate and solve subproblem p

> let $\underline{z}_p$ and x_p be the optimal value and solution
> if x_p is integral **then**
> $z = \min\{z, \underline{z}_p\}$, $x = x_p$
> update z and x if necessary
> **push**(p, P)
> **for** $j = 1, \ldots, \#$ of processors **do**
>   **branch-and-bound**(j, x, z)
> **pop**(P)
> * {*End of Recursive Procedure*}

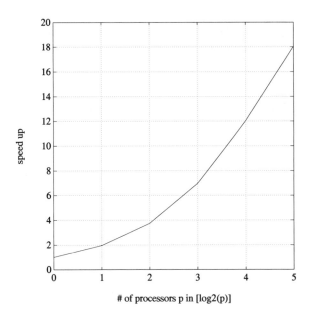

# of processors p in [log2(p)]

FIG. 25.2. *Speedup obtained with varying numbers of processors.*

25.4. Remarks

Computational experiments have been conducted with a parallel implementation of the subgradient algorithm on a 512-node nCube located at the California Institute of Technology. The program is written in Express C, an extended C language for distributed-memory parallel computers. In addition to standard C, the language provides a variety of message-passing functions.

Small problems (up to 64 dimensions) have been tested. Preliminary results show that most of the test problems can be solved by the subgradient algorithm in only a few, at most $\mathcal{O}(n)$, iterations, if proper initial guesses are used. For a test problem of dimension n there are 2^n 0–1 integer points. In the worst case, therefore, the algorithm might need to run 2^n iterations, which cannot

be done in a reasonable time, even for an average n, say 32, for which there are
$4,294,967,296$ 0–1 integer points in total. In reality, however, most of the test
problems can be solved more efficiently than by exhausting all possible integer
points. For test problems of dimension 32 or 64, with some initial guesses, only
several iterations were taken.

For the parallel implementation of the program on the nCube, if the number
of processors p $(p \leq n)$ is doubled, the total computation time can often be
reduced by almost half. Figure 25.2 shows the speedup that can be obtained for
a 32-dimension test problem when different numbers of processors are used.
The greatest speedup is about 18, which can be improved by testing larger
problems.

References

[1] P. ANGLARD AND P. DAVID, *Hierarchical steady–state optimization of very
large gas pipelines*, presented at Pipeline Simulation Interest Group Annual Meeting,
Toronto, Ontario, Canada, 1988.

[2] R. BIXBY, J. DENNIS, AND Z. WU, *A Subgradient Algorithm for Nonlinear
Integer Programming and Its Parallel Implementation*, Technical Report, Center for
Research on Parallel Computation, Rice University, Houston, TX, 1991.

[3] P. HANSEN, B. JAUMARD, AND V. MATHON, *Constrained Nonlinear 0–1
Programming*, RRR #47–89, RUTCOR, Rutgers University, New Brunswick, NJ,
1989.

[4] P.B. PERCELL, *Steady–state optimization of gas pipeline network operation*, in
Proc. Pipeline Simulation Interest Group Annual Meeting, Tulsa, OK, 1987.

Chapter 26

Parallelizing FDTD Methods for Solving Electromagnetic Scattering Problems

Sandy Nguyen
Brian Zook
Xiaodong Zhang

Editorial preface

Owing to increased use of electronics in many forms of industrial and commodity products, the study and characterization of electromagnetic scattering is becoming a mainstream area of study for industry. The solution to the electromagnetic scattering problem using the finite-difference time-domain (FDTD) method via a parallel virtual machine (PVM) for networks of workstations is presented.

This article originally appeared in *SIAM News*, Vol. 28, No. 4, April 1995. It appears in its original form owing to its recent appearance in *SIAM News*.

The study of electromagnetic scattering involves the modeling, simulation, and analysis of the electromagnetic responses of complex systems to various electromagnetic stimuli [4]. Knowledge of a system's responses is used in the design and modification of the system. The finite-difference time-domain (FDTD) method, one of the many computational approaches to this problem, is applicable to a spectrum of problems [2]. The problem domain of the simulation, consisting of the scatterer surrounded by some amount of free space, is gridded along Cartesian coordinates into cells and is truncated at planar boundaries. The scatterer can be a collection of objects of various shapes and sizes, composed of a variety of materials. The FDTD method explicitly follows the evolution of an incident electromagnetic wave as it impinges on the modeled system, calculating the values of the electromagnetic field throughout the volume of the problem space.

This application is a good arena for the study of algorithms and implementations in parallel and distributed computing environments. Electromagnetic scattering is a computationally intensive task: the problem domain can become very large as the number of cells or timesteps is increased for computing precision. Furthermore, large problems require large amounts of memory, making a distributed-memory environment desirable. Another consideration is the computational pattern of the problem. The mathematical foundation of an FDTD simulation is the solution of the two Maxwell curl equations in the time domain:

$$(26.1) \qquad \epsilon \frac{\partial \vec{E}}{\partial t} = \nabla \times \vec{H} - \sigma \vec{E},$$

$$(26.2) \qquad \mu \frac{\partial \vec{H}}{\partial t} = -\nabla \times \vec{E}.$$

These coupled partial differential equations are discretized by means of central finite differencing. Because the fields are updated at each timestep, this computation has the special feature of involving only nearest-neighbor interactions. This minimization of global communications makes the FDTD method well suited to a distributed computing implementation.

We began this work with a complete sequential production code. Then, using PVM, we developed a parallel implementation on a distributed workstation network. By first implementing the application across a local area network (LAN), we avoided some costly developmental time on an MPP system while still producing efficient distributed algorithms. A cluster of workstations is a useful environment for intermediate-sized simulations—those that are too intensive for a single workstation but do not warrant the use of an MPP system. In addition to PVM, we studied message-passing models for FDTD methods and their performance on the Intel iPSC/860 and the CM-5 multicomputers.

26.1. The FDTD Method

The FDTD code we used to solve the electromagnetic scattering problem was originally developed in Fortran by Luebbers [2]. This program uses a marching-in-time scheme to follow the evolution of the electric (E) and magnetic (H) fields in the problem space. The computational domain is defined as a set of Cartesian cells on which the electric and magnetic field components are to be computed. A sample cell is illustrated in Figure 26.1.

The electric field components are computed at the midpoints of the cell edges; the magnetic field components are computed at the centers of the cell faces. Only the component of the electric field parallel to the cell edge is computed, and only the component of the magnetic field perpendicular to the cell face is computed. Offsetting the field components within each cell results in spatially centered differencing. Time-centered derivatives are obtained by computing the electric field values at integer timesteps and the magnetic field values at half-integer timesteps. The basic algorithm for the FDTD method is as follows.

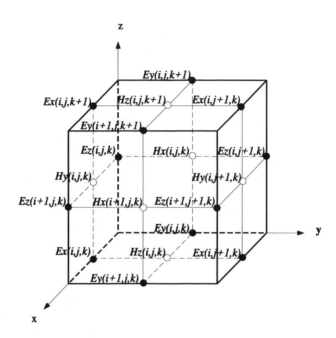

FIG. 26.1.   *A sample FDTD cell showing the points for computation of the electric (E) and magnetic (H) fields.*

```
     Initialize problem space.
Do:
   Update E-field with neighboring H-field.
   Increment timestep by 1/2.
   Update H-field with neighboring E-field.
   Apply H-field boundary condition.
   Save near-field values.
   Calculate and accumulate far-field.
   Increment timestep by 1/2.
Until predetermined timestep is reached.
Save far-field values.
```

The field-component algorithms calculate the response of a component from its value at the previous timestep and from the values for its nearest neighbors, according to the material type at that component location. For updating the E-field components, the neighboring H-field components must be available for the previous half timestep. Similarly, updating the H-field components requires the neighboring E-field components at the previous half timestep.

Since many of the desired modeling structures are situated in free space, the scattered fields should propagate into boundless space. The FDTD computational space must be bounded, however, and so for each timestep, an outgoing-radiation boundary condition is applied to absorb the scattered field at the outermost portion of the computational domain and thus prevent artificial reflections. The boundary conditions need to be applied only to either the tangential E-field or H-field components.

26.2. Distributed Method and Implementation

We have developed distributed FDTD algorithms by applying a divide-and-conquer strategy. Our principal objectives for the distributed algorithms are to minimize the communication cost and balance the execution workload.

26.2.1. Domain Decomposition via Recursive Bisection.
The domain decomposition scheme we used to parallelize the FDTD program is a bisection algorithm. The computational domain is cut into two equal parts along its longest edge, and these two parts are in turn cut along their longest edges. Cutting of the subboxes continues until the number of boxes equals the number of nodes available [1]. These subboxes are then assigned to the nodes. Because of the nature of the recursive bisection, the number of nodes that can be utilized concurrently must be a power of 2. More sophisticated decomposition algorithms can be used to get around this minor limitation as long as the common faces of adjacent partitions have the same dimensions.

For most of the cells in a subdomain, nearest-neighbor information local to the node is sufficient for updating the field at each timestep. One layer of electric and magnetic cells, however, will require field information from a neighboring node for completion of the update, as illustrated in Figure 26.2. To update E_x along the partition boundary, node 2 requires the value of H_z contained in node 1. Conversely, to update the value of H_z, node 1 requires the value of E_x contained in node 2. For communication of the required grid points to the neighboring nodes, the cells along the partition boundaries of the subboxes are duplicated when the computational box is partitioned. Each duplicated cell, called a "guard grid," is computed by one node and then sent to the neighboring node.

26.2.2. Communicating Guard Grids.
After the electric field grids have been updated, each partition needs to acquire the E-field guard grids from the neighboring partition in order to do the magnetic field updates. Then, once all the H-field grids have been updated, the H-field guard grids have to be duplicated by the neighboring partitions so that the E-field grids can be updated at the next iteration.

A strategy used by designers of effective algorithms for distributed computing is to minimize the ratio of communication to computation and to maximize the size of messages. We have implemented two algorithms for communicating the guard grids. The straightforward method is to send all the grid points directly to the corresponding guard grids of neighboring nodes and wait to

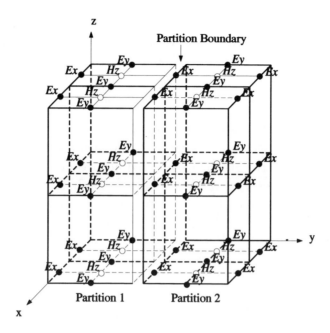

FIG. 26.2. *The partition boundary between two cells showing the relationship of the discretization points for the electric and magnetic fields in the guard grid.*

receive the local guard grids from other neighboring nodes. The guard grids along the edges and at the corners are sent to multiple nodes. This algorithm requires exchanging at most seven messages: three to adjacent partitions that intersect at boundary planes, three to the partitions that intersect only along the edges, and one to the partition that is common at the extreme corner, at either the lowest or the highest partition indices. Because the messages from sending nodes do not arrive in a predetermined sequence, the receiving node must recognize the origin of each message in order to unpack it properly.

The second communication pattern interleaves the send and receive messages by having all partitions simultaneously communicate in one direction before starting the next direction. All the partitions send and receive first along the XY plane, then along the YZ plane, and finally along the ZX plane. Unlike the first pattern, the second pattern sends duplicate guard grids along the edges and at the corner indirectly, by combining them with other partition boundaries. Use of the second method will reduce the number of messages to at most three packets, while increasing the size of the message packets. Although we would expect the second pattern to be more efficient, we have not found this to be the case, as discussed below.

26.2.3. Outer Partitions. Following the electric and magnetic field updates, the outer boundary condition must be applied to the magnetic field

values along the outer surfaces of the computation box. Without any attempt
to balance the load among all the nodes, we make the outer nodes responsible
for applying the boundary condition to the magnetic fields in their partitions.
In addition, the nodes containing cells along the boundary of the closed surface
are used to complete far-field values. We do not discuss the far-field transformation in this article; details can be found in [3].

26.2.4. The FDTD Distributed Algorithm.   Either the master/slave or
the single-program multiple-data (SPMD) model with static load balancing can
be used to implement the FDTD distributed algorithm. In the master/slave
model, the master is in charge of partitioning the problem space and assigning
a subset of the problem to each slave. Each slave then applies the FDTD code
to its local problem space and communicates the guard grids to the master
when necessary. With the SPMD model, each node concurrently partitions
the same problem space and selects the partition to which it will apply the
FDTD distributed code, depending on its node number. The concurrent FDTD
algorithm proceeds as follows.

```
    Initialize local problem space.
Do:
    Update local E-field with local H-field.
    Communicate local E-field guard grids.
    Increment timestep by 1/2.
    Update local H-field with local E-field.
    If outer partition,
       Apply H-field boundary condition.
    Communicate local H-field guard grids.
    Save local near-field values.
    If outer partition,
       Calculate and accumulate local far-field.
    Increment timestep by 1/2.
Until predetermined timestep is reached.
Global sum of far-field.
If node 0,
    Save far-field values.
```

The application starts by determining the number of nodes available for
the computation. Using the bisection algorithm, it partitions the problem
space and assigns a partition to each node. Depending on the model, the
partitioning of the problem space can be executed either by one node or by all
nodes concurrently. Once the partitions have been defined, each node applies
the distributed FDTD code. The outer boundary condition and the far-zone
calculations apply only to the partitions that contain the outer boundaries of
the problem space.

26.3. FDTD on a Workstation Network

The concurrent FDTD code was first implemented on a network of workstations. The message-passing library used to support the distributed computing is PVM version 3.2.

For evaluating the distributed FDTD code, the parallel virtual machine (PVM) environment that would most closely emulate an MPP is a cluster of homogeneous workstations in a closed LAN, which minimizes contention from other sources. Since this was not possible, our alternative was to run the code in a heterogeneous environment during periods of light network traffic. Because of the heterogeneity of the network, we used the master/slave model, with the fastest workstation running the master program.

The workstation network used for this computation is composed of four SUN SPARCstation 2 and four IBM RS/6000 machines. The four IBMs are clustered in one LAN, and the four SUN workstations are spread across various LANs, all located at the Southwest Research Institute. To compare the efficiency of the two communication algorithms on a network of workstations, we treated the heterogeneous cluster as a homogeneous cluster of workstations of the slowest type, using the homogeneous partitioning algorithm.

As mentioned earlier, we expected communication pattern 2 to be more efficient than pattern 1. To compare the two communication patterns, we ran two sets of tests. By running FDTD programs with the two communication patterns, we were able to compare the effects of the communication overhead. Figure 26.3 plots the execution times for a fixed problem size, $64 \times 64 \times 64$, running on one, two, four, and eight computers. The performance results indicate that the first communication pattern used with the FDTD program outperformed the second one when more than two processor nodes were used. Figure 26.4 provides further insight, showing significant differences in the communication overheads for the two programs when we increased both the size of the problem and the number of nodes in the network to keep the parallel computing efficiency roughly constant.

As shown in Figure 26.4, communication pattern 1 is more efficient than pattern 2. The use of pattern 2 produces extra delays because the nodes have to wait for the first message to arrive. Some information from the first message is then sent on with the second message, and a similar delay occurs with the communication of the third message. With communication pattern 1, the transmission delay is not as apparent. All seven messages are packed and sent at the beginning; by the time the nodes are ready to receive, the first few messages have already arrived at their destinations and are ready to be unpacked. Because the actual data transfer takes place during the packing and unpacking of other messages, the waiting time is avoided.

Both programs running on the eight-processor virtual machine achieved reasonably good speedups (5.3–6.1) and efficiencies (67–76%). We expect further gains for the FDTD program with the first communication pattern, as compared with the second, when a larger problem is solved on a larger system.

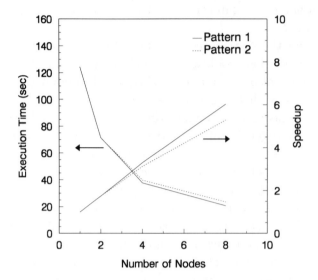

FIG. 26.3. *Execution timings for the two communication patterns with PVM on the LAN-based workstation cluster.*

26.4. FDTD on the iPSC/860 and the CM-5

To port the distributed FDTD code from PVM to the iPSC/860 and the CM-5, we had to change three major sections of the code to suit the specific architecture of the target machine. The iPSC/860 NX and CM-5 CMMD message-passing routines replaced the PVM routines. Whereas the master/slave model was used to implement the distributed FDTD code in PVM, the SPMD model, sometimes called a "hostless program," was used on the iPSC/860 and the CM-5. In the SPMD model, a single node program runs on all the nodes. Knowing its own node number, each node can determine its partition of the problem space. Since each architecture has different message-passing mechanisms, different algorithms were used to communicate the guard grids on the various machines in order to minimize communication overhead.

To study the load balance and to evaluate the performance of the two communication patterns on the iPSC/860 and the CM-5, we executed tests similar to those performed on each machine under PVM. For the speedup test on the iPSC/860, we used a constant problem size of $(128 \times 128 \times 64)$, running on one, two, four, eight, 16, 32, and 64 nodes. Figure 26.5 shows the performance results for the two communication patterns. As the number of nodes was increased from one to 32, the speedup achieved was nearly perfect. Beginning at 32 nodes, the speedup began to decline, although the computation and

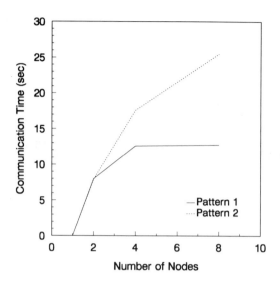

FIG. 26.4. *Communication overheads for the two communication patterns on the LAN-based workstation cluster when the problem size was increased to keep the parallel efficiency approximately constant.*

communication times continued to decrease. This is because as the number of processors is increased, the startup time for reading the configuration files and initializing the local problem space increases while the partition size decreases. Since startup time occurs only once, at the beginning of the simulation, this overhead becomes less significant when the number of timesteps is increased.

On the CM-5, the speedup test used a constant problem size of $(256 \times 256 \times 128)$, running on 32, 64, 128, 256, and 512 nodes. In calculating speedup, the base value used for comparison was the total execution time from the 32-node run. Figure 26.6 reports performance results for patterns 1 and 2.

Even though the iPSC/860 statistics are not as dramatic as those for PVM, communication pattern 1 still proved to be more efficient than pattern 2. The transmission rate on the iPSC/860 was faster than that on a LAN. The difference in the results for the CM-5 is even less pronounced: overall, the time spent communicating for pattern 1 was only slightly less than that for pattern 2.

26.5. Summary

Our experiments on a PVM LAN network and on the Intel iPSC/860 and the CM-5 show that our message-passing models for distributed FDTD algorithms are highly effective and scalable. The choice of network architectures is seen to have a direct effect on the computation's scalability. In general, the

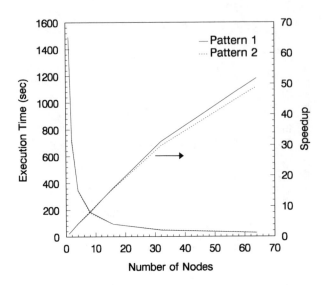

FIG. 26.5.   *Timings for the two communication patterns on the iPSC/860.*

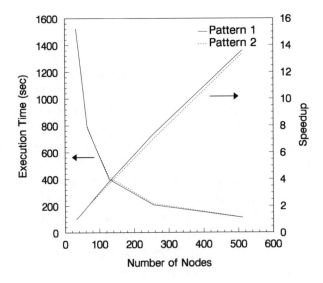

FIG. 26.6.   *Timings for the two communication patterns on the CM-5.*

interconnect latency increases proportionally as the iPSC/860 architecture
is scaled. The interconnect latency of the CM-5 data network decreases as
the system is scaled. A major design and performance evaluation issue for
a scalable multicomputer system is the interconnect latency limit and its

changing patterns in a computation as the system is scaled. In this study, what appeared to be a superior communication pattern was shown to be inferior.

References

[1] T. CWIK, *Parallel decomposition methods for the solution of electromagnetic scattering problems*, Electromagnetics, 12(1992), pp. 343–357.

[2] K.S. KUNZ AND R.J. LUEBBERS, *The Finite Difference Time Domain Method for Electromagnetics*, CRC Press Inc., Boca Raton, FL, 1993.

[3] S.T. NGUYEN, B.J. ZOOK, AND X. ZHANG, *Distributed computation of electromagnetic scattering problems using finite-difference time-domain decompositions*, in Proc. Third IEEE International Symposium on High Performance Distributed Computing, IEEE Computer Society Press, August 1994, pp. 85–93.

[4] K.S. YEE, *Numerical solution of initial boundary value problems involving Maxwell's equations in isotropic media*, IEEE Trans. Antennas and Propagation, AP-14(1966), pp. 302–307.

Chapter 27

Using a Workstation Cluster for Physical Mapping of Chromosomes

Steven W. White
David C. Torney

Editorial preface

The mapping of chromosomes, the "Human Genome Project," is a nontrivial computational task, which has recently received significant press. Fortunately the task has inherent parallelism. This parallelism is exploited and the application's concurrency achieved on networks of workstations is described. What makes the implementation approach unique is the use of the UNIX mail facility as the communication and synchronization mechanism.

This article originally appeared in *SIAM News*, Vol. 26, No. 2, March 1993. It was updated during the summer/fall of 1995.

Human heredity involves the DNA sequences of 24 large molecules called chromosomes (1 through 22, X and Y). These sequences can be thought of as long strings made up of beads of four colors, corresponding to the four molecules (bases) adenine, cytosine, guanine, and thymine. The objective of the "Human Genome Project" is to determine the sequence of the approximately 3×10^9 of these bases contained in human chromosomes [7, 3].

With the ability to clone fragments of chromosomal DNA, and then to make as many copies of the cloned fragments as desired, researchers can generate the "raw materials" for subsequent experiments. A physical map, a minimal set of clones spanning a chromosome, would be a useful resource. With such a map it would be possible, for example, to generate a chromosome's sequence by sequencing the clones of the spanning set.

Ordering of the clones into a minimal spanning set has been accomplished primarily through computational analysis of "fingerprint" experiments and unique sequence (sequence-tagged site) content. Fingerprint experiments fall into three categories: restriction enzyme digestion, probe sequence hybridization, and partial sequencing. (A fingerprint experiment can include more than one of the basic types.) Each type of fingerprint experiment extracts part of the sequence information, as shown in this article.

Restriction enzymes cut DNA whenever a specific sequence (typically 4–8 bases in length) occurs. Therefore, if two clones are overlapping and share many of these sequences, a restriction enzyme digestion of the two clones followed by the size determination for the resulting fragments will result in many pairs of "matching" fragments. Probe sequence hybridization is used to detect the presence in the clone of sequences closely similar to the probe sequence. If the clone is cut into restriction fragments, we can ask which of the fragments contain sequences similar to the probe sequence. This provides additional information: "matched" pairs of fragments must be comparable in size and, in the absence of experimental error, have the same hybridization probe results. In experiments being done at Los Alamos National Laboratory to map human chromosome 16, two different restriction enzymes are used separately and jointly to generate three sets of restriction fragments for each clone. Los Alamos researchers have determined the size (by electrophoresis) and the presence or absence of several repeated DNA sequences (by probe sequences) for every fragment.

Initially, approximately 2000 clones from chromosome 16 were fingerprinted at Los Alamos [4]. There are about 40 fragments per clone and two hybridization values per fragment, resulting in 10^5 pieces of data. To construct an overlapping clone physical map, we begin by determining the probability of overlap for all clone pairs; 2×10^6 clone pairs must be examined. To a first approximation, our clones are all the same length, 4×10^4 bases, and have equal probability of originating at each possible base in the target DNA sequence. The computer program that calculates overlap probabilities for pairs of clones attempts to match as many fragment pairs as possible separately for each restriction digest of a pair. This calculation results in a number closely related to the ratio of the likelihood of the data and overlap to the likelihood of the data and nonoverlap for an applicable statistical model [1].

In the Los Alamos data, the average number of fragments for each of the three restriction enzyme digests is between 10 and 20; the position of the restriction enzyme fragments within the clone is not ordinarily determined in these experiments. Because the order is not known, the fragments must be permuted. Fortunately, exhaustive permutation of the n_1 fragments in a digest of the first clone and the n_2 fragments in the same digest of the second clone is not necessary.

Using sorted lists of fragment sizes and a cutoff for accepting matrix elements, we ordinarily compute only a small fraction of the $n_1 \times n_2$ elements of a comparison matrix. The matrix is effectively sparse because of the parameters of the experiment. To compute the likelihood ratio, we then sum the (nonzero) elements, sum all possible products of two elements, and so on, up to products of $\min(n_1, n_2)$ elements. The products are restricted so that no fragment is matched more than once. The number of nonzero elements is typically quite small, particularly for nonoverlapping clones, and the dominant part of the calculation is to determine which matrix elements are worth computing and then to compute them. Because the a priori overlap probability is approximately 10^{-3} for a chromosome of 10^8 bases, the nonoverlapping clone pairs dominate the calculation and shape the algorithms used. We collect a three-dimensional histogram of the three likelihood ratios (one for each digest) computed for simulated fingerprint data of nonoverlapping and of overlapping clones to determine the overlap probability for a pair of clones. The details of the computer algorithm are discussed elsewhere [6, 9].

The overlapping set(s) of clones can be constructed from the overlapping pairs by finding the connected components of the graph whose vertices are clones and whose edges between vertices exist when the corresponding overlap probability is above some threshold [5]. Currently, the physical map of chromosome 16 has about 4000 4×10^4 base-long clones and several hundred longer clones, resulting in about 450 clone islands—groups of overlapping clones. A similar approach has been applied to the mapping of all human chromosomes with a library of megabase (10^6 base-long) clones; more than 20,000 clones have been fingerprinted [2].

Although current experiments typically involve fewer than 10^4 clones (originally 2000 for chromosome 16), to perform a statistically complete analysis, we typically simulate considerably more clones (say 20,000) and do all of the pairwise comparisons. Because the comparison process is time consuming (more than 10 hours of CPU time on an IBM RISC System/6000 model 530 for 20,000 clones) and each pair comparison can be done independently, the use of parallel processing substantially increases the number of simulated experiments that a scientist can perform in a given time. A larger number of simulated experiments increases the accuracy of the statistical analysis performed on the data [9].

27.1. Inherent Parallelism

To allow algorithm evaluation for data sets of different sizes, we create simulated data sets with characteristics similar to the experimental data. The routine SIMCLS, which generates the simulated clone data (fragment sizes and six probe sequence hybridization arrays), represents a negligible amount of the overall execution time. This code segment runs serially on the *master* processor before any parallel work is started.

For the analysis of nonoverlapping clones, the main body (PAIRS) of the program consists of $(n_c - 1)(n_c/2)$ pairwise comparisons of n_c unique data sets, where n_c is the number of clones. The main outer loop selects a data set while the inner loop compares this data set with all data sets logically to the right of it. For each pair, a complex function is evaluated to select one bin in a three-dimensional histogram, which is then incremented. Because the selection of the appropriate bin for a particular pair is independent of the computation for any other pair, iterations of the inner loop could be run in parallel. Exploiting the parallelism at the outer loop is more efficient, however, and still allows sufficient granularity to provide excellent load-balancing characteristics. Parallelization at such a high level also allows the mathematical algorithms that select a bin to be easily modified without substantially affecting the parallel performance traits.

A feature of outer-loop parallelism in this application is the natural load balancing resulting from the fact that later iterations of the outer loop perform fewer iterations of the inner loop. Each of the early chunks of parallel work (i.e., groups of iterations of the outer loop) requires an enormous amount of compute time, and the relative cost for synchronization will be almost zero. Toward the end of the computation, each chunk becomes very small, and the total waiting time of the other processors for the last-to-complete processor should be minimal. For 20,000 clones, the first outer-loop iteration should take about four seconds, while the last should require only 200 microseconds. To ensure that chunk-distribution overhead is amortized efficiently, chunks consist of multiple outer-loop iterations.

In the serial version of the code, the array containing the histogram information is the only array modified during the **PAIRS** phase of execution. Because it is independently updated by each iteration of the inner loop and it is only 200,000 bytes, we give each worker its own private copy to update. This eliminates the need for placing the histogram update code in a critical region. The merging of the workers' histogram arrays is extremely quick and is performed as each task completes; therefore, because the workers are not expected to finish simultaneously, the majority of the histogram merge is complete by the time the last iteration of the outer loop is finished.

The amount of computation required for a clone pair depends on the degree of similarity of the two data sets. As a result, vectorization has not provided even marginal performance improvement. The code's resistance to vectorization and its well-behaved stride-1 storage-accessing patterns map well to superscalar machines, such as the RISC System/6000. This code, because of its coarse-grain nature and minimal communication requirements, is an ideal candidate for a distributed-memory environment. The following section describes the problem partitioning, synchronization, and other factors that allow codes of this type to perform well on a cluster of workstations. It is important to realize that the class of codes that exhibit this level of computational independence would not be suitable for the majority of applications.

27.2. Loosely Coupled Cluster Environment

A serial version of the code, developed at Los Alamos, was submitted to the PERFECT club to represent the grand challenge area often referred to as the "Human Genome Project." A shared-memory parallel version that had been written for IBM ES/3090 systems [8] contained a hint of distributed computing in that it exploited all 12 processors of two six-processor mainframes using IBM

Clustered Fortran. This article focuses on a purely distributed-memory version for networked workstations. The distributed-memory cluster programming model uses a *master* process and n nearly identical *worker* processes. One of the worker processes is unique in that it resides on the same system as the master process, which has load-balancing implications.

The main goal of this work is to illustrate the acceleration of a well-suited real application that can be accomplished by utilizing otherwise wasted cycles. Workstations can perform useful work for others while their owners are attending meetings, sleeping, or doing both simultaneously. To maximize the number of available workers, a key part of our goal is to place as few requirements as possible on the workers. The code should run on most systems without change to the code or system. Given a common data format, such as IEEE-754, the code should run on a heterogeneous collection of AIX/UNIX boxes.

The major functional problems of communication and synchronization between parallel processes across a network can be summarized in the following questions: Has the data been written for me to read? Has the file been updated since I last read it? Has all of the file been written? One solution uses NFS remotely mounted files to simulate shared disks. A predefined set of shared files, with a distinct writer per file, can be used to implement a disk version of software locks. Alternatively, a variable in a separate control file can be monotonically updated to indicate that the data file is again ready to be read. Because the updates are defined to be monotonic, a pair of these control files can be used alternately to signal "the data file has been updated" and "the last update has been read."

The extensive system caching of files made synchronization via lock/control files difficult because the order of file updates did not necessarily match the order seen by all readers. (We did not flush the files or use NFS locks.) We also encountered problems with spurious end-of-file problems when one process tried to read a shared file that was also being updated by another process. Remotely mounted files also resulted in a separate directory for each worker, which made administration awkward for a continually varying set of workers. Multiple directories with multiple files also make it more difficult to determine which file to open. Furthermore, any approach that uses remotely mounted files initially requires root authority for each worker to allow the mount.

Although these problems are solvable, we wanted a communication/synchronization scheme that would not require much effort from the system administrators (system changes should be kept to a minimum when accounts are received from 20 different machine owners) and is readily available on most systems. Due to the very coarse-grain nature of this application, we decided to try a rather novel approach to parallel processing synchronization across the workstations—use of the AIX/UNIX mail facility. Each mail message is like a token. The receiver can read each message only once. This makes it easy to answer the question: Has the data been sent? Multiple workers can simultaneously send messages to the master without requiring locks—they exploit the synchronization that exists in the normal mail handler. Likewise, the master can send out messages without checking to see whether previous messages have been received. Because the control information (MBOX value) is embedded with the data (usually specifying the columns to be operated on next), synchronization is provided by the mail facility.

Use of the standard mail facility also simplifies the administration of the systems. When requesting use of a machine as a worker, the only root authority required is that to create the login ID and ensure that mail is running. No special permissions are required.

Use of a general facility such as mail requires a mechanism for coping with "unexpected" messages from existing users. Ideally, the master and workers use special login IDs reserved for parallel processing. If no one sends mail to these phantom users, no spurious mail messages arrive while the job is executing. Because we use existing login IDs, however, we include tag information in the message headers as a mechanism for detecting (and ignoring) unexpected messages.

The bulk of the data transferred during a run is from the SIMCLS portion of the program, the simulated clone data. To diminish the communication delays for this large block of data, which must be transferred before any worker begins, we made an exception to our "communicate by mail" strategy. Transferring the data in binary form reduces the traffic and latency. Binary files also minimize the parallel overhead—each of the n workers has to read the file, and unformatted I/O is more efficient. Unfortunately, mail does not easily handle binary files (without ENCODE/DECODE). Therefore, for transferring the clone data, we use remote copy, rcp, which allows more compact binary files.

The master program, written in Fortran, uses open/write/close statements to create files with the appropriate mail headers and mail data. It uses "call system()" to issue AIX/UNIX commands, such as sendmail and inc/show (receive). A common set of parameterized routines (send mail, start mail, check mail) is used by both the master and worker routines. A template of the master and worker code with these mail routines, as well as the application code itself, is available as unsupported code.

27.3. Loosely Coupled Cluster Program Structure

Figure 27.1 illustrates the relations between tasks. To begin, each worker requests work by sending a mail message (with a variable, MBOX, set to zero) to the master. The worker then checks for mail and performs the work as requested by the master. This approach is convenient for several reasons. First, the master process does not need a list of worker machines prior to the run. The master inspects the "from" portion of each message header to identify available workers. When new requests (MBOX=0) are received, the name of the worker is added to a dynamically created list. As a result, workers can be added at any time during the run. Finally, the code and files require no

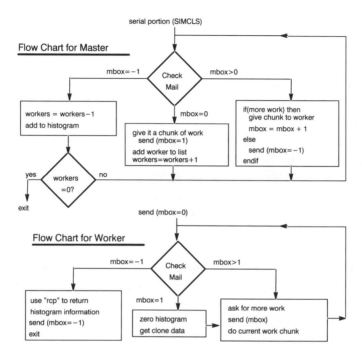

FIG. 27.1. *Parallelism at the task level.*

changes as the number of available processors fluctuates.

Before checking for mail from potential workers, the master process generates the simulated clone data and writes the data to a local file. To allow the workers to start as soon as possible, the master sends each worker a chunk of work (a unique range of outer-loop iterations) along with the initial reply that the clone data is available (MBOX=1). It is important that the worker does not have to send a message to get the initial chunk of work as the master processor is probably busy processing the other workers' initial requests. When MBOX=1, the worker knows to initialize its private histogram array and to get the read-only clone data arrays from the master using the rcp command. A message requesting additional work is sent to the master, and the worker then starts on the current chunk.

To make it easier for all workers to finish at about the same time, we use the classic approach of breaking the work into chunks that are far more numerous than workers, making the later chunks smaller, and dynamically allocating chunks as workers request more work. This is usually better than static load balancing because there are many sources of differences in execution rates, such as cache behavior, paging, and multiprogramming in a nondedicated environment. (In this application, furthermore, the amount of work in each of the parallel tasks varies with the clone data, even when the chunk sizes are the same.)

When the master sees a worker's request for more work (MBOX> 0), it mails a message to the worker with the next range of outer-loop iterations; MBOX is then set to one more than the previous value sent to that worker. The worker handles the block of iterations given to it and updates its private histogram array accordingly. The master task keeps track of the consecutively numbered MBOX values for each worker. Because the worker's reply echoes the MBOX value from the master's message, both the worker and the master can verify that work messages have not been lost.

To decrease the probability that a worker will have to wait for the next chunk of work, we choose to let the worker ask for more work before starting its current chunk. For this coarse-grain application, most of the chunks of work take significantly longer than the latency of mail messages. Therefore, it is quite likely that when a worker finishes a chunk of work, a message containing more work will be waiting to be read.

At some point, all work will have been allocated when a worker requests more work. The master will then send a message with MBOX=−1, signaling that this is the last dispatch. This mail message also contains a unique file name selected by the master process. The worker writes its local histogram array to a file and uses `rcp` to copy the file to the specified location on the master processor. It is only on this final transfer that the worker histogram information is transferred to the master to be merged into the final histogram results. The worker sends a final message, echoing MBOX=−1 and the unique file name, to notify the master that the local histogram data exists in the specified file on the master processor. The worker process then terminates.

When the master receives the message (with MBOX=−1), it reads the file name from the message, gets the data from the file, and merges it with the histogram data from the other workers. As workers are added (messages with MBOX=0), their names are appended to the internal list. As workers complete (MBOX=−1), their names are deleted from the list. When the last worker is removed, the job has been completed.

27.4. Results

Table 27.1 shows the performance measurements for problems of various sizes. Speedup is defined as the ratio of the execution times for the best serial version to the best p-processor version; n_c is the number of clones; $n_c = 20,000$ is a typical production run. Most of our measurements are done at night and on weekends, when the systems are reasonably idle, so that the potential speedups can be illustrated. Dedicated systems are not required, however; any extra cycles that a scientist can "steal" from a moderately loaded system will contribute to a reduced turnaround time for the job to be done.

The runs were made on a collection of RISC System/6000 model 530, machines. Some runs, in addition to those shown in Table 27.1, used workers more than 1200 miles away, but the speedups attained were within 2% of those listed here. The only common threads were that all machines were

TABLE 27.1

Speedup ratios for problems of various sizes. When the code is partitioned into two units (master/worker), arrays and code sections that are not referenced in one unit are not included in that unit. As a result, the order of data and code in storage varies slightly from the serial version. In the particular case of $n_c = 20,000$, the $p = 1$ row shows that the rearrangement is slightly (1%) favorable.

	Speedup		
p	$n_c = 5000$	$n_c = 10,000$	$n_c = 20,000$
1	0.98	0.99	1.01
2	1.89	1.95	2.00
3	2.72	2.88	2.99
4	3.46	3.75	3.96
5	4.20	4.64	4.90
6	4.87	5.46	5.82
7	5.50	6.23	6.75
8	6.15	7.06	7.68
9	6.60	7.75	8.53
10	6.97	8.42	9.43
11	7.35	9.17	10.23
12	7.63	9.75	11.08
12[†]	8.20	10.26	11.39

[†]Run for which the rcp times are not included.

model 530s (to make the interpretation of speedups easier) and that mail was running. Because this approach to distributed computing uses generally available facilities and requires no special privileges, this application could easily have been run on a heterogeneous collection of AIX/UNIX machines. (If the collection includes machines with differing data types, ASCII data sets or XDR may be required.)

The data transfers included the read-only clone data (about 180 n_c bytes per copy) and the histogram arrays (200 Kbyte each). The master process ran on a machine connected to the network via a slow (4 Mbit/second) token ring connection. As the number of workers increases, the average time spent waiting for the initial rcp (before a worker can start) increases due to contention. Therefore, although communication is almost negligible, the use of a faster (16 Mbit/second) token ring would cut this overhead to one-fourth of the current level, resulting in a move slightly closer to the ideal linear speedup. The last row of Table 27.1 shows the speedup for 12 processors if the rcp overhead is completely eliminated. (The rcp overhead is eliminated if the simulated (or measured) data is transferred once prior to many runs.) Without the `rcp` calls, for $n_c = 20,000$, this corresponds to 3299 seconds. A linear speedup of 12.0, over the $p = 1$ case, corresponds to 3096 seconds. Most of the 203-second difference is accounted for by the 100 seconds of serial time before the workers are started. Imperfect load balancing is the next most significant contributor.

TABLE 27.2

Average time per clone pair (in microseconds).

RISC System/6000 model 530	IBM ES/3090 600E	Cray Y-MP	RISC System/6000 model 560
185	180	140	95

On the RISC System/6000 model 530, the serial $n_c = 20,000$ production run takes more than 10 hours. With 10 model 530s, the time is just over an hour. The time for moderate runs ($n_c = 10,000$) drops from more than two and a half hours to less than 20 minutes with nine processors. Table 27.2 shows the average time per clone pair required for several platforms. In a processor-to-processor comparison, the model 530 is almost as fast as the IBM ES/3090 model 600E. A newer RISC System/6000, the model 560, is approximately twice as fast as either the ES/3090 or the model 530, and it is 50% faster than the Cray Y-MP.

27.5. Summary

Construction of an ordered-clone physical map of a human chromosome benefits substantially from the computational capabilities described in this article. To do justice to the data, ordered-clone physical mapping requires a considerable amount of computational time in preparation for and during the pairwise comparisons. The performance of the loosely coupled implementation of this algorithm for detecting clone overlap allows scientists to simulate experiments substantially larger than those that would be feasible on a single machine. Today's scientific/engineering workstations are extremely competitive with mainframes for this application, as it is difficult to exploit the vector hardware that often makes mainframes advantageous. The granularity of this application efficiently supports a rather novel approach to parallel processing synchronization—use of the AIX/UNIX mail facility. The efficiency of parallel speedups of greater than 11 on a 12-processor cluster makes parallel processing across clusters a practical alternative to 10-hour waits or expensive shared-memory mainframes. The coding structure developed for this application to exploit a cluster of RISC System/6000 systems could be considered a starting framework for applications with similar characteristics.

27.6. Acknowledgments

Doyce Nix, IBM Santa Fe, helped with the initial Fortran program on an IBM ES/3090 model 600E. Clive Whittaker, IBM Santa Fe, made considerable algorithm improvements in the overall application and ported the serial program to the RISC System/6000 platform. Maurice Franklin and John Iacoletti, IBM Austin, provided system administration education that was key to this article. Finally, we thank the system owners/administrators who contributed a large number of cycles to the project and the scientists in the C.H.G.S. at Los Alamos, in particular Norman Doggett. This work was partially supported by U.S. DOE/OHER.

References

[1] D.J. BALDING AND D.C. TORNEY, *Statistical analysis of fingerprint data for ordered clone physical mapping of human chromosomes*, Bull. Math. Biol., 53(1991), pp. 853–879.

[2] C. BELLANNE-CHANTELOT ET AL., *Mapping the whole human genome by fingerprinting yeast artificial chromosomes*, Cell, 70(1992), pp.1059–1068.

[3] C.R. CANTOR, *Orchestrating the human genome project*, Science, 248(1990), pp. 49–51.

[4] R. STALLINGS, D.C. TORNEY, C.E. HILDEBRAND, J. LONGMIRE, L. DEAVEN, J. JETT, N. DOGGETT, AND R. MOYZIS, *Physical mapping of human chromosomes by repetitive sequence fingerprinting*, Proceedings National Academy of Sciences, 87(1990), pp. 6218–6222.

[5] R.E. TARJAN, *An improved algorithm for hierarchical clustering using strong components*, Inform. Proc. Lett., 17(1983), pp. 37–41.

[6] D.C. TORNEY, C.C. WHITTAKER, S.W. WHITE, AND K. SCHENK, *Computational methods for physical mapping of chromosomes*, in Proc. First International Conference on Electrophoresis, Supercomputing, and the Human Genome, C.R. Cantor and H.A. Lim, eds., World Scientific Publishing, New York, 1991, pp. 268–278.

[7] J.D. WATSON, *The human genome project: Past, present, and future*, Science, 248(1990), pp. 44–49.

[8] S.W. WHITE, D.C. TORNEY, AND C.C. WHITTAKER, *A parallel computational approach using a cluster of IBM ES/3090 600Js for physical mapping of chromosomes*, in Proc. Supercomputing '90: IEEE Computing Society and ACM SIGARCH, New York, NY, Nov. 12–16, 1990, ACM Press, New York, pp. 112–121.

[9] R.L. DOUGHERTY, R.L. STALLINGS, S.W. WHITE, D.C. TORNEY, C.C. WHITTAKER, M.O. MUNDT, V. FABER, D.J. BALDING, *Computations for mapping genomes with clones*, Internat. J. Genome Res., 1(1993), pp. 195–226.

Chapter 28

Massively Parallel Computations in Finance

Stavros A. Zenios

Editorial preface

Advanced architectures are often mistakenly associated with solutions to problems in the physical sciences; this chapter presents their application to financial modeling. Specifically, the valuation of mortgage-backed securities is discussed, a field that is growing in popularity. A large portfolio is solved on the CM-2, the solution of which presents a distinct advantage to large financial institutions.

This article originally appeared in *SIAM News*, Vol. 24, No. 2, March 1991. It was updated during the summer/fall of 1995.

It is difficult to find an area of the fixed income markets that does not rely on analytic techniques for the pricing of financial instruments. The valuation of the complex instruments that appears in the banking and insurance industries requires simulations of their cashflow behavior in a volatile interest rate environment. These simulations are complex and computationally intensive; their use, thus far, has been limited to interday analysis and planning.

The emergence of advanced architecture computers has opened new possibilities for the use of these simulation models. Applications that would take several minutes on mainframes can now be executed in fractions of a second on a massively parallel computer. Hence, models that were once restricted to secondary support roles can now be moved to the trading desk. Analyzing a large portfolio of these instruments—a task that would take a couple of days on a mainframe—can now be completed in less than one hour. As a result, it is possible not only to build a risk profile for a given portfolio, but also to build portfolios with better risk profiles.

This article describes research carried out at the HERMES laboratory of the Wharton School, with collaborators from Thinking Machines Corporation and the Union Bank of Switzerland on the valuation of *mortgage-backed securities* (MBS). Details on the material presented here can be found in the paper by Hutchinson and Zenios [3].

28.1. Understanding MBS

MBS are created when mortgages from individual homeowners are pooled together. Investors purchase interest in the pool and receive prorated shares of the pool's cashflows. The issuing institution handles the transfer of funds and retains a service fee. These instruments facilitate the flow of funds from the ultimate lenders in the capital markets to the mortgage borrower.

MBS emerged in the early 1980s. As of the second quarter of 1988, outstanding mortgage debt in the U.S. was approximately $3.5 trillion, nearly 70% of which was in residential mortgages. Some 25% of the outstanding residential debt has been securitized via the issuance of MBS, which represent the fastest growing segment of the debt markets. Interest in MBS is not restricted to the U.S. Almost 90% of all residential debt in Denmark has been securitized. In Canada, where the first issue took place in January 1987, there are more than 70 issues backed by more than $500 million in mortgage loans.

MBS, in that they embody features of both bonds and options, are complex and difficult to value. The ability of the homeowner to prepay outstanding principal represents a *call* option on the underlying mortgage. For any specific mortgage within a pool, it is uncertain whether this call option will be exercised and, if so, when. Many factors outside the characteristics of the pool may affect the option's value. Among them are the level, structure, and history of interest rates, the market perception of future interest rates, and total and disposable consumer income. Adding to the complexity of early MBS has been the constant stream of innovative new derivative securities whose risk and return characteristics can bear little resemblance to the original MBS.

Determining a "fair" price for MBS is a complex process that relates the possible future paths of interest rates to the cashflows generated by the security. Such cashflows take into account both payment of principal and interest and prepayment of mortgages (i.e., exercise of the underlying call option by some homeowners). In the analysis, simulations are used to generate paths of interest rates, usually in monthly intervals for a period of 30 years. The state space from which simulations are drawn could be enormous: for a binomial lattice model of interest rates, there are 2^{360} paths from which a sample can be drawn.

28.2. The Valuation Methodology

The general framework of the valuation analysis has three phases:

Phase I: Generate arbitrage-free interest rate scenarios that are consistent with the prevailing term structure of interest rates. Here, we use the Cox–Ingersoll–Ross model [2], which assumes that interest rates follow a Markovian diffusion process:

$$dr(t) = \kappa(\mu - r(t))dt + \sigma\sqrt{r(t)}d\omega(t),$$

where $d\omega(t)$ is a Wiener process, i.e., $d\omega(t) = \lim_{\Delta t \to 0} \sqrt{\Delta t}z$, and z is a standard, normal random variable. The discrete approximation to the diffusion process, which is used recursively in a Monte Carlo simulation, is

$$r_{t+1} = r_t \exp\left[\kappa\left(\frac{\mu}{r_t} - 1\right) + \sigma\sqrt{r_t}z_t\right].$$

An alternative to the simulation of a diffusion process—also used in our models—is to assume that interest rate movements can be approximated by a discrete binomial process represented by a lattice (Figure 28.1).

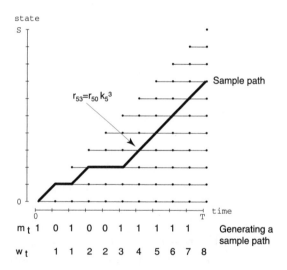

FIG. 28.1. *Binomial lattice of interest rates and the generation of a sample path.*

Discrete points in time are marked on the horizontal axis, and the nodes of the lattice represent possible states of interest rates at every point in time. The rates can move to one of two possible states between successive points in time—conveniently called the "up" and "down" states. The lattice is connected in the sense that an "up,down" and a "down,up" path starting from the same state will lead to the same state after two periods. After t time periods from the origin, the lattice has t possible states. Each of these states can be reached through 2^t possible paths. Short-term forward rates at the nodes of the lattice are computed on the basis of market data in such a way that the arbitrage-free property is satisfied. In our work we use the Black–Derman–Toy model [1] for calibrating the lattice.

Once the binomial lattice has been fitted to the current term structure—in itself a difficult and compute-intensive process—we can represent the short-term rate at time period t and at state ω by the relation

$$(28.1) \qquad\qquad r_{t\omega} = r_{t0} k_t^{\omega}.$$

The quantities r_{t0} and k_t, for $t = 1, 2, 3, \ldots, 360$, represent the 0th (i.e., ground) state and the volatility of short-term rates at period t, respectively. These parameters are estimated by the Black–Derman–Toy model. We compute a large number of interest rate scenarios by using equation (28.1) to sample paths from the binomial lattice.

Phase II: Generate cashflows along each interest rate scenario. The cashflows generated by MBS at each month have three components:

1. *Interest payment* is the portion of cashflow that reflects interest. It depends on the mortgage contract interest rate and the outstanding mortgage balance, net any servicing fee that is kept by the lending institution.

2. *Scheduled principal payment* is the scheduled payment of outstanding principal.

3. *Projected principal prepayment* represents any cashflow generated by the exercise of the call option by some homeowners. Due to sale of the property or refinancing of the loan, some mortgages in the pool will pay a lump sum equal to the outstanding mortgage balance. Principal prepayment is estimated on the basis of projected monthly *prepayment rates* that indicate the fraction of outstanding balance that is prepaid. The primary factor that affects prepayments is the interest rate path, as generated in Phase I. If interest rates are low, then prepayment rates are high as homeowners refinance their mortgages. A model for estimating prepayment activities is described in Kang and Zenios [4].

Phase III: Use the cashflows and the short-term interest rates along each path to compute an *option-adjusted spread* (oas) over the yield curve.

The option-adjusted spread is the incremental spread over the short-term rates that equates the expected present value of the cashflows under all scenarios with the market price. If P denotes the market price, the option-adjusted spread is obtained by solving for *oas* the nonlinear equation

$$P = \frac{1}{|S|} \sum_{s \in S} \left\{ \sum_{t=1}^{360} cf_t^s \prod_{\tau=1}^{t} \frac{1}{(1 + r_\tau^s + oas)} \right\},$$

where S is the sample of interest rate scenarios, and cf_t^s denotes the cashflow at time period t under short-term rate r_t^s, for each scenario $s \in S$.

28.3. Massively Parallel Designs

The simulation procedure parallelizes nicely if each processor carries out all
the computations for a single interest rate path. Multiple processors can then
execute in parallel multiple simulations. Communication across processors is
required only in computing statistics across all simulations. On a system like
the Connection Machine, however, we want to exploit the massive parallelism
not only in simulating multiple paths simultaneously but also in performing the
calculations for each path. Otherwise, a large number of processing elements
will remain unused, and the performance of the program will fall far short of
the typical performance of the hardware.

The key to our implementation is the configuration of the CM-2's virtual
processors into a two-dimensional NEWS grid. One dimension of the grid is
equal to the desired number of simulations, which for computational efficiency
is taken to be a power of 2. The second dimension of the grid is equal to the
number of time periods, rounded up to the next integer that is a power of 2.
A 1024×512 NEWS grid suffices. Each of the 1024 rows of virtual processors
carries out the calculations for a single path. The first 360 virtual processors
in each row execute the path-dependent calculations (the remaining 152 are
idle).

With this configuration of the CM-2, it is possible to implement efficiently
all the components of the model. However, substantial reformulations of the
model were needed to bring it into a form that would allow us to use the
parallel primitive operators of the CM-2. These primitives facilitate parallel
computations at distinct processors and efficient communication of results
among processors. For example, the `scan-add` primitive accepts as input data
a_j from processor $j = 1, 2, 3, \ldots, P$ and returns to processor $i = 1, 2, 3, \ldots, P$
the value $\beta_i = \sum_{j=1}^{i} \alpha_j$. This point is illustrated for the sampling of a binomial
lattice.

To generate sample paths from the binomial lattice, we need to determine
the state of each path at each point in time. Once the state ω of the sth path
in the binomial lattice is specified at the virtual processor with NEWS address
(s, t), the short-term rate can be computed by a simple application of equation
(28.1). Of course, the paths must be continuous: the state of the lattice at
instance t must be attainable by either an "up" or a "down" step from the
state at instance $t - 1$.

Such a sequence of states is produced on the CM-2 as follows: a random
bit, $m_t \in \{0, 1\}$, is first generated at each virtual processor. A `scan-add`
operation along the time axis on these bits generates an index $(\omega_t = \sum_{\tau}^{t} m_{\tau})$
indicating the state of the virtual processor (i.e., its distance from the ground
state r_{t0}). Clearly, the distance from the ground state at instance t differs at
most by one unit from the distance at instance $t - 1$. Once the distance ω_t
is determined, equation (28.1) can be evaluated simultaneously by all virtual
processors. Figure 28.1 illustrates the use of this procedure to sample a path

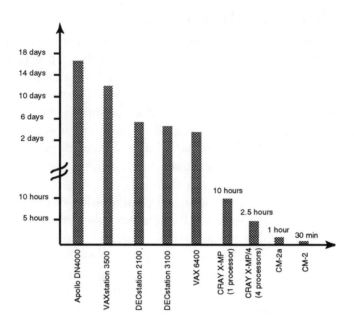

FIG. 28.2. *Timing the option-adjusted analysis for a portfolio of* 3000 *MBS.*

from a binomial lattice.

A complete model was implemented on a Connection Machine CM-2a, with 4K processing elements, using the language C/Paris. The various components of the model achieve computing rates up to 120 Mflops. Figure 28.2 illustrates the performance of the model in evaluating a portfolio of 3000 MBS. It is compared with the performance of an identical system developed by us and run on a variety of computers. The analysis that would take two weeks on a workstation and two days on a mainframe can be completed in less than one hour on the smallest CM-2 model.

28.4. New Avenues: Stochastic Optimization

The massively parallel designs described in the preceding section bring quantitative improvements in the use of mortgage-related models.   For example, we can now observe the duration (i.e., first derivative of price) and convexity (i.e., second derivative of price) of MBS with different prepayment characteristics under a host of interest rate scenarios. The response time for the analysis of a single security (1–2 seconds) makes the system usable for real-time applications.

The quantitative improvements in performance are now leading us to seek qualitative improvements in the modeling process. In particular we can now attempt answers to the following question: "Can we construct a portfolio of MBS with a total cashflow stream that is independent of the interest rate

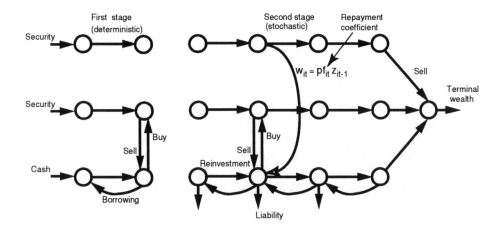

FIG.   28.3.   *Two-stage stochastic network optimization model for planning
under uncertainty with MBS. The nodes along the vertical axis represent financial
instruments (this example includes only a security i and cash). Time evolves along the
horizontal axis. Arcs pointing forward in time represent investment decisions, while
reverse arcs from $t+1$ to t represent borrowing decisions. Multipliers on forward arcs
represent reinvestment interest rates; multipliers on reverse arcs represent borrowing
rates. Of course, all multipliers after period 1 are stochastic. The key coefficients
pf_{it}^s represent the cashflow of security i to the interest rates that prevail at period t
under scenario s. They are generated by the simulation program and, together with
the holdings z_{it-1} of security i, determine the total cashflow. L_t indicates the level
of liability at time period t. At the end of the time horizon, the total net wealth is
accounted for, and some measure of utility of this wealth is optimized.*

environment or that, at least, varies less with interest rate changes than
the cashflow components of individual securities?"   Such questions can be
formally posed as two-stage stochastic network optimization programs.   A
simple multiperiod model, with only one security and a riskless asset, is
illustrated in Figure 28.3.

Using the CM-2 it is now feasible to generate the data for this model
using several hundred securities.   However, the model itself can grow very
quickly.   Constructing a portfolio with 100 securities for 30 yearly intervals
under 500 scenarios, for example, gives rise to an optimization problem with
approximately 1.5 million variables and 50,000 constraints. We have designed
massively parallel algorithms [5] that solve problems of this size in 10–15
minutes of elapsed time on the 32K CM-2.   These developments open new
possibilities for financial planning under uncertainty.   Without the use of
advanced computer architectures, such models would have been very difficult
to build and impossible to solve.

References

[1] F. BLACK, E. DERMAN, AND W. TOY, *A One-factor Model of Interest Rates and its Application to Treasury Bond Options*, Discussion Paper 1, Goldman Sachs, June 1988.

[2] J. COX, J. INGERSOLL, AND S. ROSS, *A theory of the term structure of interest rates*, Econometrica, 53(1985), pp. 385–407.

[3] J.M. HUTCHINSON AND S.A. ZENIOS, *Financial simulations on a massively parallel connection machine*, Internat. J. Supercomputer Appl., 5(1991), pp. 27–45.

[4] P. KANG AND S.A. ZENIOS, *Complete prepayment models for mortgage-backed securities*, Management Science, 38(1992), pp. 1665–1685.

[5] S. NIELSEN AND S.A. ZENIOS, *A massively parallel algorithms for nonlinear stochastic network problems*, Oper. Res., 41(1993), pp. 319–337.

Chapter 29

Robust Optimization on PC Supercomputers

John M. Mulvey

Editorial preface

Robust optimization is an enhancement to the classic linear programming method that allows for the treatment of "noisy" input data or uncertainty in the input data. Fortunately, the structure of the larger system that is needed to accomplish the robustness is such that parallelism and vectorization are easily applied.

Three examples are presented to illustrate the application and origin of robust optimization problems. Performance results are offered on both a network of workstations and on classic vector computer hardware.

This article originally appeared in *SIAM News*, Vol. 26, No. 7, November 1993. It was updated during the summer/fall of 1995.

Robust optimization is a practical approach for handling noisy data and uncertainty within the context of optimization models. The approach broadens the range of applicability for linear and nonlinear programs.

The key idea is to define a spectrum of plausible model representations—depicted by the usual data coefficients—as a set of scenarios. The resulting large-scale optimization problem considers the original objective, e.g., profit, cost, market share, and a new objective called the "robustness" term. The purpose of the latter objective is to ensure that the model recommendations are close to optimal, regardless of which scenario occurs. When this situation arises, the solution is said to be "robust over the scenario universe." Model robustness is an indication that the recommended course of action is insensitive to potential modifications to the data coefficients.

While the robust optimization approach does not address all of the issues associated with the presence of uncertainty within a linear program (LP) or nonlinear program (NLP), we have found that it provides a useful framework for a number of important problem areas. The three examples presented in this article illustrate the range of applicability for the approach.

The size of a robust optimization model is much larger than the corresponding linear program. Nevertheless, such models possess a special structure that can be taken advantage of by optimization algorithms tailored to these problems. Substantial progress has been made in solving problems in this class by means of clusters of desktop computers connected via a broadband communication network such as Ethernet. These distributed algorithms can also take advantage of classic supercomputer vector capability, if available.

The increasing use of robust optimization will be accelerated by several new developments. First, several software packages have been developed to link to the most popular spreadsheet packages, e.g., Microsoft's Excel and Borland's Quattro Pro, in a simple and intuitive manner. These products will ease the pain of generating robust optimization models. In addition, the arrival of the Windows NT operating system will bring much more powerful hardware to bear on the solution of the resulting spreadsheet-generated optimization problems, via desktop 64-bit processors. The new PC/workstations, when employed effectively in a distributed computing environment, will generate power almost at the supercomputer level for applications that can take advantage of distributed computations.

The trend to bring a high level of computational performance to a large number of users should have a profound effect on the types of problems that are addressed via LPs and NLPs. In the 1940s, George Dantzig had the foresight to realize that solving large systems of equations (hundreds or thousands) would someday be feasible and important. Today, we are on the verge of being able to solve, simultaneously, a large number of linear or nonlinear programs. This power should expand the range of appropriate optimization problems. Robust optimization is a natural area for application of this expanded linear and nonlinear program solving capability.

29.1. The Robust Optimization Framework

The models used in robust optimization have two distinct components: (1) a *structural* component that is fixed and free of any noise in its input data and (2) a *control* component that is subjected to noisy input data. To define the appropriate model, we introduce two sets of variables. $x \in \Re^{n_1}$ denotes the vector of decision variables that depend only on the fixed, structural constraints. The optimal value of these *design* variables is independent of any realization of the uncertain parameters. $y \in \Re^{n_2}$ denotes the vector of *control* decision variables that are subjected to adjustment once the uncertain parameters are observed. Their optimal value depends both on the realization of uncertain parameters and on the optimal value of the design variables.

Design variables determine the structure of the system and the size of production modules. Control variables are used to adjust the mode and level of production in response to disruptions in the system, changes in demand or production yield, and so on. Viewed in the context of portfolio management, for example, the design variables specify the composition of the portfolio.

Control variables determine the decisions about borrowing, reinvesting, or rebalancing that are made once the returns are observed.

With these variables, the optimization model we are interested in has the following structure:

$$\text{minimize} \quad c^T x + d^T y, \qquad x \in \Re^{n_1}, y \in \Re^{n_2}$$

(29.1)   subject to $Ax = b,$

(29.2) $Bx + Cy = e,$

$$x, y \geq 0.$$

Here (29.1) denotes the *structural constraints*, which are fixed and free of noise. (29.2) denotes the *control constraints*, the coefficients of which are subject to noise.

We now introduce a set of scenarios, $\Omega = \{1, 2, 3, \dots, S\}$. With each scenario $s \in \Omega$, we associate control variables, $\{y_1, y_2, \dots, y_S\}$, and constraints with coefficients $\{d_s, B_s, C_s, e_s\}$. The probability of scenario s is equal to p_s ($\sum_{s=1}^{S} p_s = 1$). The optimal solution of the mathematical program (29.1)–(29.2) will be robust with respect to optimality if it remains "close" to optimal for any realization of the scenario $s \in \Omega$. It is then termed *solution robust*. The solution is also robust with respect to feasibility if it remains "almost" feasible for any realization of s. It is then termed *model robust*. The meanings of "close" and "almost" are made precise through the choice of norms. The robust optimization approach is most appropriate when a moderate number of scenarios (currently ≤ 1000) is adequate to cover the range of possibilities. The generation of representative scenarios is a current research topic [8].

Of course, it is unlikely that any solution to the mathematical program will remain both feasible and optimal for all realizations of s. If substantial redundancies are built into the system being modeled, then it might be possible to find solutions that remain both feasible and optimal. Otherwise, we need a model that will allow us to measure the tradeoff between solution and model robustness.

We have proposed a model that formalizes a way to measure this tradeoff. We define a set $\{z_1, z_2, \dots, z_S\}$ of error vectors that will measure the infeasibility allowed in the control constraints under scenario s. The following is the *compact formulation* of the robust optimization model:

(29.3)   minimize $\sigma \ (x, y_1, \dots, y_S) + \omega \rho(z_1, \dots, z_S)$

subject to $Ax = b,$

$B_s x + C_s y_s + z_s = e_s \qquad \forall s \in \Omega,$

$x \geq 0, \ y_s \geq 0 \qquad\qquad \forall s \in \Omega.$

With multiple scenarios, the objective function $\xi = c^T x + d^T y$ becomes a random variable taking the value $\xi_s = c^T x + d_s^T y_s$ with probability p_s. Hence, there is no longer a single choice for an aggregate objective. We could use the mean value:

$$\sigma(\cdot) = \sum_{s \in \Omega} p_s \xi_s.$$

This is precisely the function used in stochastic programs. In worst-case analysis, the model minimizes the maximum value, and the objective function is defined by

$$\sigma(\cdot) = \max_{s \in \Omega} \xi_s.$$

In (29.3), the second term in the objective function, $\rho(z_1, \ldots, z_S)$, is a feasibility penalty function. It penalizes violations of the control constraints; these constraints are called "soft" and may be conditional under some of the scenarios. This proposed model takes a multicriteria objective form. The first term measures optimality robustness, whereas the penalty term is a measure of model robustness. The goal programming weight ω is used to derive a spectrum of answers that trade solution for model robustness. The introduction of the penalty function distinguishes the robust optimization model from existing approaches for dealing with noisy data.

Two popular approaches for handling moderate- to high-risk decisions in the context of robust optimization are (1) mean/variance models and (2) von Neumann–Morgenstern (VM) expected utility. For the former, risk is equated with the variance of the outcomes. A high variance for ξ_s means that the outcome is much in doubt. Given outcome variance as a surrogate for risk, we are naturally led to the minimization of the expected outcome for a given level of risk. In this case an appropriate choice for $\sigma(\cdot)$ would be the mean plus a constant (say, λ) multiplied by the variance:

$$\sigma(x, y_1, y_2, \ldots, y_S) = \sum_{s \in \mathcal{S}} p_s \xi_s + \lambda \sum_{s \in \mathcal{S}} p_s \left(\xi_s - \sum_{s' \in \mathcal{S}} p'_s \xi'_s \right)^2.$$

An efficient frontier can be readily constructed by parametrizing the tradeoff between risk and expected outcome. This approach requires that the distribution of the random variable ξ_s is symmetric around its mean. Third and higher moments are simply ignored.

The derivation of the efficient frontier gives the user an opportunity to achieve a *robust* recommendation, which is not possible by means of traditional sensitivity analysis. The slope of the risk-return curve, as shown in Figure 29.1, provides a true measure of the robustness of the solution. This curve becomes a surface in three dimensions when we add the feasibility robustness objective. Alternatively, a series of curves can be drawn for various levels of feasibility attainment. Regardless, the robust optimization approach should replace the usual parametric curve.

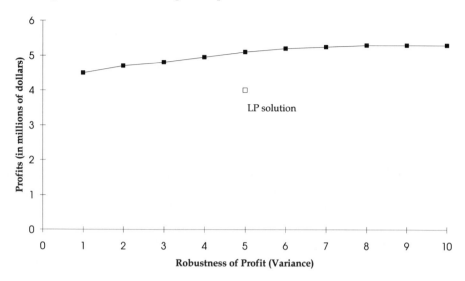

FIG. 29.1.   *The efficient frontier. The slope of the risk-return curve provides a true measure of the robustness of the solution.*

An alternative approach to handling risk is based on the use of VM utility curves [4] via the concept of certainty equivalence. The result is, for risk-averse decision makers, a concave utility function $U(\cdot)$. A decision maker displays consistent behavior by maximizing expected utility. In this situation we define

$$\sigma(\cdot) = -\sum_{s\in\Omega} p_s U(\xi_s).$$

The primary advantage of the expected-utility model over the mean/variance approach is that asymmetries in the distribution of ξ_s are also captured. A consistent and repeatable decision process can also be implemented, given a time-invariant utility function. Of course, an additional information burden is placed on the user, who must decide on an appropriate level of risk tolerance.

As compared with alternative approaches for dealing with uncertainty, robust optimization has several advantages, while it is not without its shortcomings. Sensitivity analysis (SA), a reactive approach to controlling uncertainty, just measures the sensitivity of a solution to changes in the input data. It provides no mechanism by which this sensitivity can be controlled.

Stochastic programming (SP), like robust optimization, is a constructive approach. Both SP and robust optimization are superior to SA. With stochastic programming models, the decision maker is afforded the flexibility of *recourse* variables. These variables are identical to the control variables of robust optimization and provide the mechanism for adjusting the model recommendations to account for the data realizations.

The SP model, however, optimizes only the first moment of the distribution of the objective value ξ_s. It ignores higher moments and the decision maker's preferences with respect to risk. These aspects are particularly important for asymmetric distributions, and for risk-averse decision makers.

Furthermore, striving for expected-value optimization implicitly assumes an active management style whereby the control (i.e., recourse) variables are easily adjusted as scenarios unfold. Large changes in ξ_s may be observed among the different scenarios, but their average value will be optimal. The robust optimization model minimizes higher moments as well, e.g., the variance of the distribution of ξ_s, and, hence, assumes a less aggressive management style. Since the value of ξ_s will not differ substantially among different scenarios, little or no adjustment of the control variables will be needed.

This important distinction between robust optimization and SP defines the domains of applicability for the two approaches. Applied to personnel planning, for example, an SP solution will design a workforce that can be adjusted (by hiring or layoffs) to meet demand at the least expected cost. The important consideration of maintaining stability of employment cannot be captured. The robust optimization model, on the other hand, will design a workforce that will need few adjustments to cope with demand for all scenarios. However, this cost will be higher than the cost of the SP solution.

The importance of controlling variability of the solution (as opposed to just optimizing its first moment) is, of course, well recognized in portfolio management applications due to the work of Markowitz [5]. It has been ignored in most other applications of mathematical programming. The robust optimization framework allows modelers to address this issue directly.

As mentioned earlier, another important distinction between robust optimization and SP is the handling of the constraints. Stochastic programs seek to find the design variable x such that for each realized scenario there is a control variable setting, y_s, that satisfies the constraints. For systems with substantial redundancy, such a solution might always be possible. Indeed, the SP literature even allows for the notion of *complete recourse*, whereby a feasible solution y_s exists for all scenarios and for any value of x that satisfies the control constraints. What happens in cases where no feasible pair (x, y_s) is possible for every scenario? The SP model is declared infeasible. Robust optimization explicitly allows this possibility. In engineering applications (e.g., image restoration), such situations inevitably arise due to measurement errors. Multiple measurements of the same quantity may be inconsistent with each other. Hence, even if the underlying physical system has a solution it will not satisfy all the measurements. The robust optimization model, through the use of error terms $\{z_s\}$ and the penalty function $\rho(\cdot)$, will find a solution that violates the constraints by the least amount. Such an approach is fairly standard in medical imaging (see Elfving [2] and Herman et al. [3]).

29.2. Three Robust Optimization Examples

The concepts of robust optimization are perhaps best explained in the context of some real-world applications. Certainly, the idea that data coefficients are not always accurate with machine precision is unassailable for actual linear and nonlinear programming problems. The following three examples are from the area of planning and scheduling. Further details of these and other applications are available [6]. In each example the goal is to locate recommendations that are relatively immune to errors or mis-specifications in the data coefficients.

29.2.1. Air Force STORM.
The Air Mobility Command (AMC, formerly MAC) is responsible for managing all U.S. military airplanes and flights worldwide. It is this group that planned the massive airlift for the 1991 operation in Iraq. One of AMC's planning systems, called STORM, assigns military aircraft to routes in such a way that costs are minimized and anticipated travel plans are met for each of the services (Army, Navy, Air Force, and Marines). In this domain, the planners assign military planes before leasing any civilian aircraft in order to reduce the overall costs.

Each service generates its demand for cargo and other movements roughly one month ahead of time. Constraints specify capacity for each aircraft in terms of size, weight, and cargo type. For instance, tanks can be moved only by the wide-body planes (e.g., the C5A). The resulting linear programs are of moderate size—500 to 5000 equations and 2000 to 20,000 variables—depending on the geographic area under study.

Several difficulties arise in the solution of the deterministic STORM LP. First, the cargo demands must be modified as the month unfolds; delays occur, and new crises and other unplanned events arise. Thus, the minimum-cost solution as defined by the LP may not be very accurate. A second difficulty involves the disruptions in the detailed scheduling of individual aircraft to time-sequenced routes. These disruptions cause problems for the staff who are responsible for scheduling of specific airplanes. The goal of a robust solution to the STORM model is a set of core routes and contingency plans that are cost effective over a range of scenarios and cause minimum disruption in scheduling. The robust optimization framework handles this problem in a natural fashion.

29.2.2. Telecommunication Networks.
Before deregulation of long-distance telephone service, telephone networks were designed with a great deal of redundancies, such as multiple links between every major pair of cities (nodes in the network), and overcapacities. In contrast, today's telecommunication networks are designed to be adaptive, via new switching technology and other measures (e.g., Digital Cross Connect), and cost effective. Thus, there is a tension between supplying adequate capacities under a variety of traffic scenarios, e.g., Mother's Day, or a major disaster, such as an earthquake in Los Angeles, and the costs of building or leasing extra transmission lines.

For this problem a robust solution is a communication network that remains reliable and stable for a spectrum of scenarios, such as loss of a single node, and that is relatively inexpensive to build and operate. Clearly, there is a tradeoff between short-run profit and system reliability. The robust optimization model provides a systematic approach for handling this difficult planning problem.

29.2.3. International Investment Strategies. Diversification of risks is a key ingredient in financial planning. As any seasoned investor knows, the goal of diversification is to provide stable profits regardless of the scenario that actually occurs. In this respect, a well-diversified portfolio equals a robust solution. Investments that are relatively uncorrelated with each other are prized because they can be employed with great advantage to reduce risks. International investments are important, for example, because of their relative lack of correlation with U.S. markets.

A robust solution for an investor depends on that investor's circumstances. For instance, people close to retirement should have more conservative risk profiles than people just starting their careers. In addition to the relevance of the horizon, the users of the moneys must be considered. Savings for a home purchase should be treated differently than savings for a child's college tuition. The notions of robust optimization are ideally suited to the problem of personal investment strategies.

29.3. Solution Strategies

Solution of the robust optimization problems that arise in these examples is a formidable computational task. The total number of decision variables and equations can quickly exceed the capacity of most computers. To give an example, the LOQO linear programming system [9, 10] solved a version of the STORM model with a nonlinear expected-utility function. The size for this deterministic LP is 585 constraints and 1380 variables. Figure 29.2 depicts the size of the robust optimization problems and the resulting execution time as a function of the number of scenarios. Even for these problems, the solution time hits a wall at about 20 scenarios on the Silicon Graphics Indigo workstations for the LOQO direct solver, primarily due to the large size of the convex robust optimization problems.

Specialized solvers are more successful than direct optimization methods for handling the structure of the robust optimization model. As an example, the diagonal quadratic approximation (DQA) method [7] handled much larger problems, as detailed in Figure 29.2. Methods of this type parcel out the work to computers on a network, each according to its capacity, with more powerful machines getting larger subproblems. The DQA algorithm takes advantage of the distributed computational resources and attains a substantial performance level.

The examples shown in Figure 29.2 were solved at Princeton University

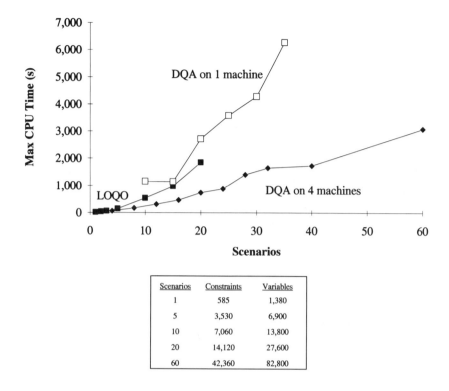

FIG. 29.2. *CPU execution time as a function of the number of scenarios for the U.S. Air Mobility Command's STORM aircraft planning system. The DQA algorithm can handle much larger problems than direct optimization methods by parceling out the work to the computers on a network, each according to its capacity.*

using the PVM software package in conjunction with a standard Ethernet connection. By breaking the problem into smaller digestible pieces, we were able to handle a 35-scenario STORM model using a single SGI workstation (SGI R4000 Indigo). Moving to a four-machine configuration, we solved up to 60 scenarios. Finally, with a configuration of 20 SGI machines, we solved a 220-scenario problem in 5374 CPU seconds. These NLPs are quite large: 300,000 variables and 130,000 constraints. Their solvability can be attributed to the collective resources of our workstation network.

In addition, DQA is able to take advantage of the vector capabilities of classic supercomputers, as shown in Figure 29.3. Here, since each scenario has an identical sparsity pattern, with different numerical values, the computationally expensive Cholesky factorization can be recast. Normally, the factorization's inner loop is quite short, and indirect addressing must be used; these features are the bane of high performance on a vector computer. When using DQA, we can recast the factorization so that the inner, vectorized loop is going "across" the scenarios. The inner loop is then as long as the number of scenarios, and there is no indirect addressing.

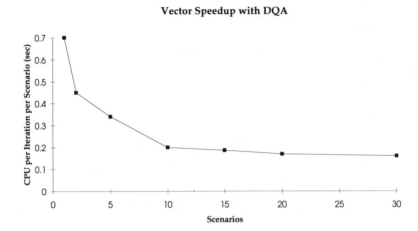

Fig. 29.3. *CPU execution time per iteration as a function of the number of STORM scenarios, showing that the DQA algorithm can also take advantage of the vector capabilities of traditional supercomputers. The inner loop of the computationally expensive Cholesky factorization, which is normally short, goes "across" the scenarios with DQA, and there is no indirect addressing.*

Another technique that shows great promise is to specialize interior-point algorithms for the structure of robust optimization and multistage stochastic programming models. The approach is called tree dissection [1]; it orders the problem matrix such that its sparse properties are maintained during factorization. Empirical tests have shown that the run time of problems with this method grows linearly in the number of scenarios. Alternative ordering heuristics, such as minimum degree ordering, do not take advantage of the special structure of the multistage stochastic program. These techniques can result in dense factorizations and longer running times.

Tree dissection was tested on a financial optimization problem. In this problem, the decision maker is an individual who wishes to decide how to allocate assets among seven different asset categories in each future time period. The decision maker must also decide whether to pay back loans early. The constraints model the flow of funds between asset categories as well as the flow of funds outside the system, such as transaction costs and interest payments. The uncertainties in this problem are the returns on the assets and the adjustable interest rate on the debt. Table 29.1 shows the size of the model, which depends on the number of scenarios and the structure of the scenario tree. The basic single scenario model has 96 constraints, 156 variables, and six time periods. The structure of the tree depends on the number of splits that occur at each level of the tree ("branchings") and the way linking constraints are handled in the model ("compact form" and "split variable form").

These problems were run on a Silicon Graphics Power Challenge workstation with a 100 MHz R8000/R8010 chip and 4GB of main memory. Berger et al.
incorporated tree dissection into LOQO version 2.07 and compared the results
with the same problems running on an unmodified version of LOQO running
using its default settings (LOQO default) as well as using LOQO settings that
attempt to resolve the dual problem first (LOQO "dual first"). The results are
summarized in Table 29.2. These results indicate that small problems can be
solved faster with tree dissection, and problems formerly unsolvable can now
easily be handled with tree dissection.

TABLE 29.1

Problem size for tree dissection tests, rounded to nearest thousand.

Scenarios	Branchings	Compact form		Split variable form	
		Constraints	Variables	Constraints	Variables
1,024	16×4×4×2×2	65,000	109,000	146,000	160,000
2,048	16×8×4×2×2	131,000	218,000	299,000	319,000
4.096	32×8×4×2×2	263,000	435,000	597,000	639,000
8,192	32×16×4×2×2	525,000	869,000	1,195,000	1,278,000
16,384	64×16×4×2×2	1,050,000	1,738,000	2,390,000	2,556,000

TABLE 29.2

*Runtimes (in minutes) until solution. Final relative optimality gap $< 10^{-5}$. NS
denotes problem not solved due to excessive memory requirements.*

Scenarios	LOQO run times		Tree dissection	
	Default settings	"dual first"	Run time	Iterations
512	423	18	7.5	31
1,024	NS	45	24	49
2,048	NS	431	86	73
4,096	NS	1776	167	70
8,192	NS	NS	330	57
16,384	NS	NS	600	51

In the future, algorithmic designers will need to match the hardware
characteristics with the available software to solve the robust optimization
problem under study. Specialization, unfortunately, may become more
common in this environment. Due to their enormous sizes, robust optimization
problems will require efficient use of advanced computer architectures in order
to achieve solutions within practical time limits.

29.4. Research Challenges

The future of robust optimization will depend on the attainment of three goals. The first, and the main, challenge is to develop realistic approaches for generating scenarios that can be readily understood by optimization modelers and the user community. Representing noisy data can be quite onerous and dependent on the actual problem environment. Second, research must continue on the design of efficient parallel algorithms for robust optimization. Third, graphical interfaces for helping users select an acceptable level of robustness for their problems will be essential. Much progress on each of these goals has been made.

The popular scenario managers in spreadsheets will help in the modeling of noisy data for robust optimization. The upcoming spreadsheet implementations will open the topic of robust optimization to millions of potential users. Still, the modeling public will be able to use the robust optimization concepts effectively only if further progress is made on the aforementioned goals.

References

[1] A.J. BERGER, J.M. MULVEY, E. ROTHBERG, AND R.J. VANDERBEI, *Solving Multistage Stochastic Programs Using Tree Dissection*, Report SOR 95–07, Princeton University, Princeton, NJ, 1995.

[2] T. ELFVING, *An algorithm for maximum entropy image reconstruction from noisy data*, Math. Comput. Modelling, 12(1989), pp. 729–745.

[3] G.T. HERMAN, D. ODHNER, K. TOENNIES, AND S.A. ZENIOS, *A parallelized algorithm for image reconstruction from noisy projections*, in Large Scale Numerical Optimization, T. Coleman and Y. Li, eds., SIAM, Philadelphia, 1990, pp. 3–21.

[4] R.L. KEENEY AND H. RAIFFA, *Decisions with Multiple Objectives: Preferences and Value Tradeoffs*, John Wiley, New York, 1976.

[5] H. MARKOWITZ, *Portfolio Selection, Efficiency Diversification of Investments*, Cowles Foundation Monograph 16, Yale University Press, 1959; 2nd edition, Vasil Blackwell, Cambridge, 1991.

[6] J.M. MULVEY, R.J. VANDERBEI, AND S.A. ZENIOS, *Robust optimization of large-scale systems*, Oper. Res., 43(1995), pp. 264–281.

[7] J.M. MULVEY AND A. RUSZCZYNSKI, *A new scenario decomposition method for large-scale stochastic optimization*, Oper. Res., 43(1995), pp. 477–490.

[8] J.M. MULVEY, *Generating scenarios for the Towers Perrin investment system*, Interfaces, 26(1996), pp. 1–15.

[9] R.J. VANDERBEI, *LOQO User's Manual*, Report SOR 92–5, Princeton University, Princeton, NJ, 1992.

[10] ——, *LOQO: An Interior Point Code for Quadratic Programming*, Report SOR 94–15, Princeton University, Princeton, NJ, 1995.

Chapter 30

History Matching of Multiphase Reservoir Models on Hypercubes

Jianping Zhu

Editorial preface

The inverse problem associated with modeling the behavior of oil reservoirs offers a challenging simulation problem. This chapter details the formulation of the problem occurring when fluid is injected into an underground reservoir to force the oil to the extraction well(s). The algorithm, which is run on the Intel hypercube, uses multilevel grid methods. The results indicate a highly parallel solution method.

This article originally appeared in *SIAM News*, Vol. 25, No. 2, March 1992. It was updated during the summer/fall of 1995.

Scientific study shows that only a small portion (15–25%) of the oil in a reservoir can be recovered by natural means—through reservoir pressure forcing the oil to the surface. As removal of the oil continues, the well output declines due to the diminishing pressure gradient and increasing viscosity. To maximize the oil production, various secondary and enhanced oil recovery processes have to be used. If they are properly designed and implemented, the use of these processes can result in the production of at least 10–15% more oil from the reservoirs.

These processes involve primarily the injection of water or chemicals into the reservoir to build up the pressure gradient or to increase the mobility of the oil. The moving fluid front between the injected fluid and the oil should move in such a way as to drive the oil out of the production wells. The stability of this front is crucial to the success of these processes. If the front becomes unstable, it will no longer displace oil toward the production wells.

In the simple two-dimensional reservoir oil field shown in Figure 30.1, the injection well is located at the lower left corner and the production well is located at the upper right corner; the contour lines represent the front between the injected fluid and the oil at different times. The injected fluid is supposed to displace oil toward the production well. However, when the front becomes

FIG. 30.1. *The fingering phenomenon. The long narrow "fingers" represent the unstable injected fluid front bypassing the oil.*

unstable as indicated by the contour lines in the figure, the injected fluid can move much faster than oil, producing the so-called "fingering phenomenon" [5]. What comes out from the production wells can then be the injected fluid, rather than oil. To prevent this from happening and to make enhanced oil recovery methods productive, we must be able to monitor and predict the fluid front movement during the injection process.

Because reservoirs are usually deep under the ground, the movement of the fluid front between the injected fluid and the oil is not directly observable. Therefore, mathematical models are very important for keeping track of the changes in reservoir pressure and the movement of the fluid front.

A commonly used two-phase black oil reservoir model describing the flow of slightly compressible fluids through porous media is given by the following system of nonlinear partial differential equations (PDEs) [8]:

$$(30.1) \qquad \nabla \cdot [\lambda_o K(\underline{x})(\nabla P_o - \rho_o \mathbf{g})] = \frac{1}{\rho_o} \frac{\partial(\Phi \rho_o S_o)}{\partial t} + q_o,$$

$$(30.2) \qquad \nabla \cdot [\lambda_w K(\underline{x})(\nabla P_w - \rho_w \mathbf{g})] = \frac{1}{\rho_w} \frac{\partial(\Phi \rho_w S_w)}{\partial t} + q_w,$$

$$(30.3) \qquad S_o + S_w = 1, \qquad P_{cow} = P_o - P_w = f(S_w),$$

for $t > 0$, $\underline{x} = (x, y, z) \in \Omega$, with the initial conditions

$$(30.4) \qquad P_l(x, y, z, 0) = P_l^0(x, y, z), \qquad l = o, w, \qquad \underline{x} \in \Omega$$

and the boundary conditions

(30.5) $$\frac{\partial P_l}{\partial \underline{n}} = 0, \qquad l = o, w, \quad \underline{x} \in \partial\Omega, \quad t > 0,$$

where $K(\underline{x})$ is the absolute permeability distribution; the subscripts o and w represent phases of oil and water, respectively; λ_l is the phase mobility, which depends on the phase saturation, S_l; P_l is the phase pressure; ρ_l is the phase density; Φ is the rock porosity; q_l is the production (injection) rate; $\underline{n}$ is the normal vector to the boundary $\partial\Omega$; and $\mathbf{g}$ is the gravitational acceleration. These equations were derived from conservation of mass of all phases. If all the parameters are known, we can solve (30.1)–(30.5) and get the information on pressure changes and fluid front movement (in terms of saturations) that is needed to facilitate the oil recovery process.

Unfortunately, some parameters in the above equations, like the absolute permeability distribution $K(\underline{x})$, are usually not known in advance. Because of the high cost of drilling (at least \$100,000 for a well), it is unrealistic to drill many wells and get rock samples to measure the values of the permeability distribution. Therefore, a numerical algorithm for estimating the underground reservoir structural parameters by computer simulations is necessary for the reservoir simulations.

Although the true permeability distribution can not be measured directly in general, the values of reservoir pressure P_l corresponding to $K(\underline{x})$ are known from measurements made at observation wells (including both production and injection wells) during the production period. This means that we can obtain partial information about the solution of (30.1)–(30.5) without actually solving them. The task of identifying the unknown parameter distributions by using the available knowledge of the solution of a given PDE system is called an inverse problem. (This is opposed to the forward problem in which all parameters are known and the major task is to solve the PDEs to get the solution.) In particular, the purpose of solving inverse problems in reservoir modeling is to identify the reservoir structural parameters by using the available pressure, saturation, and other production data obtained at wells.

30.1. History Matching

The process of identifying the unknown permeability distribution usually starts from an initial guess, K^0, at the true permeability distribution K. By solving (30.1)–(30.5) with K^0, we can get the pressure distribution P_l^0 that corresponds to K^0. If K^0 happens to be the same as K, meaning that we had a perfect initial guess, then P_l^0 should also be the same as, or very close to, the measured P_l, which is supposed to be the solution corresponding to the true permeability distribution K. Such a coincidence is unlikely in practice and the initial guess, K^0, is in general different from K, which means that P_l^0 is also different from P_l. The differences $\delta P_l = P_l - P_l^0$ can then be used to calculate a correction term δK^0, so that $K^1 = K^0 + \delta K^0$ will be a better approximation to K. The entire iterative process can be described as follows:

STEP1: Solve (30.1)–(30.5) with K^n to get the corresponding pressure distribution P_l^n. For $n = 0$, K^0 is the initial guess of the absolute permeability distribution. We can solve (30.1)–(30.2) by linearization and a finite difference scheme. The discretized form of (30.1)–(30.2) is

$$(30.6) \qquad\qquad \mathbf{Ap} = \mathbf{f},$$

where $\mathbf{A}$ is normally sparse and structured.

STEP2: Compare P_l^n with P_l. If the computed pressure $P_l^n \cdot$matches the measured pressure data P_l, go to **STEP4**; K^n is the true permeability distribution. Otherwise, go to **STEP3**.

STEP3: Set $\delta P_l^n = P_l - P_l^n$ and $\delta K^n = K - K^n$, where P_l^n is the computed pressure that corresponds to the permeability distribution K^n, and the measured pressure P_l corresponds to the true permeability distribution K. We then have

$$(30.7) \qquad\qquad P_l = P_l^n + \delta P_l^n, \qquad K = K^n + \delta K^n.$$

P_l and K should satisfy (30.1)–(30.5), as should P_l^n and K^n. If we substitute P_l and K into (30.1)–(30.5) and utilize the fact that P_l^n and K^n should also satisfy (30.1)–(30.5), we will have a PDE system that relates δK^n to δP_l^n [18]. By using the finite difference scheme to discretize this PDE system and putting together those equations at the observation wells (so that δP_l^n is known), we will have the following algebraic equations:

$$\mathbf{B}\delta \mathbf{k}^n = \mathbf{b},$$

where $\mathbf{B}$ is a dense rectangular matrix and $\mathbf{b}$ represents the influence of δP_l^n. This equation can be solved by the least squares method. Because of the illposedness of inverse problems, Tikhnov's regularization method [12] is usually used to solve a better conditioned system

$$(30.8) \qquad\qquad (\mathbf{B}^T\mathbf{B} + \alpha\mathbf{I})\delta\mathbf{k}^n = \mathbf{B}^T\mathbf{b}.$$

Here α is a nonnegative parameter that is usually called the regularization factor. The key idea is to solve a better conditioned system whose solution can be taken as a reasonable approximation to the solution of the original ill-conditioned system. The value of α is problem dependent and sometimes needs to be adjusted dynamically during the computations [18]. After solving this system, K^n can be updated by $K^{n+1} = K^n + \delta K^n$, and the process returns to **STEP1**.

STEP4: Stop.

Because the key issue involved here is to match the computed pressure P^n with those measured history pressure data P by adjusting the unknown permeability distribution $K(\underline{x})$, the process is called history matching in the oil industry.

Since the early 1970s when Chen et al. [2] proposed the use of optimal control methods for automatic history matching, there have been many improvements in automatic history matching [1, 4, 6, 10, 13, 14]. However, two major difficulties continue to impede further development of efficient history matching algorithms for multiphase multidimensional nonlinear reservoir models:

- The sensitive dependence of the identified parameter distribution on the initial guess used to start the iteration.

- The large amounts of numerical computation involved. As can be seen from **STEP1** and **STEP3**, the nonlinear reservoir models of (30.1)–(30.5) must be solved repeatedly at many timesteps and a dense regularized least squares problem needs to be solved at each iteration to compute δK^n from δP_l^n.

Some recent efforts to solve these problems are presented in the next two sections.

30.2. Multilevel Grid Method

It is the experience of many researchers that the identified permeability distribution is very sensitive to the initial guess. If the initial guess K^0 is not close enough to the true distribution, the identified permeability distribution K^n might not resemble the true distribution K at all, even if the iterative process converges in the sense that the computed pressure P^n matches the measured pressure history P. In practical applications, however, it is unlikely that a very accurate initial guess will be available for history matching. Therefore, a good history matching algorithm must be capable of identifying the true permeability distribution from a fairly general initial guess.

Chen and Zhang [3] first used multigrid strategy to reduce the computational complexity of a parameter identification algorithm (the GPST algorithm) for linear diffusion equations. McCormick and Wade [7] applied the multigrid method to the inverse problem in electrical impedance tomography. Recently, we used the multilevel grid method to reduce the sensitive dependence of the identified permeability distribution on the initial guess [17]. The multilevel grid method was chosen based on the finding that, as the number of unknown parameters (grid points) increases, the history matching process becomes more sensitive to the initial guess K^0. If the reservoir model is discretized by a relatively coarse grid system with fewer unknown parameters to be identified, the parameter estimation process will be less sensitive to the initial guess. However, the resolution of the identified permeability distribution on the coarse grid is usually not good enough for large-scale reservoir simulations (which can involve thousands of grid points). This was the motivation for the use of the multilevel grid method in history matching.

We begin the process by constructing a hierarchy of grids $G_0 \subset G_1 \subset G_2 \cdots \cdots \subset G_q$, where G_0 is just fine enough to accommodate all the wells and some auxiliary points added to keep the grid regular. G_i can be generated from G_{i-1} in many ways. We used the bisection method for its simplicity.

The history matching process starts on grid G_0 with an initial guess $K_{G_0}^0$. The identified permeability distribution $K_{G_0}^n$ is then interpolated to the finer grid G_1 and used as the initial guess $K_{G_1}^0$ on G_1 to start the history matching process. Because history matching on the coarse gird is less sensitive to the initial guess, $K_{G_0}^n$ should be a reasonable approximation to the true permeability distribution on the coarse grid G_0. Therefore, the interpolation of $K_{G_0}^n$ to $K_{G_1}^0$ could be a very good initial guess of the true permeability distribution on the finer grid G_1. Although the history matching process becomes more sensitive to the initial guess as the number of grid points increases, the use of the multilevel grid technique ensures that the history matching process has a better initial guess $K_{G_1}^0$ to start the iteration on the finer grid G_1. This process can be repeated until the desired resolution has been achieved.

30.3. Implementation on Hypercubes

Advanced parallel computers have been used to cope with the large amounts of numerical computation involved in the history matching process with multiphase multidimensional reservoir models [16, 18, 19]. We have used the iPSC/2 hypercube at the Cornell Theory Center and the iPSC/860 hypercube at the NSF Engineering Research Center of Mississippi State University in our investigation of parallel processing. The code optimizations used to achieve large speedups have been concentrated in the following two aspects:

- dividing the whole solution process into smaller tasks that can be executed in parallel,

- reducing the internode communication as much as possible.

Timing analysis of the sequential program shows that the solution of (30.6) and (30.8) is the most time consuming part of the process, taking more than 95% of the total CPU time when several thousand grid points are involved in the computation. The rest of the CPU time is used for processing data and generating (30.6) and (30.8).

Equation (30.6) is a sparse system that can be solved by iterative methods. We find that the block SOR method with red-black ordering is very efficient for the following reasons:

- Simplicity for parallelization. With red-black ordering, all unknowns of the same color can be solved in parallel.

- Satisfactory convergence rate. Since the initial reservoir pressure distribution is a very good initial guess to start the iteration, and the pressure distribution calculated at the current timestep can be used as a good initial guess for the iteration at the next timestep, the successive over-relaxation (SOR) iteration for (30.6) always has good initial guesses. Furthermore, we find that the number of SOR iterations needed at each timestep does not increase significantly as the number of grid points increases.

Internode communication is required on two occasions in the solution of (30.6).

1. Because the domain is decomposed into subdomains, which are assigned to different nodes, each node needs to communicate with nodes in adjacent subdomains at every SOR iteration. For simplicity, we discuss rectangular domains; irregular domains can be mapped into rectangular domains by numerical grid generation schemes [11]. For rectangular domains, the subdomains can be cut into strips and mapped to nodes by using the binary reflected gray code so that the internode communications always occur between the nearest neighbors.

   A simple example can explain this. Suppose we have eight subdomains D_i, $i = 0, \ldots, 7$, and each subdomain D_i needs to exchange information with D_{i-1} and D_{i+1}, except for D_0 and D_7, which exchange information with D_1 and D_6, respectively. If we map D_i directly to node i on a hypercube, the communication may go beyond the nearest neighbors. If we use the binary reflected gray code to map the domains onto the hypercube, however, communication will always be between the nearest neighbors. The code is a sequence of n-bit binary numbers such that any two successive numbers in the sequence differ in only one bit and all binary numbers with n bits are represented. The one-bit binary code is $C_1 = \{0, 1\}$. The n-bit code can be built recursively from the $(n-1)$-bit code by prepending a 0 to each element in the $(n-1)$-bit code and then prepending a 1 to the same $(n-1)$-bit code elements taken in a reversed order. Since there are eight (2^3) subdomains, a three-bit gray code is needed:

$$C_3 = \{000, 001, 011, 010, 110, 111, 101, 100\}.$$

   The elements in C_3 gives the node numbers to which the corresponding subdomains

$$\{D_0, D_1, D_2, D_3, D_4, D_5, D_6, D_7\}$$

   should be assigned. For example, D_7 should be assigned to node 4, rather than directly to node 7.

2. Verification of the stopping criterion requires that each node compute the error e_i associated with its assignment of solution vectors and that

$\max(e_i) \leq \epsilon$, where ϵ is the given tolerance. We used two different schemes to accomplish this. Because the host program of the cube does not do any numerical computation, we first let all nodes send e_i's to the host for comparison; the host then broadcasts the signal to all nodes for continuation or termination of the iteration.

In the second scheme one node (not the host) is selected as the master node. All other nodes send the errors e_i to the master node, which then determines whether the criterion $\max(e_i) \leq \epsilon$ has been satisfied and broadcasts a signal to all other nodes for the continuation or termination of the iteration process. We find the second method faster, which is an indication that communication between the host and nodes should be avoided on architectures of this type.

Equation (30.8) is in the form of the normal equation for a least squares problem. For numerical stability, the best way to solve this system is to use QR decomposition, rather than forming the normal equation explicitly. We can rewrite (30.8) as

$$(30.9) \qquad (\mathbf{B}^T, \sqrt{\alpha}\mathbf{I}) \begin{pmatrix} \mathbf{B} \\ \sqrt{\alpha}\mathbf{I} \end{pmatrix} \delta\mathbf{k} = (B^T, \sqrt{\alpha}\mathbf{I}) \begin{pmatrix} \mathbf{b} \\ 0 \end{pmatrix}$$

or

$$(30.10) \qquad \mathbf{C}^T\mathbf{C}\delta\mathbf{k} = \mathbf{C}^T \begin{pmatrix} \mathbf{b} \\ 0 \end{pmatrix}.$$

The main task is to carry out QR factorization on the matrix $\mathbf{C}$, which has the structure

$$(30.11) \qquad \mathbf{C} = \begin{bmatrix} b_{11} & b_{12} & \cdots & b_{1n} \\ \vdots & \vdots & \vdots & \vdots \\ b_{m1} & b_{m2} & \cdots & b_{mn} \\ \sqrt{\alpha} & & & \\ & \sqrt{\alpha} & & \\ & & \ddots & \\ & & & \sqrt{\alpha} \end{bmatrix}.$$

This is neither a complete full matrix nor a sparse matrix. The upper part is full and the lower part is sparse (in diagonal form). Many QR factorization schemes for general dense matrices were implemented and tested for comparison [9, 15]. Timing results show that the column-oriented Householder transform is more efficient than other algorithms for matrices of this special type. In this algorithm, the matrix is distributed to different nodes and transformed by columns. For an $m \times n$ equation system (assume $m > n$), n steps are required to annihilate all nonzero elements below the main diagonal. In the ith step the node holding the pivoting column $\mathbf{c}_i$ broadcasts the pivoting column to all other nodes. After receiving the pivoting column, each node can proceed to update its assignment of columns.

The efficiency of the algorithm has been further improved by the following:

1. Take the special structure of matrix (30.11) into consideration. In the first step of the transformation, for example, only the elements in the submatrix

$$
\begin{bmatrix}
b_{11} & b_{12} & \cdots & b_{1n} \\
\vdots & \vdots & \vdots & \vdots \\
b_{m1} & b_{m2} & \cdots & b_{mn} \\
\sqrt{\alpha} & & &
\end{bmatrix}
$$

will be affected, rather than the entire matrix. Similarly, in each of the n transformation steps, the transformation will be applied to only part of the original matrix.

2. Schedule the node holding the next pivoting column c_{i+1} to broadcast the column to all nodes as soon as it finishes updating c_{i+1}, thereby reducing the waiting time of all other nodes for the arrival of the next pivoting column. Given the matrix in (30.11), if n is even and two nodes are available on the hypercube, the columns can be assigned to the two nodes in a wrap-around fashion for better load balancing. Node 1 is assigned columns $1, 3, \ldots, n-1$ and node 2 is assigned columns $2, 4, \ldots, n$. In the first step, node 1 sends column 1 to node 2 and updates all the columns it has; node 2 updates its columns by using the pivoting column received from node 1. The new pivoting column for the next step will be column 2, which is held by node 2. Node 2 should send column 2 to node 1 as soon as it finishes updating the column and before it starts updating columns $4, \ldots, n$ in the first transformation step. In this way, node 1 gets the new pivoting column sooner for the second transformation step.

30.4. Numerical Experiments

The algorithm has been tested by extensive numerical computations in two and three dimensions. Using the multilevel grid in history matching significantly reduces the dependence of the identified permeability distribution on the initial guess K^0 of the true permeability distribution. With the multilevel grid, the algorithm discussed here can be used to identify the absolute permeability distributions varying by an order of magnitude over the domain, with an initial guess deviating from the true distribution by two orders of magnitude.

Timing results for this example on iPSC/2 and iPSC/860 hypercubes (32 nodes), with two different final grids, are given in Figure 30.2. The final grids are 32×32 for the first model and 64×64 for the second model. The curves in the figure represent speedup versus the number of nodes on the iPSC/2 and iPSC/860 hypercubes. We can see clearly the linear speedup on iPSC/2 with the 64×64 final grid, while the speedup for the 32×32 grid slows down as the number of nodes increases.

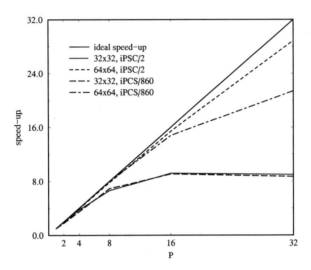

FIG. 30.2.   *Speedup curves on iPSC/2 and iPSC/860 hypercubes. P is the number of processors.*

The same program runs about 25 to 30 times faster on the iPSC/860 than on the iPSC/2. By incorporating routines from the Basic Linear Algebraic Subroutine package into the program, we achieved a sustained rate of about 20 Mflops on a single node. There is, however, a communication problem to be solved in order to fully utilize the computing power of large number of processors. It is obvious from the figure that the speedup for iPSC/860 is not as good as that of iPSC/2. The imbalance between the significantly improved computation speed and the relatively slow communication rate on the iPSC/860 is responsible for this situation. For the history matching problems discussed here, we find that the sustained numerical computation rate in Fortran is about 40 times faster on i860 nodes than on the 386-based iPSC/2 nodes, while communication is only three to five times faster. Therefore, as the number of nodes increases, the performance degradation due to internode communication is more significant on the iPSC/860 than on the iPSC/2.

30.5. Conclusions

History matching is very important to oil reservoir simulations. A successful history matching algorithm can save millions of dollars by helping engineers identify the underground reservoir structural parameters without drilling additional wells. These parameters can then be used for reservoir simulations to make the enhanced oil recovery process more productive and generate even greater economic benefit. This could involve billions of dollars.

When the reservoir model is large enough, the efficiency of the history matching algorithm can be well above 90% with linear speedup. A 128-node iPSC/860 hypercube can deliver Gflop performance, which is comparable with that of the Cray Y-MP, and the performance/price ratio is at least an order of magnitude higher for the former. Considering that distributed-memory parallel computers can be scaled to a large number of processors, they are very promising for large-scale computational problems in reservoir modeling.

References

[1] G. CHAVEN, M. DUPUY, AND P. LEMONNIER, *History matching by use of optimal control theory*, Soc. Pet. Eng. J., 15(1975), pp. 74–86.

[2] W.H. CHEN, G.R. GAVALAS, J.H. SEINFELD, AND M.L. WASSERMAN, *A new algorithm for automatic history matching*, Soc. Pet. Eng. J., 14(1974), pp. 593–608.

[3] Y.M. CHEN AND F.G. ZHANG, *Multigrid strategy for efficiency improvement of GPST inversion algorithm*, Appl. Numer. Math., 6(1989), pp. 431–446.

[4] G.R. GAVALAS, P.C. SHAH, AND J.H. SEINFELD, *Reservoir history matching by Bayesian estimation*, Soc. Pet. Eng. J., 16(1976), pp. 337–350.

[5] J. GLIMM AND O.A. MCBRYAN *The mathematics of reservoir simulation*, in Inverse Problems in Partial Differential Equations, R.E. Ewing, ed., SIAM, Philadelphia, 1983, pp. 107–160.

[6] T. LEE AND J.H. SEINFELD, *Estimation of two phase petroleum reservoir properties by regularization*, J. Comput. Phys., 69(1987), pp. 397–419.

[7] S.F. MCCORMICK AND J.G. WADE, *Multigrid solution of a linearized, regularized least-squares problem in electrical impedance tomography*, Inverse Problems, 9(1993), pp. 697–713.

[8] D.W. PEACEMAN, *Fundamentals of Numerical Reservoir Simulation*, Elsevier–North Holland, Amsterdam, 1977.

[9] A. POTHEN AND P. RAGHAVAN, *Distributed orthogonal factorization: Givens and Householder algorithms*, SIAM J. Sci. Statist. Comput., 10(1989), pp. 1113–1135.

[10] Y.N. TANG, Y.M. CHEN AND W.H. CHEN, *Generalized pulse-spectrum technique for 2-D and 2-phase history matching*, Appl. Numer. Math., 5(1989), pp. 529–541.

[11] J.F. THOMPSON, Z. U.A. WARSI, AND C.W. MASTIN, *Numerical Grid Generation*, North–Holland, New York, 1985.

[12] A.N. TIKHONOV AND V. ARSENIN, *Solutions of Ill-Posed Problems*, John Wiley & Sons, New York, 1977.

[13] M.L. WASSERMAN, A.S. EMANUEL, AND J.H. SEINFELD, *Practical applications of optimal control theory to history matching multiphase simulator models*, Soc. Pet. Eng. J., 15(1975), pp. 347–355.

[14] A.T. WATSON, *Sensitivity analysis of two phase history matching*, SPE Reservoir Engineering, 4(1989), pp. 319–324.

[15] J.P. ZHU, *QR factorization for the regularized least squares problem on hypercubes*, Parallel Computing, 19(1993), pp. 939–948.

[16] J.P. ZHU AND Y.M. CHEN, *Parameter estimation for multiphase reservoir models on hypercubes*, Impact Comput. Sci. Engrg., 4(1992), pp. 97–123.

[17] J.P. ZHU AND Y.M. CHEN, *On the application of the multilevel grid to the solution of inverse problems of reservoir simulations*, Appl. Numer. Math., 10(1992), pp. 159–174.

[18] ———, *History matching multi-phase multi-dimensional oil reservoir models on IBM supercomputer*, Scientific Excellence in Supercomputing: The 1990 IBM Prize Papers, Baldwin Press, Athens, GA, 1992.

[19] ———, *History matching for multiphase reservoir models on shared memory supercomputers*, Internat. J. Supercomputer Appl., 6(1992), pp. 193–206.

A Retrospective on the "Applications on Advanced Architecture Computers" *SIAM News* Column

Greg Astfalk

Editorial preface

This brief article appeared shortly before the *SIAM News* column on "Applications on Advanced Architecture Computers" reached its fifth anniversary. In answer to questions about how the column originated, I gave a somewhat personal and casual tale of its origin. At the time, as mentioned in the appendix, the column was enjoying a milestone, its first repeat author!

This article originally appeared in *SIAM News*, Vol. 27, No. 9, November 1994.

With the publication of this issue's "Applications on Advanced Architecture Computers" column, *SIAM News* editor Gail Corbett and I thought it was an appropriate time to look back over the years since the column first appeared in March 1990. It was the author of this issue's article, Craig Douglas, who unknowingly prompted our look to the past.

Craig, whose article begins on page 1 [*SIAM News*, Vol. 27, No. 9, November 1994], is our first repeat author. His previous column article, which appeared in the May 1992 issue of *SIAM News*, was actually something of a departure for the column in that it focused on a method—multigrid—rather than an application. Multigrid methods resurface in his current article on the solution of two-dimensional models of laminar flames. Gail and I felt that having a repeat author gave the column a certain maturity. (In the very near future we will also have our first two-part article.)

As a matter of history, the column got started when I happened to be in the Philadelphia area on business and paid SIAM managing director Ed Block a visit. Over lunch, Ed talked about the nature of *SIAM News* and asked if I had any ideas for increasing interest and improving circulation. At that point I unwittingly decided my fate by telling Ed that *SIAM News* could use more technical articles.

Ed and I had been talking earlier about the growing union between mathematicians and computers. "Ahhh, a series of articles, or better yet, a column, would be just thing," Ed said, looking directly at me. I was volunteered (an Ed Block euphemism for enlisted), and that was the beginning. At least he paid for lunch!

The March 1990 article was a brief survey of high-performance computing architectures that I wrote. At that point, justifiably concerned about how a column would be received (and whether it would be possible to acquire enough articles to keep it going), we didn't formally identify it as a column. The second article, in which Mike Mascagni detailed how the use of a SIMD architecture led to a "backward" random walk algorithm for elliptic boundary value problems, appeared two issues later (July 1990). Mike's article was still not formally identified as a column article. It was the article in the next issue (September 1990), Liam Healy's symbolic and parallel approach to celestial mechanics, that officially kicked off the column.

Over the three and a half years since the column began, we have worked with a group of high-quality, enthusiastic and incredibly cooperative authors to produce articles on a regular basis. Craig's article in this issue is the 25th to appear. Since the first article appeared, in fact, we have missed only eight issues. The complete set of articles is 306 pages in 12pt LaTeX.

For readers who are new to *SIAM News*, or for those who haven't saved (or read) back issues, we include a complete list of the articles.[16] Reprints of the articles are available (send requests to `siamnews@siam.org`). If you would like to offer feedback or feedforward (i.e., recommendations) for the column contact Gail (`corbett@siam.org`).

As to the future, we are *always* looking for contributors for the column. The prerequisites are simple: the articles should describe some real application with a formulation that has some mathematical flavor and a solution done on an "exotic" architecture of any form. Anyone interested in submitting an article can contact me (`astfalk@rsn.hp.com`).

---

[16]We have eliminated the list in this book since the articles themselves appear.

Index